Advances in Genomics

Advances in Genomics

Advances in Genomics

Edited by **Samuel D'costa**

R CALLISTO
REFERENCE

New York

Published by Callisto Reference,
106 Park Avenue, Suite 200,
New York, NY 10016, USA
www.callistoreference.com

Advances in Genomics
Edited by Samuel D'costa

International Standard Book Number: 978-1-63239-045-5 (Hardback)

Contents

Preface

The origins of the term "genomics" can be traced to Dr. Tom Roderick, a geneticist at the Jackson Laboratory (Bar Harbor, Maine). It is the subset of genetics that pertains to recombinant DNA, DNA sequencing systems, bioinformatics, and other similar such fascinating disciplines. The applications of genomics can be found in many industries. Genomics forms the basis of molecular biology and genetic-mapping. Research based on intragenomic phenomena such as heterosis, epistasis, pleiotropy and other interactions between loci and alleles within the genome are also covered under this subject.

In 1941, after the confirmation of helical structure by Rosalind Franklin, genetics as a field of study took on mammoth proportions. Stellar work has been done by biologists and scientists like Frederick Sanger and Walter Gilbert. Researchers and scientists across the globe have propelled the growth of genomics through its nascence in the 1970s and 80s. Before them were trailblazers like James D. Watson and Francis Crick.

Clubbed with the recent developments in genomics and the intense work by the forefathers of the field, it has made us capable of tracking genetic codes and data for numerous viruses, bacteria and fungi, aiding the development of the necessary prevention measures. Genomics includes analysis and synthesis of genomes which involves various kinds of sequencing like shotgun sequencing, high throughput sequencing and illumines sequencing.

This book takes a look at the outstanding advancements made in this field. Experts have lucidly put their knowledge here for easy understanding. It's aimed to be a book that will help not just students and scientists, but also those who are professionally involved in genome studies for further advancements.

Editor

Network Analysis of Functional Genomics Data: Application to Avian Sex-Biased Gene Expression

Oliver Frings,[1,2] **Judith E. Mank,**[3] **Andrey Alexeyenko,**[1,4] **and Erik L. L. Sonnhammer**[1,2,5]

[1] *Stockholm Bioinformatics Centre, Science for Life Laboratory, Box 1031, SE-171 21 Solna, Sweden*
[2] *Department of Biochemistry and Biophysics, Stockholm University, SE-106 91 Stockholm, Sweden*
[3] *Department of Genetics, Evolution and the Environment, University College London, WC1E 6BT, UK*
[4] *School of Biotechnology, Royal Institute of Technology, SE-171 65 Solna, Sweden*
[5] *Swedish eScience Research Center, SE-100 44 Stockholm, Sweden*

Correspondence should be addressed to Erik L. L. Sonnhammer, erik.sonnhammer@sbc.su.se

Academic Editors: R. Jiang, W. Tian, J. Wan, and X. Zhao

Gene expression analysis is often used to investigate the molecular and functional underpinnings of a phenotype. However, differential expression of individual genes is limited in that it does not consider how the genes interact with each other in networks. To address this shortcoming we propose a number of network-based analyses that give additional functional insights into the studied process. These were applied to a dataset of sex-specific gene expression in the chicken gonad and brain at different developmental stages. We first constructed a global chicken interaction network. Combining the network with the expression data showed that most sex-biased genes tend to have lower network connectivity, that is, act within local network environments, although some interesting exceptions were found. Genes of the same sex bias were generally more strongly connected with each other than expected. We further studied the fates of duplicated sex-biased genes and found that there is a significant trend to keep the same pattern of sex bias after duplication. We also identified sex-biased modules in the network, which reveal pathways or complexes involved in sex-specific processes. Altogether, this work integrates evolutionary genomics with systems biology in a novel way, offering new insights into the modular nature of sex-biased genes.

1. Introduction

Although primary sex determining genes are responsible for the initial sex determining cues in the gonad, most of the heritable differences in morphology, behavior, and life history between males and females are the result of different expression levels of genes present in both sexes [1, 2]. Sex-biased genes, which comprise up to 50% of metazoan transcriptomes [3–7], are the product of sexually antagonistic selection for different male and female optima [8, 9]. This antagonism is resolved with the emergence of sex-specific transcriptional regulatory elements that decouple expression between the sexes, thereby allowing separate female and male phenotypes to emerge from a shared genome. Sex-biased genes behave according to the evolutionary predictions for sexually selected and sexually antagonistic traits [10–18], and the study of sex-biased gene expression is emerging as a method to connect sex-specific selection pressures, which act on the whole organism, to the encoding loci.

This connection between sex-biased genes and sexually dimorphic traits offers a way to study the complex interactions between the phenotype to the underlying genome. Most studies of sex-biased gene expression treat individual genes as independent units, ignoring correlated expression that results from the interactive nature of genetic pathways and networks. This simplification compresses the multidimensional nature of the transcriptome. However, because many sexually dimorphic phenotypes are complex amalgams of numerous genes [5, 19, 20], we require a way to study the interactions of the genes underlying them if we wish to understand the constraints acting on these traits and how they respond to selection. In addition to this complexity, many genes contribute to more than one phenotype, pathway, or subnetwork. This pleiotropy is likely an important factor in the evolution of sex-biased gene expression which may ameliorate intralocus sexual conflict acting on a given gene [9].

For genes with high levels of pleiotropy, the many functions of a single locus result in strong evolutionary constraints hindering change due to selection pressure for any single function [21]. This is important for studies of sex-biased genes, as sex-biased gene expression patterns resulting from sexually antagonistic selection for any single function may be detrimental in other functionalities [22]. This would suggest that genes with many pathway connections, though not necessarily less likely to experience sexually antagonistic selection, are less likely to resolve that antagonism through sex-biased expression, as this could result in detrimental effects in other phenotypes encoded by the same loci [23]. More simply stated, the resolution of sexually antagonistic selection may be more common for genes with fewer network interactions. This prediction suggests that (1) pleiotropic genes may contain relatively high levels of unresolved sexual conflict and (2) sexually dimorphic phenotypes are more often encoded by genes with few other functions. This has important implications for evolutionary models of sexual selection which typically assume single functionalities and simple inheritance patterns.

Here we test the relationship between network interaction and sex-biased gene expression with a newly developed gene interaction atlas of the chicken. Previously, we have shown that sex-biased expression is prevalent in chicken [23] and that sex-biased genes in chicken exhibit evolutionary patterns consistent with sexual selection and sexual conflict [16, 18, 24]. In this analysis, we created a functional coupling network from data integration [25] of chicken and incorporated sex-biased expression data into it in order to analyze the connectivity of sex-biased and unbiased genes in both the gonad and soma. Overall, our goal was to better understand the relationship between sexually dimorphic phenotypes, the sexually antagonistic selection pressures shaping them, and the genes encoding them.

2. Materials and Methods

2.1. Network. The chicken network was generated using the FunCoup framework [25, 26]. This framework reconstructs global large-scale networks of functional coupling by Bayesian integration of diverse high-throughput datasets. More specifically raw scores of various types of functional coupling are turned into probabilistic estimates that are then integrated across different types of data and model organisms. The different types of evidence comprised: protein-protein interactions, mRNA coexpression, subcellular colocalization, phylogenetic profile similarity, cotargeting by either miRNA or transcription factors, protein co-expression, and domain-domain interactions. The integration of data from different sources enabled more comprehensive network reconstruction with higher quality. Furthermore, data from other eukaryotic species were transferred via orthologs. Ortholog assignments for cross-species mapping were obtained from the InParanoid database [27]. Signaling and metabolic pathways from KEGG as well as both pathway types combined were used as gold standard for Bayesian training. The network has consequently three different kinds of links: metabolic, signaling, and combined.

The network was predicted using seven chicken-specific microarray expression datasets (see Table S1 in Supplementary Materials available online at doi:10.1100/2012/130491), phylogenetic profile similarity across eukaryotes, and information transferred from other species via orthologs. The use of ortholog transfer was of special importance in this case, as it allows us to overcome the lack of chicken-specific interaction data.

2.2. Microarray Expression Datasets. The network was studied in the context of sex bias under different conditions. We used three different Affymetrix chicken expression datasets from the embryonic gonad, the adult gonad, and the adult brain (previously described in Mank et al. [16], Mank and Ellegren [28], and Mank et al. [24]). Each tissue/time-point array hybridization was based on three replicate nonoverlapping within-sex pools of 3–5 individual samples from male and female embryonic and adult chickens. All datasets were normalized using the MAS5 algorithm from the Affy Bioconductor package.

2.3. Differential Gene Expression. There are several different ways to define differential gene expression. Traditionally genes that are meant to be over- or underrepresented in one condition compared to a second condition have been identified by fold-change. Although this method is still widely used, it might be biased in multiple ways. A high fold-change can be caused by a single flawed sample or by negligible differences in expression level just above the detection limit. In other words it ignores if the differences in expression change are statistically significant or not. Different methods have been proposed to assess the significance of changes in gene expression. The Student's t-test and Welch test are commonly used to estimate the significance of differential gene expression. However, the reliability of those methods strongly depends on the sample variance and the number of samples for each condition. Besides numerous statistical packages have been developed that account for differential gene expression, for example, SAM, EBAM, and so forth.

It also has been widely recognized that using different methods might result in rather distinct sets of differential expressed genes. We approached the problem by using the R MWT-package to determine significant differential gene expression [29]. The MWT method is essentially a moderated Welch test that aims to circumvent the problem of a low sample number by pooling the variance over the whole probe set. To adjust for multiple testing, all P values related to differential expression were corrected using the Benjamini-Hochberg method [30] that is rendered into false discovery rate (FDR) values.

2.4. Network Randomization. To determine the significance of the level of connectivity between a predefined set of genes and a second set (or itself) we used the CrossTalkZ network randomization package (http://sonnhammer.sbc.su.se/download/software/CrossTalkZ/). The method compares the number of observed connections between two gene sets to the number of connections in a randomized version of the network. In the course of network randomization, links between genes are swapped so that the original connectivity of a gene is conserved. The randomization was repeated 100 times, and all results were averaged. For each gene set a number of statistics were calculated including a z-score, a P value, and a Benjamini-Hochberg corrected FDR value.

2.5. Functional Gene Modules.

2.5. Functional Gene Modules. To identify gene modules that are relevant to different developmental stages and sexes we compiled for each condition networks of male or female-biased genes separately. In addition these networks contained other genes strongly connected to those sex-biased genes. We used the hypergeometric test to identify such genes, and genes with a Bonferroni corrected P value of less than 10% were included in such networks.

A large number of network clustering techniques exist to infer modules, but it is not obvious which ones are most robust, that is, perform well under many different circumstances. From a benchmark study of 8 popular methods we selected the two overall top performing methods, MGclus (http://sonnhammer.sbc.su.se/download/software/MGclus/) and MCL [31]. The latter was used with an inflation parameter of 3.5. The significance of the derived modules was evaluated by comparing the number of enriched GO terms per module to the expected number of enriched GO terms given a set of genes of that size. The expected number of GO terms per module was estimated by 500 times randomly picking n genes from the parental subnetwork, where n equals the number of genes for a module. Based on the distributions of the expected numbers of enriched GO terms, a z-score was calculated for enrichment of GO terms per clustering.

3. Results

3.1. The Chicken Network. With the FunCoup tool and dataset collection, we derived a global chicken gene interaction network. FunCoup can be used to determine confidence values regarding the value of observed functional coupling links, and the chicken network has roughly 1.8 million links at a confidence cutoff $(c) > 0.02$ and about 58,000 at $c > 0.75$ (Table 1). The network was trained on three different categories: metabolic, signaling, and both metabolic and signaling combined. In the following we used a $c > 0.25$ as it represents a reasonable tradeoff between accuracy and coverage.

The proteins with the highest connectivity are mainly related to fundamental cellular processes such as protein synthesis and degradation, translation, and transcription (Table S2). Many of them are involved in multiple processes. The most connected protein in our chicken network is the RA-related nuclear protein (RAN). Due to its various functions in nuclear transport and cell cycle regulation, it acts as a major hub with a host of other proteins. Interestingly, RAN is highly differentially expressed between male and female chicken (i.e., sex biased) in the gonad (FDR $P < 10^{-4}$ in the adult), which is actually less common for hubs as we show in the following.

3.2. Sex Bias Depends on Network Connectivity. Is there a dependency between sex bias and network connectivity? To answer this question, we first grouped the genes in three sex bias categories: male biased, female biased, and unbiased. For this we used the MWT statistic of differential expression with an FDR P value cutoff of 0.1. This was done for all four tissue/stage conditions: the embryonic and adult gonad and brain. The number of sex-biased genes in the network for each category is shown in Table 2. Remarkably, the embryonic brain contained almost no sex-biased genes and was therefore left out of this analysis. The adult brain had more sex-biased genes, but these still represented only 3% of the genes in the network. In contrast, the gonad abounded with sex-biased genes in the network: 43% in the embryo and 82% in the adult.

Sex-biased hub genes were thus frequent in the gonad, but not in the brain, and this may be due to the fact that the male and female gonads have extensive sex-specific functions, while the brain consists of many different tissues of which only small fractions of our microarray samples may be affected by the sex. The sex-specific expression signal in the brain will therefore be diluted by the nonaffected tissues until it is no longer statistically significant. Finer-scale analysis of specific brain tissues might reveal more dimorphism in gene expression, particularly those regions related to vocalization differences between male and female birds [32] or reproductive behavior [33].

We calculated Spearman's rank correlation coefficient between FDR values from differential expression analysis and node degree (i.e., the number of connections a gene has in the network), for each tissue/stage combination. As can be seen in Table 3, all but one of the sex-biased categories in the gonads had a significant positive correlation at FDR $P < 0.1$, indicating a tendency for fewer network connections as sex-bias increased. The exception was male biased genes in the adult gonad, but when lowering the cutoff to 0.001 these gave a weak but significant positive correlation ($r = 0.1, P < 0.05$). Unbiased genes were not significantly correlated with connectivity, nor were the brain genes, as may be expected given the dilution problem of brain expression mentioned above. This trend also held true when using fold-change as a measure for sex-bias. In other words, sex biased pathways seemed to generally affect local components of the network, except for the ones overexpressed in the male adult gonad, which tends to act more often in global components.

3.3. Sex-Biased Hub Genes. From the previous section it is clear that the level of sex was a function of both by tissue/stage condition as well as connectivity. We demonstrated that low connectivity genes tend to be more sex biased than high connectivity genes, yet some hub genes have strong sex bias. To focus on such sex-biased hubs, we first ranked each gene according to sex bias or connectivity separately and then reranked them according to the sum of both ranks. The highest ranked genes thus represent the most sex-biased hub genes. Table S3 shows the twenty top ranked sex-biased hubs in each condition.

TABLE 1: Number of links (first number) and unique genes (second number) at different FBS (final Bayesian score) cutoffs in FunCoup, where c is the corresponding confidence value of functional coupling.

	Metabolic	Signaling	Combined	Total
FBS > 3 (c > 0.02)	1375931/10555	601101/10569	1152763/10549	1809810/11311
FBS > 5.9 (c > 0.25)	171490/5520	33934/4861	124616/5383	199120/6748
FBS > 7 (c > 0.5)	89818/4132	13401/2990	62707/3885	100887/4902
FBS > 8 (c > 0.75)	52285/3175	6365/1869	35821/2930	57690/3673

TABLE 2: Number of sex-biased and unbiased genes separated by the cut-off FDR < 0.1 according to MWT.

	Male	Female	Unbiased
Embryonic gonad	1304	1128	3244
Adult gonad	1934	2693	1049
Embryonic brain	0	2	5675
Adult brain	87	92	5497

TABLE 3: Sex-biased genes tend not to be hubs. This is evidenced by Spearman's correlation coefficient between differential expression (measured as MWT FDR) and network connectivity which was significantly positive in most cases. Sex-biased genes were separated from unbiased genes using a cutoff of FDR = 0.1. The first number is the correlation, and in brackets is the corresponding P value. Significant correlations ($P < 0.01$) are marked in bold.

	Male	Female	Unbiased
Embryonic, gonad	**0.12 (7.98e − 06)**	**0.16 (1.55e − 07)**	−0.03 (0.12)
Adult, gonad	−0.04 (0.10)	**0.06 (9.91e − 04)**	−0.003 (0.93)
Adult, brain	0.12 (0.29)	−0.06 (0.60)	0.03 (0.02)

Among the highly connected hubs in Table S3, tubulin alpha-3e (TUBA3E), ranked first and second in the embryonic and adult gonads, has 426 links. It is female biased in the embryonic gonad but male biased in the adult gonad. Other highly connected tubulins are also in the list. This indicates that sexual differentiation and sex-specific function are partly orchestrated via sex-specific tubulin assembly. Some of the proteins in Table S3 are directly implicated in sex determination, for example, the testis-specific tubulin alpha-2 (TUBAL2, connectivity 94), the meiotic recombination SPO11 (connectivity 37), or the NASP the nuclear autoantigenic sperm protein (connectivity 97). A major hub in the embryonic as well as the adult gonad is CDK3, cell division protein kinase 3 (connectivity 189). CDK3 is further linked to the KEGG pathways oocyte meiosis as well as progesterone-mediated oocyte maturation. Intriguingly, CDK3 was strongly female biased in the embryonic gonad but strongly male biased in the adult gonad. Overall this suggests that a major difference between the sexes results from a complex interplay between components of the cell division and development systems.

3.4. Interconnectivity of Sex-Biased Genes. To answer the question if sex-biased genes are stronger connected to genes of the same bias, we compared the connection frequencies between the different categories to what is expected by chance alone. For topology-preserving randomization of the network we used the CrossTalkZ program (see Section 2) and performed 100 randomization runs. The results for the embryonic and adult gonads are shown in Figure 1.

In the gonad we found genes of the same sex bias (e.g., male versus male) to be more frequently connected to each other than to genes of a different sex bias or unbiased genes. It is striking that both in the embryonic and adult gonads, male-biased genes have significantly fewer connections to female-biased genes than expected by chance. In the brain, we did not observe a significant crosstalk between male- or female-biased genes, probably due to the dilution problem mentioned above. Separate female- and male-specific networks are thus common throughout the chicken network in the gonad, and these sex-specific networks function to encode dimorphic processes in this tissue.

Sex-biased genes on the Z-chromosome are shown as separate nodes in Figure 1. The Z chromosome had to be treated separately from the autosomes due to the lack of complete dosage compensation in birds which results in a pervasive male bias for nondosage sensitive genes [34]. Sex-biased genes on the Z-chromosome are shown as separate nodes in Figure 1. Genes of the same sex bias located on the Z-chromosome were more connected to each other than expected by chance and were significantly enriched in links to genes of the same sex bias on other chromosomes. Connections between female and male genes on the Z-chromosome were about as frequent as expected, but there were significantly fewer connections than expected between whole-genome male and Z-chromosome female-biased genes and vice versa. These results show that the reconstructed chicken network is largely made up of male-specific and female-specific modules.

TABLE 4: Results of the inparalog group analysis showing the number of groups in the various categories. In total we found 69 groups with at least two inparalogs in chicken. However, only 59 could be processed since expression data were not available for all genes in ten of the groups. The number in brackets in the mixed cluster field is the number of groups that contain both male- and female-biased genes. A significant difference between the observed number of inparalogs and what is expected by chance is indicated by *($P < 0.05$), + indicates a number higher than expected by chance, and − a number lower than expected.

	Gonad adult	Gonad embryo	Brain adult
All-male	15+*	7+*	0
All-female	18+*	8+*	0
All-unbiased	6+*	25	58+
Mixed	20 (5)−*	19 (1)−*	1 (0)−

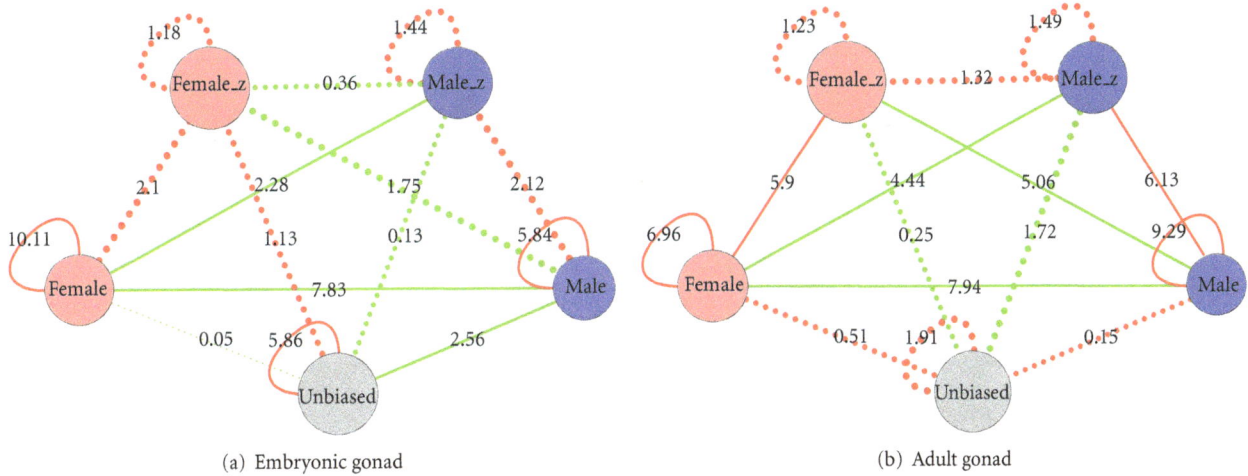

(a) Embryonic gonad

(b) Adult gonad

FIGURE 1: Network of crosstalk, that is, enrichment or depletion of links, between sex-biased and unbiased genes. Positive crosstalk (i.e., enrichment of links) is shown in red and depletion in green. Solid lines indicate significant crosstalk with FDR < 0.05. Edge width and label show the z-score of the crosstalk analysis.

3.5. Duplicated Sex-Biased Genes. Gene duplication is a mechanism for creating new functions, and such a functional niche could be associated with a particular sex bias. Previous work has shown that duplicates of unbiased genes often develop sex-biased expression [35]. However, it is not yet clear if sex-biased genes that were recently duplicated tend to maintain the same pattern of expression bias. To answer this question, we restricted the analysis to orthologs. Orthologous genes are known to retain identical or closely related biological function more often than other types of homologs [36–39]. Two genes in one species are considered as inparalogs with respect to another species if the gene duplication occurred after the respective speciation event. In order to clarify if inparalog genes in chicken would more often have the same sex bias or are biased towards the opposite sex, we selected all inparalogs between chicken and human from the InParanoid database [27]. An ortholog group was only analyzed if it had at least two alternative inparalogs in chicken and if expression data were available. Roughly half of the groups could thus be analyzed. The group was then evaluated for differentially expressed genes using a FDR cutoff of 10%.

The results of this analysis can be seen in Table 4 (and Table S4). In the gonad, the numbers of male- and female-biased groups were similar, while in the brain none of the groups were biased towards one of the sexes. A big

fraction of the groups is however a mix between sex-biased and unbiased genes. Remarkably, five ortholog groups in the adult gonad contained both male- and female-biased genes. An example of such a group contained female-biased glutathione S-transferase 2 (GSTA2) and male-biased glutathione S-transferase 3 (GSTA3). These inparalogs were connected to each other in the FunCoup network as well as to a set of other sex-biased genes (see Figure 2). However, 75% of GSTA2 links and 48% of GSTA3 links were not in common. At a cutoff of 0.5 only 2 links were shared between the two genes. It thus represents a likely example of subfunctionalization driven by sex differentiation.

To evaluate significance of these findings we compared the obtained numbers of inparalogs with the same bias to the distribution expected by chance (see Table S4). To this end, we randomly sampled genes of each ortholog group from the complete expression dataset. This procedure was repeated 1000 times, and the obtained numbers of groups with the same sex bias were compared to the observed values. For both the embryonic and adult gonadsthe original number of inparalogs with the same sex bias significantly exceeded the number of what would be expected by chance alone ($P < 0.05$). In the brain however there was no clear trend. We conclude that inparalogs that emerged after the mammal-bird speciation generally preserved sex bias, although a few exceptions exist.

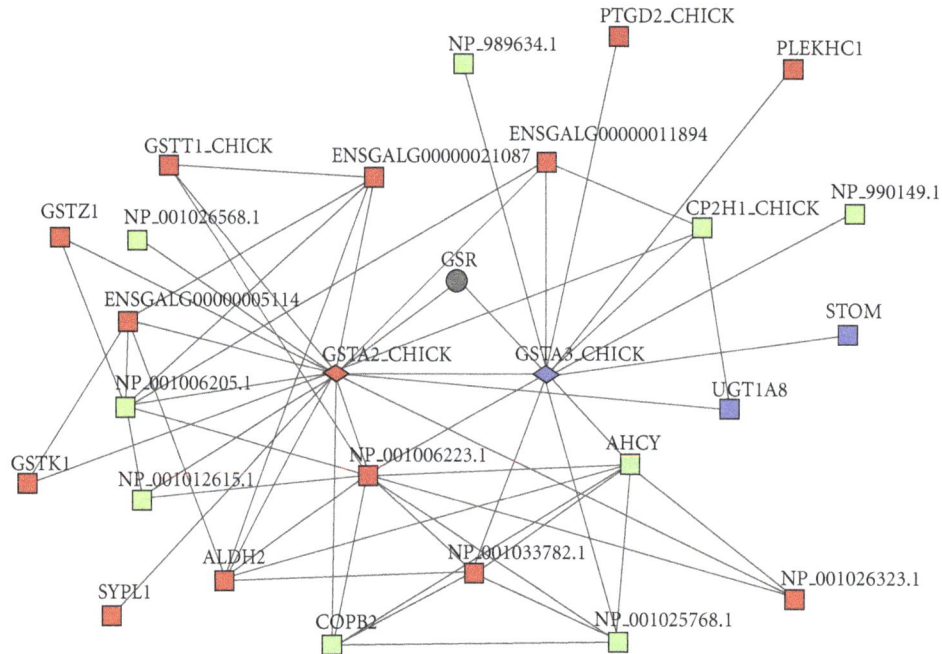

FIGURE 2: Example of sex-differentiation-driven subfunctionalization. The chicken genes GSTA2 and GSTA3 (glutathione S-transferases 2 and 3; shown as diamonds) originate from a duplication that happened after the divergence from human, making them inparalogs. GSTA2 is male biased, but GSTA3 is female biased in the adult gonad. Their interaction partners in the chicken network are shown with sex bias. Male-biased genes are shown in blue, female-biased in red, unbiased in green, and unknown in grey.

3.6. Sex-Biased Network Modules.

Network modules, or clusters, can be useful to find groups of functionally related genes. Such modules may represent parts of pathways or complexes that can be discerned as cliques of genes that are strongly linked to each other in the network. To identify functional modules of sex-biased genes, we calculated for each condition a set of male- or female-biased modules. We derived a network of sex-biased genes as well as genes which were strongly connected to them for each condition (see Section 2). Different clustering methods have different advantages and disadvantages and might as well result in relatively different sets of modules. We used two different methods to derive functional modules, MCL and MGclus. In the following we contrast the results of both methods as well as discuss the significance of the derived modules based on a few selected examples.

MCL is a global clustering approach that simulates random walks in the underlying interaction network. MGclus tries to identify clusters of strongly mutually linked genes using a scoring function that additionally accounts for shared neighbors. Thus nodes in the same cluster are thus likely to share a large fraction of shared neighbors, which increases cohesiveness within the cluster.

The overall outcomes of the MGclus and MCL clusterings are shown in Table 5. In all of the cases the clusters were significant, that is, had at least one enriched GO term, which was assigned to more than one gene in the cluster. Further, in all cases except for the adult brain, all clusters had on average significantly more enriched GO terms than random modules of the same size. The adult brain might however have too few sex-biased genes to see this. It is also worth noting that

the MGclus clusters had on average more enriched GO terms than the MCL clusters.

How different are MCL and MGclus clusters? The overlap strongly depended on the size of the input network. While the overlap was notable for smaller networks (e.g., the embryonic gonad or brain), it was limited for larger networks. To illustrate the overlap between the different clusterings we calculated UPGMA trees based on the fraction of the intersection relative to the union (Jaccard index) of genes in MCL and MGclus clusters. The same was done for enriched GO terms. For the male adult gonad, a few MGclus and MCL clusters overlap to a high degree, but most of them do not have a counterpart with more than 30% overlap (Figure S2). On the other hand, for the male embryonic gonad, which had much fewer differential expressed genes, most of them find a counterpart with more than 60% overlap (Figure S3), indicating that these clusters are relatively reliable. Unsurprisingly, gene and GO term overlap trees were very similar.

One module identified from the embryonic gonad contained eight male-biased genes and one female-biased gene (Figure 3(a)). The female-biased gene was included because it was significantly enriched in connections to the male-biased genes. This module was functionally related to cell growth and development. It contained eight enzymes with biosynthetic functions and one extracellular matrix protein Tenascin (TENA_CHICK) which is important in tissue development. Two of the enzymes (AL1A1_CHICK, ADH1_CHICK) are important for retinoic acid (RA) synthesis [40]. RA is known to be crucial for embryonic development, growth, and reproduction. Four of the genes

TABLE 5: Number of MGclus and MCL clusters, number of clusters with significant GO term enrichment, and the level of significant GO term enrichment compared to random. A z-score above 2 corresponds to a significance level of $P < 0.05$.

	Clusters	Significant	Avg. sig. terms	Random avg. sig. terms
MGclus				
Gonad embryo male	6	6	43.50	18.72
Gonad adult male	24	24	20.58	9.56
Brain adult male	3	3	16.67	13.01
Gonad embryo female	6	6	31.17	16.09
Gonad adult female	22	22	62.23	23.58
Brain adult female	3	3	25.67	20.94
MCL				
Gonad embryo male	5	5	30.80	16.87
Gonad adult male	31	29	18.52	6.92
Brain adult male	6	6	9.00	6.94
Gonad embryo female	3	3	16.67	10.85
Gonad adult female	64	64	27.97	10.62
Brain adult female	3	3	24.33	16.42

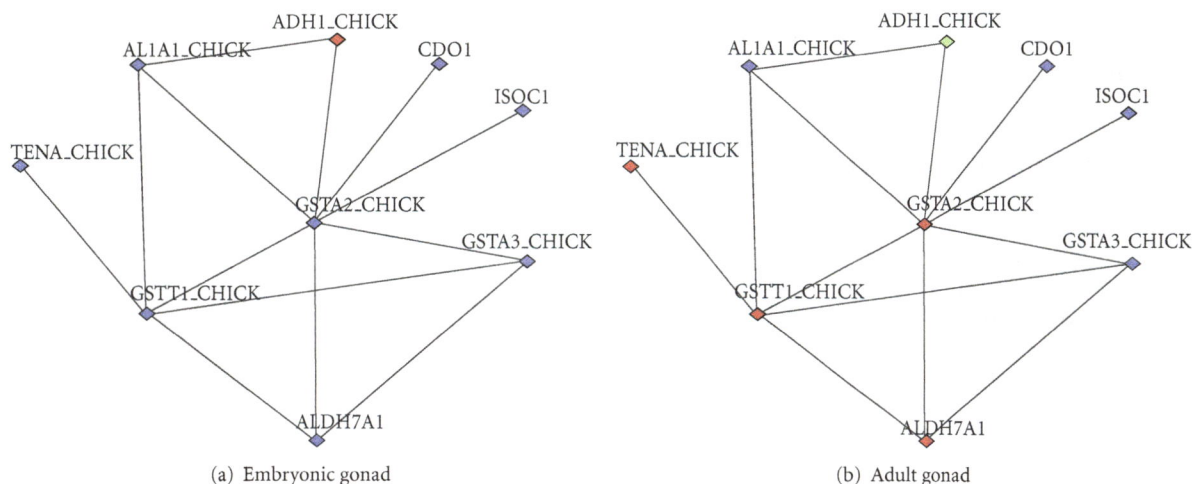

(a) Embryonic gonad   (b) Adult gonad

FIGURE 3: Example of sex bias switching between developmental stages. Shown is an MGclus cluster colored according to sex bias in the embryonic (a) and adult (b) gonads. Male-biased genes are shown in blue, female-biased in red, and unbiased in green.

change sex bias and become female-biased in the adult gonad (Figure 3(b)), indicating that this module switches its function during development depending on the sex.

4. Discussion

By analyzing sex bias within the chicken gene network, we have been able to deduce several network properties pertaining to sexual dimorphism that gives new biological insights. Our analysis suggests that network hub genes tend not to be sex biased, although with some interesting exceptions. This suggests that most sex-biased genes tend to act within local network environments, and relatively few of them interact on a more global scale. This is consistent with recent studies that show that pleiotropy, as measured by expression breadth, tends to constrain the evolution of sex-biased expression [41, 42]. This analysis extends the measure of pleiotropy to network connectivity, with broadly consistent results.

We also investigated the propensity of sex-biased genes to form network modules in several ways. First, we noted that genes of the same sex bias tend to be more connected to each other than expected. Second, recently duplicated genes, which are similar in biochemical function, tend to have the same sex bias. Finally, a set of sex-biased modules were extracted from the network, and these showed unexpected functional homogeneity. These observations support a network structure that embodies sex-biased network modules. The implication of this is that the mechanisms underlying sex-specific development can be organized according to these modules, which simplifies the study and understanding of this complex system.

This work provides the first integrated, multidimensional analysis of the network structure underlying sex-biased gene expression and, as such, offers a more realistic link between sex-biased gene expression and sexually dimorphic phenotypes. Our analysis suggests, that rather than operating as distinct entities, genes of the same sex bias often group together in network modules, potentially due to shared regulatory elements or hierarchical pathway structures. This has several evolutionary genetic implications. First, it suggests that when many genes act in concert to encode sexually dimorphic phenotypes, they may be controlled by a shared regulatory apparatus. This collective regulatory control could then be exploited by emergent sexual dimorphisms, resulting in associated phenotypic differences [43]. It also suggests that single- or oligolocus models of sexual selection evolution (e.g., [44, 45]) are appropriate for some sexually dimorphic traits, even when transcriptome analysis reveals that gene expression of those phenotypes differs for many genes between the sexes. Although genes do not operate as independent units but are rather tethered in modules in a complex network of interactions, they however often work in concerted regulatory patterns. Therefore, our analysis somewhat paradoxically suggests that the control of complex sexual dimorphism may be ultimately attributable to relatively few key regulators.

Sex chromosomes often exhibit a nonrandom distribution of sex-biased genes associated with masculinizing or feminizing selection [46, 47]. Additionally, female heterogametic sex chromosomes, including those exhibited by birds, are also predicted to be particularly associated with the evolution of certain types of sexually selected traits [45, 48, 49]. Our analysis is consistent with these predictions. The crosstalk observed in the adult gonad between sex-biased genes on the Z chromosome and sex-biased genes on the autosomes suggests that the Z chromosome, which contains a relatively modest proportion of the total avian coding content, may play a disproportionately large role in the regulation of sex-biased genes.

Previous work has shown a nonrandom distribution of sex-biased genes on the avian Z chromosome [50–52], with more male-biased and fewer female-biased genes on the Z chromosome than would be expected by chance alone. However this issue is complicated by the incomplete dosage compensation observed on the avian Z chromosome. Studies in a range of bird species have shown a persistent male bias on the Z chromosome due to the fact that males have two copies of every locus and females just one [34, 53, 54]. It has therefore been difficult to disentangle the effects of masculinizing selection for gene expression from incomplete dosage compensation [18]. Our analysis does not suffer from this type of conflation, as the crosstalk enrichment takes the relative abundances of different biases into account. This should minimize any effects of incomplete dosage compensation on our network.

In conclusion, our results suggest that network approaches to the study of sex-biased gene expression can offer new insights into the programming and genetic basis of sexual differentiation. Current transcriptome profiling produces massive datasets measuring relative gene expression, but this approach alone results in the false perception that each locus is independent of all others. Gene network approaches such as the one described here make it possible to consider a more multidimensional and integrated view of genome regulation which is particularly insightful for complex phenotypes.

Acknowledgments

O. Frings was supported by a grant from the Swedish Research Council. J. E. Mank is supported by the BBSRC and ERC (Grant AGREEMENT 260233).

References

[1] H. Ellegren and J. Parsch, "The evolution of sex-biased genes and sex-biased gene expression," *Nature Reviews Genetics*, vol. 8, no. 9, pp. 689–698, 2007.

[2] J. E. Mank, "Sex chromosomes and the evolution of sexual dimorphism: lessons from the genome," *The American Naturalist*, vol. 173, no. 2, pp. 141–150, 2009.

[3] J. M. Ranz, C. I. Castillo-Davis, C. D. Meiklejohn, and D. L. Hartl, "Sex-dependent gene expression and evolution of the Drosophila transcriptome," *Science*, vol. 300, no. 5626, pp. 1742–1745, 2003.

[4] P. Khaitovich, I. Hellmann, W. Enard et al., "Evolution: parallel patterns of evolution in the genomes and transcriptomes of humans and chimpanzees," *Science*, vol. 309, no. 5742, pp. 1850–1854, 2005.

[5] X. Yang, E. E. Schadt, S. Wang et al., "Tissue-specific expression and regulation of sexually dimorphic genes in mice," *Genome Research*, vol. 16, no. 8, pp. 995–1004, 2006.

[6] Y. Zhang, D. Sturgill, M. Parisi, S. Kumar, and B. Oliver, "Constraint and turnover in sex-biased gene expression in the genus Drosophila," *Nature*, vol. 450, no. 7167, pp. 233–237, 2007.

[7] B. Reinius, P. Saetre, J. A. Leonard et al., "An evolutionarily conserved sexual signature in the primate brain," *PLoS Genetics*, vol. 4, no. 6, Article ID e1000100, 2008.

[8] T. Connallon and L. L. Knowles, "Intergenomic conflict revealed by patterns of sex-biased gene expression," *Trends in Genetics*, vol. 21, no. 9, pp. 495–499, 2005.

[9] J. E. Mank and H. Ellegren, "Are sex-biased genes more dispensable?" *Biology Letters*, vol. 5, no. 3, pp. 409–412, 2009.

[10] C. D. Meiklejohn, J. Parsch, J. M. Ranz, and D. L. Hartl, "Rapid evolution of male-biased gene expression in Drosophila," *Proceedings of the National Academy of Sciences of the United States of America*, vol. 100, no. 17, pp. 9894–9899, 2003.

[11] M. Parisi, R. Nuttall, D. Naiman et al., "Paucity of genes on the Drosophila X chromosome showing male-biased expression," *Science*, vol. 299, no. 5607, pp. 697–700, 2003.

[12] P. P. Khil, N. A. Smirnova, P. J. Romanienko, and R. D. Camerini-Otero, "The mouse X chromosome is enriched for sex-biased genes not subject to selection by meiotic sex chromosome inactivation," *Nature Genetics*, vol. 36, no. 6, pp. 642–646, 2004.

[13] V. Reinke, I. S. Gil, S. Ward, and K. Kazmer, "Genome-wide germline-enriched and sex-biased expression profiles in Caenorhabditis elegans," *Development*, vol. 131, no. 2, pp. 311–323, 2004.

[14] Z. Zhang, T. M. Hambuch, and J. Parsch, "Molecular evolution of sex-biased genes in Drosophila," *Molecular Biology and Evolution*, vol. 21, no. 11, pp. 2130–2139, 2004.

[15] M. Pröschel, Z. Zhang, and J. Parsch, "Widespread adaptive evolution of Drosophila genes with sex-biased expression," *Genetics*, vol. 174, no. 2, pp. 893–900, 2006.

[16] J. E. Mank, L. Hultin-Rosenberg, E. Axelsson, and H. Ellegren, "Rapid evolution of female-biased, but not male-biased, genes expressed in the avian brain," *Molecular Biology and Evolution*, vol. 24, no. 12, pp. 2698–2706, 2007.

[17] D. Sturgill, Y. Zhang, M. Parisi, and B. Oliver, "Demasculin-ization of X chromosomes in the Drosophila genus," *Nature*, vol. 450, no. 7167, pp. 238–241, 2007.

[18] J. E. Mank and H. Ellegren, "Sex-linkage of sexually antagonis-tic genes is predicted by female, but not male, effects in birds," *Evolution*, vol. 63, no. 6, pp. 1464–1472, 2009.

[19] A. L. Ducrest, L. Keller, and A. Roulin, "Pleiotropy in the melanocortin system, coloration and behavioural syndromes," *Trends in Ecology & Evolution*, vol. 23, no. 9, pp. 502–510, 2008.

[20] D. Wright, S. Kerje, H. Brändström et al., "The genetic architecture of a female sexual ornament," *Evolution*, vol. 62, no. 1, pp. 86–98, 2008.

[21] J. F. Ayroles, M. A. Carbone, E. A. Stone et al., "Systems genetics of complex traits in Drosophila melanogaster," *Nature Genetics*, vol. 41, no. 3, pp. 299–307, 2009.

[22] M. J. Fitzpatrick, "Pleiotropy and the genomic location of sexually selected genes," *The American Naturalist*, vol. 163, no. 6, pp. 800–808, 2004.

[23] J. E. Mank, L. Hultin-Rosenberg, M. T. Webster, and H. Ellegren, "The unique genomic properties of sex-biased genes: insights from avian microarray data," *BMC Genomics*, vol. 9, article 148, 2008.

[24] J. E. Mank, K. Nam, B. Brunström, and H. Ellegren, "Onto-genetic complexity of sexual dimorphism and sex-specific selection," *Molecular Biology and Evolution*, vol. 27, no. 7, pp. 1570–1578, 2010.

[25] A. Alexeyenko and E. L. L. Sonnhammer, "Global networks of functional coupling in eukaryotes from comprehensive data integration," *Genome Research*, vol. 19, no. 6, pp. 1107–1116, 2009.

[26] A. Alexeyenko, T. Schmitt, A. Tjarnberg, D. Guala, and O. Frings, "Comparative interactomics with Funcoup 2.0," *Nucleic Acids Research*, vol. 40, no. 1, pp. D821–D828, 2012.

[27] G. Östlund, T. Schmitt, K. Forslund et al., "Inparanoid 7: new algorithms and tools for eukaryotic orthology analysis," *Nucleic Acids Research*, vol. 38, no. 1, Article ID gkp931, pp. D196–D203, 2009.

[28] J. E. Mank and H. Ellegren, "All dosage compensation is local: gene-by-gene regulation of sex-biased expression on the chicken Z chromosome," *Heredity*, vol. 102, no. 3, pp. 312–320, 2009.

[29] M. Demissie, M. Mascialino, S. Calza, and Y. Pawitan, "Unequal group variances in microarray data analyses," *Bioinformatics*, vol. 24, no. 9, pp. 1168–1174, 2008.

[30] Y. Benjamini and Y. Hochberg, "Controlling the false discovery rate: a practical and powerful approach to multiple testing," *Journal of the Royal Statistical Society. Series B*, vol. 57, no. 1, pp. 289–300, 1995.

[31] A. J. Enright, S. Van Dongen, and C. A. Ouzounis, "An efficient algorithm for large-scale detection of protein families," *Nucleic Acids Research*, vol. 30, no. 7, pp. 1575–1584, 2002.

[32] C. V. Mello, D. S. Vicario, and D. F. Clayton, "Song presen-tation induces gene expression in the songbird forebrain," *Proceedings of the National Academy of Sciences of the United States of America*, vol. 89, no. 15, pp. 6818–6822, 1992.

[33] D. S. Manoli, M. Foss, A. Villella, B. J. Taylor, J. C. Hall, and B. S. Baker, "Male-specific fruitless specifies the neural substrates of Drosophila courtship behaviour," *Nature*, vol. 436, no. 7049, pp. 395–400, 2005.

[34] Y. Itoh, E. Melamed, X. Yang et al., "Dosage compensation is less effective in birds than in mammals," *Journal of Biology*, vol. 6, no. 1, article 2, 2007.

[35] M. Gallach and E. Betrán, "Intralocus sexual conflict resolved through gene duplication," *Trends in Ecology & Evolution*, vol. 26, no. 5, pp. 222–228, 2011.

[36] M. E. Peterson, F. Chen, J. G. Saven, D. S. Roos, P. C. Babbitt, and A. Sali, "Evolutionary constraints on structural similarity in orthologs and paralogs," *Protein Science*, vol. 18, no. 6, pp. 1306–1315, 2009.

[37] A. Henricson, K. Forslund, and E. L. L. Sonnhammer, "Orthology confers intron position conservation," *BMC Genomics*, vol. 11, no. 1, article 412, 2010.

[38] J. Huerta-Cepas, J. Dopazo, M. A. Huynen, and T. Gabaldón, "Evidence for short-time divergence and long-time conser-vation of tissue-specific expression after gene duplication," *Briefings in Bioinformatics*, vol. 12, no. 5, pp. 442–448, 2011.

[39] S. Movahedi, Y. van De Peer, and K. Vandepoele, "Com-parative network analysis reveals that tissue specificity and gene function are important factors influencing the mode of expression evolution in arabidopsis and rice," *Plant Physiology*, vol. 156, no. 3, pp. 1316–1330, 2011.

[40] A. Molotkov and G. Duester, "Genetic evidence that retinalde-hyde dehydrogenase Raldh1 (Aldh1a1) functions downstream of alcohol dehydrogenase Adh1 in metabolism of retinol to retinoic acid," *Journal of Biological Chemistry*, vol. 278, no. 38, pp. 36085–36090, 2003.

[41] J. E. Mank, L. Hultin-Rosenberg, M. Zwahlen, and H. Ellegren, "Pleiotropic constraint hampers the resolution of sexual antagonism in vertebrate gene expression," *The American Naturalist*, vol. 171, no. 1, pp. 35–43, 2008.

[42] R. P. Meisel, "Towards a more nuanced understanding of the relationship between sex-biased gene expression and rates of protein-coding sequence evolution," *Molecular Biology and Evolution*, vol. 28, no. 6, pp. 1893–1900, 2011.

[43] S. Fukamachi, M. Kinoshita, K. Aizawa, S. Oda, A. Meyer, and H. Mitani, "Dual control by a single gene of secondary sexual characters and mating preferences in medaka," *BMC Biology*, vol. 7, article 1741, p. 64, 2009.

[44] R. Lande and S. J. Arnold, "Evolution of mating preference and sexual dimorphism," *Journal of Theoretical Biology*, vol. 117, no. 4, pp. 651–664, 1985.

[45] M. Kirkpatrick and D. W. Hall, "Sexual selection and sex linkage," *Evolution*, vol. 58, no. 4, pp. 683–691, 2004.

[46] B. Vicoso and B. Charlesworth, "Evolution on the X chro-mosome: unusual patterns and processes," *Nature Reviews Genetics*, vol. 7, no. 8, pp. 645–653, 2006.

[47] D. Bachtrog, M. Kirkpatrick, J. E. Mank, S. F. McDaniel, J. C. Pires, and W. Rice, "Are all sex chromosomes created equal?" *Trends in Genetics*, vol. 27, no. 9, pp. 350–357, 2011.

[48] W. R. Rice, "Sex chromosomes and the evolution of sexual dimorphism," *Evolution*, vol. 38, no. 4, pp. 735–742, 1984.

[49] A. Y. K. Albert and S. P. Otto, "Evolution: sexual selection can resolve sex-linked sexual antagonism," *Science*, vol. 310, no. 5745, pp. 119–121, 2005.

[50] R. Storchová and P. Divina, "Nonrandom representation of sex-biased genes on chicken Z chromosome," *Journal of Molecular Evolution*, vol. 63, no. 5, pp. 676–681, 2006.

[51] V. B. Kaiser, M. Van Tuinen, and H. Ellegren, "Insertion events of CR1 retrotransposable elements elucidate the phylogenetic branching order in galliform birds," *Molecular Biology and Evolution*, vol. 24, no. 1, pp. 338–347, 2007.

[52] H. Ellegren, "Emergence of male-biased genes on the chicken Z-chromosome: sex-chromosome contrasts between male and female heterogametic systems," *Genome Research*, vol. 21, no. 12, pp. 2082–2086, 2011.

[53] S. Naurin, B. Hansson, D. Hasselquist, Y.-H. Kim, and S. Bensch, "The sex-biased brain: sexual dimorphism in gene expression in two species of songbirds," *BMC Genomics*, vol. 12, article 37, 2011.

[54] J. B. W. Wolf and J. Bryk, "General lack of global dosage compensation in ZZ/ZW systems? Broadening the perspective with RNA-seq," *BMC Genomics*, vol. 12, article 91, 2011.

2

Conservation of Nucleosome Positions in Duplicated and Orthologous Gene Pairs

Hiromi Nishida

Agricultural Bioinformatics Research Unit, Graduate School of Agriculture and Life Sciences, The University of Tokyo, Bunkyo-ku, Tokyo 113-8657, Japan

Correspondence should be addressed to Hiromi Nishida, hnishida@iu.a.u-tokyo.ac.jp

Academic Editor: David E. Misek

Although nucleosome positions tend to be conserved in gene promoters, whether they are conserved in duplicated and orthologous genes is unknown. In order to elucidate how nucleosome positions are conserved between duplicated and orthologous gene pairs, I performed 2 comparative studies. First, I compared the nucleosome position profiles of duplicated genes in the filamentous ascomycete *Aspergillus fumigatus*. After identifying 63 duplicated gene pairs among 9630 protein-encoding genes, I compared the nucleosome position profiles of the paired genes. Although nucleosome positions are conserved more in gene promoters than in gene bodies, their profiles were diverse, suggesting evolutionary changes after gene duplication. Next, I examined the conservation of nucleosome position profiles in 347 *A. fumigatus* orthologs of *S. cerevisiae* genes that showed notably high conservation of nucleosome positions between the parent strain and 2 deletion mutants. In only 11 (3.2%) of the 347 gene pairs, the nucleosome position profile was highly conserved (Spearman's rank correlation coefficient > 0.7). The absence of nucleosome position conservation in promoters of orthologous genes suggests organismal specificity of nucleosome arrangements.

1. Introduction

Nucleosomes are histone octamers around which DNA is wrapped in 1.65 turns [1]. Neighboring nucleosomes are separated by unwrapped linker DNA. Nucleosome density is lower, and nucleosome position is more conserved in the promoters than in the bodies of genes [2–5]. It is thought that nucleosome positioning in the gene promoter plays an important role in transcriptional regulation.

Although nucleosome positions can be partially simulated using a DNA-sequence-based approach [6], these simulations are limited due to variations between species. The nucleosome positioning mechanism varies between the 2 ascomycetous yeasts, *Saccharomyces cerevisiae*, and *Schizosaccharomyces pombe* [7]. Nucleosome positioning differs even among phylogenetically close ascomycetous yeast species [5].

Gene duplication is a driving force behind gene creation, and generating novel functions in newly created genes. Approximately one-half of cellular functions have been gained through gene duplication [8]. The duplicated genes encode similar amino acid sequences and often similar protein functions. It is uncertain, however, whether duplicated genes have similar nucleosome position profiles. In this study, I compared nucleosome positions in the promoter and body regions of duplicated gene pairs in the filamentous ascomycete *Aspergillus fumigatus*.

Previous analyses have found that nucleosome positions in *A. fumigatus* are conserved more in gene promoters than in gene bodies, even after treatment with the histone deacetylase inhibitor trichostatin A [4, 9]. In addition, nucleosome positions in *S. cerevisiae* are more conserved in gene promoters than in gene bodies between the control and the histone acetyltransferase gene *ELP3* deletion mutant, and between the control and the histone deacetylase gene *HOS2* deletion mutant [10]. The proteins Elp3 and Hos2 show the highest and the third highest evolutionary conservation, respectively, among the fungal histone modification proteins [11].

How well are nucleosome positions conserved in genes of the same origins? If there is a "nucleosome position code" that regulates nucleosome positioning, common nucleosome

TABLE 1: Duplicated gene pairs in *Aspergillus fumigatus*.

Gene pair	Chromosome	Gene body region		Gene direction	Function
AFUA1G00150	1	25442	27017	+	RING finger protein
AFUA6G09370	6	2245549	2247121	+	RING finger protein
AFUA1G00420	1	135528	137781	+	Carboxypeptidase S1, putative
AFUA8G04120	8	897824	900076	−	Carboxypeptidase S1, putative
AFUA1G00440	1	138359	140093	−	DUF895 domain membrane protein
AFUA8G04110	8	895512	897246	+	DUF895 domain membrane protein
AFUA1G00450	1	143117	144466	+	N-acetylglucosamine-6-phosphate deacetylase(NagA), putative
AFUA8G04100	8	891135	892484	−	N-acetylglucosamine-6-phosphate deacetylase(NagA), putative
AFUA1G00470	1	148615	150219	+	Betaine aldehyde dehydrogenase, putative
AFUA8G04080	8	885376	886980	−	Betaine aldehyde dehydrogenase (BadH), putative
AFUA1G00530	1	164101	164771	−	Thermoresistant gluconokinase family protein
AFUA4G12050	4	3163747	3164530	−	Thermoresistant gluconokinase
AFUA1G00550	1	177114	178593	+	Hypothetical protein
AFUA1G00910	1	328785	330274	+	Hypothetical protein
AFUA1G00580	1	184790	186873	+	Acid phosphatase (PhoG), putative
AFUA8G04050	8	870757	872487	−	Acid phosphatase (PhoG), putative
AFUA1G00650	1	215584	217006	+	Alpha-1,3-glucanase, putative
AFUA7G08510	7	1973398	1974759	+	Alpha-1,3-glucanase, putative
AFUA1G00920	1	331676	332955	−	Hypothetical protein
AFUA3G06425	3	1582748	1584037	−	Hypothetical protein
AFUA1G01050	1	385302	386244	−	Hypothetical protein
AFUA8G06160	8	1465695	1466634	+	Hypothetical protein
AFUA1G02550	1	744583	746335	−	Tubulin alpha-1 subunit
AFUA2G14990	2	3947008	3948834	−	Tubulin alpha-2 subunit
AFUA1G02730	1	788249	789373	−	Mitochondrial phosphate carrier protein (Ptp), putative
AFUA1G15140	1	4070230	4071449	−	Mitochondrial phosphate carrier protein (Mir1), putative
AFUA1G05760	1	1658382	1659697	−	Arsenite efflux transporter
AFUA5G15010	5	3882425	3883746	+	Arsenite permease (ArsB), putative
AFUA1G05760	1	1658382	1659697	−	Arsenite efflux transporter
AFUA1G16100	1	4378898	4380216	+	Arsenite permease (ArsB), putative
AFUA1G10910	1	2848155	2850137	−	Tubulin beta, putative
AFUA7G00250	7	70221	71948	+	Tubulin beta-2 subunit
AFUA1G11260	1	2971529	2971889	−	Conserved hypothetical protein
AFUA6G00270	6	79542	79886	−	Conserved hypothetical protein
AFUA1G11610	1	3060058	3060510	+	3-Dehydroquinate dehydratase, type II
AFUA3G14850	3	3929721	3930173	+	3-Dehydroquinate dehydratase, type II
AFUA1G11890	1	3129447	3131489	+	Serine palmitoyltransferase 2, putative
AFUA6G00300	6	85851	87692	−	Serine palmitoyltransferase 1, putative
AFUA1G12850	1	3398837	3400611	+	Nitrate transporter (nitrate permease)
AFUA1G17470	1	4782320	4783995	+	High-affinity nitrate transporter NrtB
AFUA1G15970	1	4338407	4339712	+	Aldo-keto reductase (AKR13), putative
AFUA8G01560	8	401581	402815	−	Aldo-keto reductase (YakC), putative
AFUA1G16030	1	4358866	4360256	−	Conserved hypothetical protein
AFUA5G14930	5	3863218	3864521	−	Conserved hypothetical protein

TABLE 1: Continued.

Gene pair	Chromosome	Gene body region		Gene direction	Function
AFUA1G16040	1	4363493	4365310	+	Metalloreductase, putative
AFUA5G14940	5	3867797	3869542	+	Cell surface metalloreductase (FreA), putative
AFUA1G16050	1	4366573	4368232	+	Hypothetical protein
AFUA5G14950	5	3870635	3872465	+	Hypothetical protein
AFUA1G16070	1	4370346	4373694	+	Conserved hypothetical protein
AFUA5G14980	5	3874539	3877890	+	Conserved hypothetical protein
AFUA1G16080	1	4374579	4375298	−	Hypothetical protein
AFUA5G14990	5	3878797	3879579	−	Hypothetical protein
AFUA1G16090	1	4377842	4378249	−	Arsenate reductase (ArsC), putative
AFUA5G15000	5	3881386	3881835	−	Arsenate reductase (ArsC), putative
AFUA1G16100	1	4378898	4380216	+	Arsenite permease (ArsB), putative
AFUA5G15010	5	3882425	3883746	+	Arsenite permease (ArsB), putative
AFUA1G16110	1	4380474	4381455	−	Arsenic methyltransferase (Cyt19), putative
AFUA5G15020	5	3883929	3884993	−	Arsenic methyltransferase (Cyt19), putative
AFUA1G16120	1	4385650	4386671	−	Arsenic resistance protein (ArsH), putative
AFUA8G07150	8	1751693	1752688	+	ArsH protein
AFUA1G16120	1	4385650	4386671	−	Arsenic resistance protein (ArsH), putative
AFUA5G15030	5	3887287	3888230	−	Arsenic resistance protein (ArsH), putative
AFUA2G00800	2	178081	179459	−	PelA protein
AFUA5G10380	5	2658656	2659968	+	Pectin lyase, putative
AFUA2G00800	2	178081	179459	−	PelA protein
AFUA7G05030	7	1182104	1183794	−	Pectin lyase B
AFUA2G04010	2	1092973	1094679	−	Alpha, alpha-trehalose-phosphate synthase subunit, putative
AFUA6G12950	6	3268958	3270783	+	Alpha, alpha-trehalose-phosphate Synthase subunitTPS1, putative
AFUA2G11270	2	2897275	2904954	−	Alpha-1,3-glucan synthase, putative
AFUA3G00910	3	210186	217666	+	Alpha-1,3-glucan synthase, putative
AFUA3G00340	3	71186	72734	+	Glycosyl hydrolase, putative
AFUA4G02720	4	751306	752715	−	Glycosyl hydrolase, putative
AFUA3G00680	3	151397	153549	−	Copper amine oxidase
AFUA7G04180	7	943893	946136	+	Amine oxidase
AFUA3G01560	3	393032	394802	−	Aminoacid permease, putative
AFUA5G04260	5	1140953	1142714	+	Arginine transporter, putative
AFUA3G02420	3	597476	598286	+	ThiJ/PfpI family protein
AFUA4G01400	4	369123	369944	−	ThiJ/PfpI family protein
AFUA3G03080	3	824479	825425	+	Endo-1,3(4)-beta-glucanase, putative
AFUA6G14540	6	3702416	3703383	−	Endo-1,3(4)-beta-glucanase, putative
AFUA3G03980	3	1134609	1136404	+	Cytochrome P450 monooxygenase, putative
AFUA5G10050	5	2589301	2591081	−	Cytochrome P450 monooxygenase, putative
AFUA3G08160	3	2094223	2095759	−	Eukaryotic translation initiation Factor eIF4A, putative
AFUA5G02410	5	621886	623486	−	DEAD/DEAH box helicase, putative
AFUA3G14420	3	3831868	3834921	+	Chitin synthase G
AFUA5G00760	5	211013	213795	+	Chitin synthase C
AFUA4G00510	4	133477	135102	−	Hypothetical protein
AFUA7G08600	7	2009436	2011052	+	Hypothetical protein
AFUA4G03110	4	868474	870249	−	Monosaccharide transporter
AFUA5G10690	5	2737348	2739178	−	Monosaccharide transporter

TABLE 1: Continued.

Gene pair	Chromosome	Gene body region		Gene direction	Function
AFUA4G03680	4	1031586	1032539	–	Oxidoreductase, short-chain dehydrogenase/reductase family
AFUA6G03520	6	764308	765294	–	Short-chain dehydrogenase/reductase family protein, putative
AFUA4G09440	4	2462920	2466158	–	Sodium P-type ATPase, putative
AFUA6G03690	6	810027	813362	–	Sodium transport ATPase, putative
AFUA4G14360	4	3774307	3776166	+	Capsular associated protein, putative
AFUA5G07560	5	1889791	1891689	–	Capsular associated protein, putative
AFUA5G00145	5	15749	16327	–	Hypothetical protein
AFUA6G11710	6	2915903	2916483	+	Conserved hypothetical protein
AFUA5G00145	5	15749	16327	–	Hypothetical protein
AFUA7G08440	7	1942319	1942804	+	Hypothetical protein
AFUA5G01030	5	266294	267439	–	Glyceraldehyde 3-phosphate dehydrogenase(Ccg-7), putative
AFUA5G01970	5	503797	505194	+	Glyceraldehyde 3-phosphate dehydrogenase GpdA
AFUA5G06240	5	1494455	1495619	–	Alcohol dehydrogenase, putative
AFUA7G01010	7	270494	271675	–	Alcohol dehydrogenase, putative
AFUA5G07980	5	2019069	2020841	+	Hypothetical protein
AFUA5G14920	5	3857350	3859167	+	Hypothetical protein
AFUA5G09130	5	2345689	2346728	–	Polysaccharide deacetylase family protein
AFUA6G05030	6	1195846	1196956	+	Polysaccharide deacetylase family protein
AFUA5G15030	5	3887287	3888230	–	Arsenic resistance protein (ArsH), putative
AFUA8G07150	8	1751693	1752688	+	ArsH protein
AFUA6G06750	6	1475239	1476209	+	14-3-3 family protein
AFUA2G03290	2	867203	868250	–	14-3-3 family protein ArtA, putative
AFUA6G07070	6	1587643	1589124	+	Cellobiohydrolase D
AFUA6G11610	6	2878078	2879676	–	1,4-beta-D-glucan-cellobiohydrolyase, putative
AFUA6G11430	6	2837140	2839051	+	Aldehyde dehydrogenase, putative
AFUA7G01000	7	267518	269163	–	Aldehyde dehydrogenase, putative
AFUA6G11710	6	2915903	2916483	+	Conserved hypothetical protein
AFUA7G08440	7	1942319	1942804	+	Hypothetical protein
AFUA6G13490	6	3431912	3433612	–	Glutamate decarboxylase
AFUA8G06020	8	1428812	1430515	+	Glutamate decarboxylase
AFUA7G00360	7	102030	103064	–	UDP-galactose 4-epimerase, putative
AFUA8G00860	8	203496	204338	+	UDP-galactose 4-epimerase, putative
AFUA7G07050	7	1725686	1726402	–	Hypothetical protein
AFUA7G08300	7	1870244	1870960	–	Hypothetical protein
AFUA7G07060	7	1728876	1732992	–	Hypothetical protein
AFUA7G08310	7	1873437	1877817	–	Hypothetical protein

positions should remain in the promoters of orthologous genes across distinct species. In this study, I compared nucleosome positions in the promoters of duplicated and orthologous genes in *A. fumigatus* and *S. cerevisiae*.

2. Materials and Methods

2.1. Identification of Duplicated Gene Pairs in Aspergillus fumigatus. Protein-coding gene pairs aligned over more

than 80% of query length and more than 70% aminoacid sequence identity were selected by performing a BLAST search of 9630 *A. fumigatus* proteins at Fungal Genomes Central on NCBI (http://www.ncbi.nlm.nih.gov/projects/genome/guide/fungi/). Pairs in which the lengths differ by more than 25% were not used. Thus, we identified 63 duplicated *A. fumigatus* gene pairs (Tables 1 and 2).

2.2. Identification of Orthologous Gene Pairs in Aspergillus fumigatus and Saccharomyces cerevisiae. In a comparison

TABLE 2: Spearman's rank correlation coefficients of nucleosome position profiles in the promoter and body regions of 63 duplicated gene pairs in *Aspergillus fumigatus*.

Gene pair		Gene promoter	Gene body
AFUA1G00150	*AFUA6G09370*	0.750176913	0.215204254
AFUA1G00420	*AFUA8G04120*	0.80370974	0.723149072
AFUA1G00440	*AFUA8G04110*	0.856048561	0.225149091
AFUA1G00450	*AFUA8G04100*	0.9213602	0.94633912
AFUA1G00470	*AFUA8G04080*	0.433118045	0.818800395
AFUA1G00530	*AFUA4G12050*	0.326735006	−0.199291887
AFUA1G00550	*AFUA1G00910*	0.676910334	0.678763937
AFUA1G00580	*AFUA8G04050*	0.305291793	0.312927993
AFUA1G00650	*AFUA7G08510*	0.448171525	0.085599209
AFUA1G00920	*AFUA3G06425*	0.719748778	−0.340001527
AFUA1G01050	*AFUA8G06160*	0.635149308	0.176679577
AFUA1G02550	*AFUA2G14990*	0.139079116	−0.036644732
AFUA1G02730	*AFUA1G15140*	0.452193259	0.067186644
AFUA1G05760	*AFUA5G15010*	0.395849711	0.117006666
AFUA1G05760	*AFUA1G16100*	0.127911436	−0.397363108
AFUA1G10910	*AFUA7G00250*	0.310718577	0.665906466
AFUA1G11260	*AFUA6G00270*	0.617303556	−0.02397179
AFUA1G11610	*AFUA3G14850*	0.636281069	0.869300724
AFUA1G11890	*AFUA6G00300*	0.254487564	0.429087849
AFUA1G12850	*AFUA1G17470*	0.644447287	0.008468365
AFUA1G15970	*AFUA8G01560*	0.373850536	−0.106800689
AFUA1G16030	*AFUA5G14930*	0.618349765	0.858488283
AFUA1G16040	*AFUA5G14940*	0.481054005	0.017330056
AFUA1G16050	*AFUA5G14950*	−0.345037268	−0.392132138
AFUA1G16070	*AFUA5G14980*	0.930922124	0.902581516
AFUA1G16080	*AFUA5G14990*	0.559980787	0.136819379
AFUA1G16090	*AFUA5G15000*	0.773867924	0.338119153
AFUA1G16100	*AFUA5G15010*	0.681170098	0.70434823
AFUA1G16110	*AFUA5G15020*	−0.261864905	0.632982505
AFUA1G16120	*AFUA8G07150*	0.356869915	0.074502754
AFUA1G16120	*AFUA5G15030*	−0.135442129	−0.239074416
AFUA2G00800	*AFUA5G10380*	0.352063916	0.460712971
AFUA2G00800	*AFUA7G05030*	−0.114184443	0.457081245
AFUA2G04010	*AFUA6G12950*	0.134790545	−0.14813806
AFUA2G11270	*AFUA3G00910*	−0.004207858	0.192024608
AFUA3G00340	*AFUA4G02720*	−0.050259987	0.055757479
AFUA3G00680	*AFUA7G04180*	0.399912713	−0.044615588
AFUA3G01560	*AFUA5G04260*	0.179395067	0.125177632
AFUA3G02420	*AFUA4G01400*	0.662712481	0.646400554
AFUA3G03080	*AFUA6G14540*	0.3401707	−0.021961056
AFUA3G03980	*AFUA5G10050*	0.486726534	0.318475376
AFUA3G08160	*AFUA5G02410*	0.309645464	−0.255654021
AFUA3G14420	*AFUA5G00760*	0.917685134	0.415111431
AFUA4G00510	*AFUA7G08600*	0.578582743	0.441363777
AFUA4G03110	*AFUA5G10690*	0.721540193	0.176996881
AFUA4G03680	*AFUA6G03520*	0.662657833	−0.320417166
AFUA4G09440	*AFUA6G03690*	0.204159184	0.000609208
AFUA4G14360	*AFUA5G07560*	0.038083073	0.515103795
AFUA5G00145	*AFUA6G11710*	0.809402393	−0.741045496

TABLE 2: Continued.

Gene pair		Gene promoter	Gene body
AFUA5G00145	*AFUA7G08440*	0.019652845	−0.575600966
AFUA5G01030	*AFUA5G01970*	0.576211178	0.132479521
AFUA5G06240	*AFUA7G01010*	0.163674862	−0.552899419
AFUA5G07980	*AFUA5G14920*	0.669562689	−0.265207346
AFUA5G09130	*AFUA6G05030*	0.092452938	0.729278817
AFUA5G15030	*AFUA8G07150*	0.337482368	−0.502391454
AFUA6G06750	*AFUA2G03290*	0.333402926	−0.538389013
AFUA6G07070	*AFUA6G11610*	−0.048961745	−0.239430043
AFUA6G11430	*AFUA7G01000*	0.0145492	0.476692417
AFUA6G11710	*AFUA7G08440*	0.243362623	0.823124582
AFUA6G13490	*AFUA8G06020*	0.768186533	0.174847009
AFUA7G00360	*AFUA8G00860*	−0.115050268	0.03815086
AFUA7G07050	*AFUA7G08300*	0.70756842	0.778653567
AFUA7G07060	*AFUA7G08310*	0.906852725	0.919839072

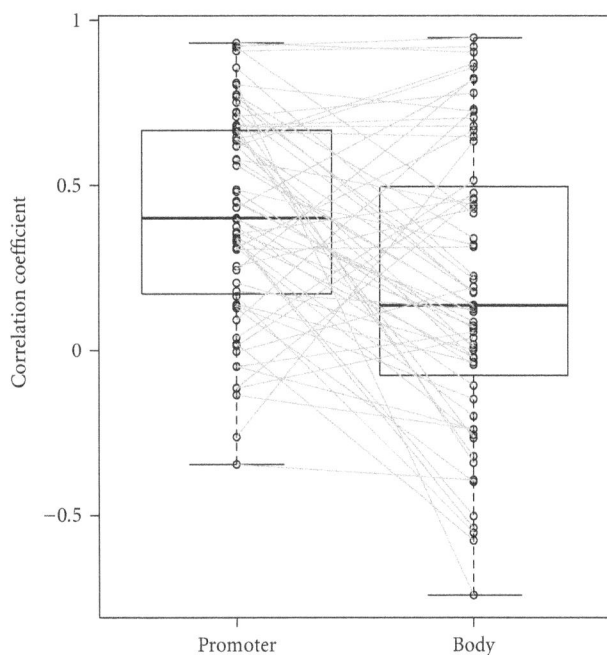

FIGURE 1: Boxplots of Spearman's rank correlation coefficients of nucleosome position profiles in the promoter and body regions of 63 duplicated gene pairs. Circles represent the correlation coefficients and values of the same genes are connected by lines.

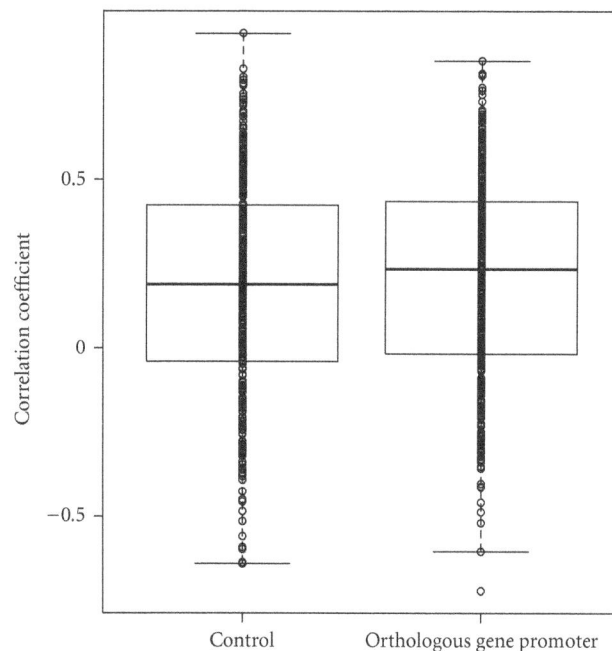

FIGURE 2: Boxplots of Spearman's rank correlation coefficients between nucleosome position profiles in the promoters of 347 orthologous gene pairs between *Aspergillus fumigatus* and *Saccharomyces cerevisiae*. The same number of gene pairs was chosen at random to serve as a control. Dots indicate correlation coefficients. The distributions of correlation coefficients did not significantly differ (*P*-value = 0.28 in Kolmogorov-Smirnov test) between the orthologous gene promoters and the controls.

of nucleosome positioning between *A. fumigatus* and *S. cerevisiae*, I focused on 466 genes (Table 3) that showed notably high conservation of nucleosome positioning in the promoters of the control and the *ELP3* and *HOS2* deletion mutants from the previous study [10].

A total of 3339 ortholog clusters were identified (See table 1 in Supplementary Material available at doi: 10.1100/2012/298174) between *A. fumigatus* and *S. cerevisiae* by ortholog cluster analysis in the Microbial Genome Database for Comparative Analysis (MBGD, http://mbgd .nibb.ac.jp/) [12]. Of these orthologous gene pairs, 347 (Table 4) are yeast genes that showed a high level of

nucleosome positioning conservation in the control and deletion mutants. I focused on these 347 orthologous pairs to compare nucleosome positioning between species. The same number of pairs of *A. fumigatus* and *S. cerevisiae* genes chosen at random were used as a control.

2.3. Nucleosome Position Profile. Nucleosome mapping numbers at each genomic position were determined [13] based

TABLE 3: Genes of *Saccharomyces cerevisiae* with highly conserved nucleosome positions in the promoters of the control and histone modification gene deletion mutants.

Chromosome	Gene	Correlation coefficient between the control and the *ELP3* deletion	Correlation coefficient between the control and the *HOS2* deletion	Translational start site	Transcription direction
chr01	YAL064W-B	0.974719955	0.965814243	12047	+
chr01	YAL056W	0.989801003	0.982157328	39260	+
chr01	YAL047C	0.950243691	0.953129232	56858	−
chr01	YAR019C	0.970115162	0.960628643	175133	−
chr01	YAR033W	0.968807875	0.987221166	188101	+
chr02	YBL111C	0.995802065	0.99455065	5009	−
chr02	YBL108C-A	0.959497013	0.969966542	7733	−
chr02	YBL101C	0.958517125	0.991324096	28299	−
chr02	YBL087C	0.95719187	0.957273203	60735	−
chr02	YBL061C	0.984960212	0.992728731	107408	−
chr02	YBL060W	0.960330289	0.99357258	107934	+
chr02	YBL051C	0.97653734	0.989165943	124762	−
chr02	YBL032W	0.956874854	0.980366627	160187	+
chr02	YBL005W-B	0.950781322	0.952679326	221333	+
chr02	YBL005W-A	0.950781322	0.952679326	221333	+
chr02	YBR023C	0.960238499	0.992906362	287925	−
chr02	YBR029C	0.959231389	0.987222377	297742	−
chr02	YBR047W	0.964564522	0.963289004	331831	+
chr02	YBR060C	0.959229871	0.974718825	362512	−
chr02	YBR084W	0.962172866	0.981530879	411048	+
chr02	YBR090C	0.967658785	0.971792497	427052	−
chr02	YBR091C	0.965037862	0.985960321	427478	−
chr02	YBR131W	0.953323249	0.969685452	497157	+
chr02	YBR136W	0.963546822	0.974854426	505662	+
chr02	YBR173C	0.978518391	0.961438459	582167	−
chr02	YBR179C	0.988489146	0.991754907	589109	−
chr02	YBR180W	0.987132212	0.975974209	589736	+
chr02	YBR204C	0.970958946	0.978433534	633376	−
chr02	YBR243C	0.972253467	0.961252065	706788	−
chr02	YBR244W	0.974202306	0.992702483	707523	+
chr02	YBR249C	0.955742682	0.976101912	717989	−
chr02	YBR250W	0.954149188	0.976352348	719028	+
chr02	YBR251W	0.96273261	0.961120273	721385	+
chr02	YBR258C	0.966601899	0.955155711	730157	−
chr02	YBR260C	0.969346019	0.970918059	734634	−
chr02	YBR279W	0.956081387	0.964626558	761253	+
chr02	YBR290W	0.981586823	0.967935658	782587	+
chr03	YCL064C	0.957150698	0.95270514	16880	−
chr03	YCL058W-A	0.969895345	0.992394829	23584	+
chr03	YCL057C-A	0.990886192	0.987170576	24325	−
chr03	YCL057W	0.963062357	0.961804623	24768	+
chr03	YCL038C	0.978210537	0.957633548	56527	−
chr03	YCL035C	0.981360705	0.973568151	61173	−
chr03	YCL019W	0.953241664	0.978688459	85102	+
chr03	YCR026C	0.96886665	0.97932414	166335	−
chr03	YCR043C	0.951764864	0.966166673	206640	−

TABLE 3: Continued.

Chromosome	Gene	Correlation coefficient between the control and the ELP3 deletion	Correlation coefficient between the control and the HOS2 deletion	Translational start site	Transcription direction
chr03	YCR053W	0.989728241	0.992623598	216693	+
chr03	YCR087C-A	0.975398919	0.993232435	264464	−
chr03	YCR090C	0.965637294	0.95018878	272860	−
chr03	YCR108C	0.985803086	0.993860549	316185	−
chr04	YDL248W	0.964701365	0.980217326	1802	+
chr04	YDL247W	0.989686199	0.997997741	5985	+
chr04	YDL233W	0.957672252	0.97857489	36798	+
chr04	YDL232W	0.979592686	0.970413477	38488	+
chr04	YDL225W	0.950053133	0.978441348	52446	+
chr04	YDL208W	0.987264893	0.987894794	87513	+
chr04	YDL189W	0.965815356	0.98561241	122217	+
chr04	YDL174C	0.956215571	0.953787348	147590	−
chr04	YDL147W	0.971755059	0.950911485	190925	+
chr04	YDL116W	0.965936123	0.976379579	251566	+
chr04	YDL110C	0.974238254	0.97295169	264964	−
chr04	YDL102W	0.957786538	0.980122821	276872	+
chr04	YDL085W	0.97887937	0.960150217	303211	+
chr04	YDL035C	0.975573212	0.977145228	392054	−
chr04	YDL025C	0.96134673	0.972526893	407203	−
chr04	YDR019C	0.95269772	0.953635877	485362	−
chr04	YDR028C	0.960937436	0.959981321	500876	−
chr04	YDR034C-C	0.967222636	0.9814374	519353	−
chr04	YDR034C-D	0.967222636	0.981437429	519353	−
chr04	YDR037W	0.951365292	0.956861294	525437	+
chr04	YDR054C	0.955928802	0.982048103	562325	−
chr04	YDR055W	0.950321012	0.952540035	563525	+
chr04	YDR062W	0.974021833	0.967688258	576471	+
chr04	YDR109C	0.954396163	0.952291074	675664	−
chr04	YDR110W	0.986081905	0.974555387	676099	+
chr04	YDR120C	0.960200845	0.969981962	693258	−
chr04	YDR162C	0.967301741	0.984347121	781097	−
chr04	YDR233C	0.967024058	0.967951444	930353	−
chr04	YDR234W	0.967415159	0.980945022	931125	+
chr04	YDR238C	0.97602642	0.973622232	940812	−
chr04	YDR261C-D	0.984848646	0.993722663	992345	−
chr04	YDR261C-C	0.984848646	0.9937227	992345	−
chr04	YDR262W	0.990989397	0.995915841	993130	+
chr04	YDR270W	0.951139826	0.96263534	1005671	+
chr04	YDR281C	0.984619329	0.959392909	1022317	−
chr04	YDR300C	0.975849644	0.971677938	1062787	−
chr04	YDR301W	0.952923341	0.977515222	1063348	+
chr04	YDR307W	0.951918886	0.970033626	1075861	+
chr04	YDR310C	0.951723931	0.969273153	1084312	−
chr04	YDR311W	0.953685357	0.980595738	1085062	+
chr04	YDR317W	0.98969221	0.991598049	1102181	+
chr04	YDR322W	0.954843272	0.97204312	1110586	+
chr04	YDR328C	0.957421993	0.983957835	1126013	−
chr04	YDR334W	0.953030288	0.963155512	1135927	+

TABLE 3: Continued.

Chromosome	Gene	Correlation coefficient between the control and the *ELP3* deletion	Correlation coefficient between the control and the *HOS2* deletion	Translational start site	Transcription direction
chr04	*YDR359C*	0.977489172	0.969606029	1194877	−
chr04	*YDR365C*	0.985073544	0.998926611	1206375	−
chr04	*YDR367W*	0.990041805	0.994689201	1212840	+
chr04	*YDR369C*	0.957655919	0.96987305	1217572	−
chr04	*YDR379W*	0.962899057	0.974939821	1230159	+
chr04	*YDR397C*	0.966298918	0.975133337	1266890	−
chr04	*YDR420W*	0.971217177	0.991886941	1306259	+
chr04	*YDR424C*	0.966881808	0.989813233	1319833	−
chr04	*YDR432W*	0.955728738	0.982171718	1328775	+
chr04	*YDR438W*	0.971626925	0.98904082	1338266	+
chr04	*YDR444W*	0.973566433	0.976502447	1350282	+
chr04	*YDR453C*	0.952661492	0.954481014	1365654	−
chr04	*YDR476C*	0.969940159	0.964679279	1411119	−
chr04	*YDR477W*	0.954743343	0.979243738	1412365	+
chr04	*YDR479C*	0.977460825	0.994501786	1416866	−
chr04	*YDR480W*	0.967727074	0.986953403	1417391	+
chr04	*YDR488C*	0.983114512	0.979187155	1430781	−
chr04	*YDR497C*	0.955855412	0.958236577	1445459	−
chr04	*YDR529C*	0.966273576	0.971812076	1496540	−
chr05	*YEL072W*	0.968131913	0.977435534	13720	+
chr05	*YEL043W*	0.979079858	0.96227706	70478	+
chr05	*YEL038W*	0.981572221	0.988369151	80462	+
chr05	*YEL021W*	0.951315475	0.951565922	116167	+
chr05	*YER004W*	0.956348638	0.987671859	159579	+
chr05	*YER026C*	0.964952897	0.993875293	208473	−
chr05	*YER076C*	0.965398244	0.971982059	313494	−
chr05	*YER083C*	0.953185048	0.981521771	327027	−
chr05	*YER094C*	0.976027049	0.982595427	349342	−
chr05	*YER095W*	0.963423619	0.988752501	349976	+
chr05	*YER107C*	0.982024941	0.98488468	374541	−
chr05	*YER109C*	0.969371754	0.986034001	377610	−
chr05	*YER173W*	0.960496827	0.952930419	536295	+
chr05	*YER188C-A*	0.991489293	0.996538184	569902	−
chr05	*YER189W*	0.997057108	0.998333831	571150	+
chr06	*YFL066C*	0.970012372	0.993098415	2615	−
chr06	*YFL065C*	0.957034943	0.97004682	3338	−
chr06	*YFL060C*	0.987757312	0.995138274	10969	−
chr06	*YFL059W*	0.976205235	0.962195117	11363	+
chr06	*YFL058W*	0.9750941	0.994474768	12929	+
chr06	*YFL028C*	0.966244522	0.959308346	80211	−
chr06	*YFL026W*	0.957755207	0.952711873	82578	+
chr06	*YFR009W*	0.979200258	0.985708335	162482	+
chr06	*YFR013W*	0.972958971	0.986214291	169914	+
chr06	*YFR037C*	0.965415915	0.972070667	229173	−
chr07	*YGL255W*	0.961366208	0.972062316	20978	+
chr07	*YGL248W*	0.971369747	0.971756841	35653	+
chr07	*YGL223C*	0.956215218	0.984994924	80364	−
chr07	*YGL215W*	0.965861339	0.95181154	87980	+
chr07	*YGL201C*	0.954296336	0.979120204	120911	−

TABLE 3: Continued.

Chromosome	Gene	Correlation coefficient between the control and the ELP3 deletion	Correlation coefficient between the control and the HOS2 deletion	Translational start site	Transcription direction
chr07	YGL180W	0.982084299	0.994334205	160071	+
chr07	YGL171W	0.969321982	0.974163208	182396	+
chr07	YGL163C	0.975825825	0.968348392	196409	−
chr07	YGL138C	0.971247145	0.983410766	249536	−
chr07	YGL120C	0.981452639	0.964320425	283943	−
chr07	YGL119W	0.986610225	0.981525372	284448	+
chr07	YGL108C	0.961840326	0.985939242	304074	−
chr07	YGL058W	0.977732494	0.975176081	393992	+
chr07	YGL056C	0.960426378	0.969119302	397624	−
chr07	YGL055W	0.964044825	0.969447792	398631	+
chr07	YGL048C	0.971402915	0.953304338	411289	−
chr07	YGL043W	0.964287661	0.976335397	417487	+
chr07	YGL028C	0.960077919	0.965512687	442914	−
chr07	YGL006W	0.950151816	0.968357705	485925	+
chr07	YGR001C	0.958621356	0.990717019	498038	−
chr07	YGR006W	0.980877592	0.97721459	506074	+
chr07	YGR027W-B	0.973585999	0.969942324	536061	+
chr07	YGR027W-A	0.973585999	0.969942324	536061	+
chr07	YGR054W	0.964434273	0.980689617	596697	+
chr07	YGR076C	0.981588154	0.958402031	637581	−
chr07	YGR082W	0.986433799	0.977608093	644048	+
chr07	YGR084C	0.952879595	0.959716787	648146	−
chr07	YGR109C	0.967037071	0.989224447	706505	−
chr07	YGR109W-B	0.964433838	0.990506548	707614	+
chr07	YGR109W-A	0.964433838	0.990506548	707614	+
chr07	YGR149W	0.957244122	0.960107682	789036	+
chr07	YGR161W-B	0.958455297	0.972873884	811743	+
chr07	YGR161W-A	0.958455297	0.9728739	811743	+
chr07	YGR161C-D	0.9711882	0.974318881	823020	−
chr07	YGR161C-C	0.9711882	0.9743189	823020	−
chr07	YGR162W	0.967152562	0.972476532	824064	+
chr07	YGR165W	0.959888658	0.951709091	829121	+
chr07	YGR166W	0.978481923	0.950610672	830520	+
chr07	YGR173W	0.978946749	0.974220242	843859	+
chr07	YGR178C	0.97338186	0.951056442	853220	−
chr07	YGR193C	0.977703185	0.9816188	885746	−
chr07	YGR198W	0.954889766	0.972754857	894698	+
chr07	YGR239C	0.968888325	0.989258725	970058	−
chr07	YGR240C	0.966313371	0.971843468	973739	−
chr07	YGR255C	0.966583058	0.991666026	1003967	−
chr07	YGR267C	0.960278778	0.969707488	1025741	−
chr07	YGR280C	0.961084505	0.952670993	1051732	−
chr07	YGR295C	0.967286944	0.964457386	1082736	−
chr07	YGR296W	0.971130914	0.991185295	1084871	+
chr08	YHL044W	0.958725508	0.965538017	13563	+
chr08	YHL029C	0.988947037	0.966760586	47966	−
chr08	YHL028W	0.983034685	0.977043411	48761	+
chr08	YHL024W	0.961645582	0.966633627	56647	+

TABLE 3: Continued.

Chromosome	Gene	Correlation coefficient between the control and the *ELP3* deletion	Correlation coefficient between the control and the *HOS2* deletion	Translational start site	Transcription direction
chr08	YHL020C	0.978013871	0.986901259	67453	−
chr08	YHL016C	0.976082816	0.988408432	74241	−
chr08	YHL007C	0.973847748	0.981293306	97933	−
chr08	YHL004W	0.97418766	0.993677693	99215	+
chr08	YHL001W	0.959021584	0.965334128	104272	+
chr08	YHR001W-A	0.991644256	0.988918444	107821	+
chr08	YHR056C	0.972836967	0.982728846	217836	−
chr08	YHR081W	0.961720079	0.960688905	267540	+
chr08	YHR091C	0.955695121	0.987437891	286772	−
chr08	YHR101C	0.984820398	0.996893941	315971	−
chr08	YHR102W	0.96127201	0.995534701	316575	+
chr08	YHR107C	0.950250496	0.982165489	328039	−
chr08	YHR118C	0.973725305	0.961118169	345631	−
chr08	YHR127W	0.95014104	0.981438554	360916	+
chr08	YHR136C	0.953476026	0.965665058	375103	−
chr08	YHR148W	0.978585198	0.950179497	393537	+
chr08	YHR153C	0.959414518	0.982780385	402685	−
chr08	YHR165C	0.958747068	0.984253541	436950	−
chr08	YHR214C-D	0.987951207	0.966688891	550941	−
chr08	YHR215W	0.987951207	0.966688891	552099	+
chr08	YHR216W	0.972948554	0.975124764	554396	+
chr09	YIL158W	0.961579117	0.957386446	46201	+
chr09	YIL154C	0.954401073	0.971892535	55021	−
chr09	YIL137C	0.976010283	0.976810775	92788	−
chr09	YIL135C	0.976089791	0.979613921	96375	−
chr09	YIL134W	0.974541094	0.979148771	97395	+
chr09	YIL129C	0.958227819	0.99048832	113237	−
chr09	YIL125W	0.963070211	0.960333013	122689	+
chr09	YIL063C	0.970956253	0.957414461	243741	−
chr09	YIL061C	0.95394965	0.989156182	245556	−
chr09	YIL046W	0.967209958	0.952996831	268650	+
chr09	YIL033C	0.989017477	0.989306501	291668	−
chr09	YIL031W	0.985799922	0.98683422	292632	+
chr09	YIL030C	0.957449018	0.97109633	300008	−
chr09	YIR022W	0.960636707	0.971998311	398730	+
chr09	YIR024C	0.961569219	0.970038905	403488	−
chr09	YIR038C	0.973108358	0.975735046	424510	−
chr10	YJL221C	0.992889333	0.995409763	18536	−
chr10	YJL219W	0.992889333	0.995409763	19497	+
chr10	YJL197W	0.963017116	0.967339021	63804	+
chr10	YJL181W	0.971567753	0.971225117	85658	+
chr10	YJL176C	0.979890865	0.967535459	94528	−
chr10	YJL174W	0.983474082	0.9822043	95090	+
chr10	YJL173C	0.986867668	0.989398406	96527	−
chr10	YJL151C	0.952976229	0.987631587	136770	−
chr10	YJL149W	0.975756648	0.990848833	137376	+
chr10	YJL118W	0.953268112	0.959468203	191638	+
chr10	YJL113W	0.982940363	0.963188993	197913	+

TABLE 3: Continued.

Chromosome	Gene	Correlation coefficient between the control and the *ELP3* deletion	Correlation coefficient between the control and the *HOS2* deletion	Translational start site	Transcription direction
chr10	YJL114W	0.982940363	0.963188993	197913	+
chr10	YJL093C	0.967675882	0.964352732	256807	−
chr10	YJL092W	0.963240719	0.958166876	257418	+
chr10	YJL066C	0.953982678	0.974252054	314867	−
chr10	YJL050W	0.950677364	0.95235067	342517	+
chr10	YJL048C	0.96132641	0.958967756	348632	−
chr10	YJL039C	0.950918953	0.970557445	373794	−
chr10	YJL034W	0.963347804	0.980266739	381322	+
chr10	YJR010W	0.971667854	0.97428271	456232	+
chr10	YJR021C	0.956762721	0.994264722	469572	−
chr10	YJR029W	0.958070547	0.974229894	478337	+
chr10	YJR028W	0.958070547	0.9742299	478337	+
chr10	YJR041C	0.976618081	0.981063923	513756	−
chr10	YJR048W	0.960606245	0.975055837	526328	+
chr10	YJR049C	0.954983464	0.986149168	528469	−
chr10	YJR055W	0.976932113	0.992763445	538765	+
chr10	YJR065C	0.975569729	0.987649101	559151	−
chr10	YJR095W	0.97047495	0.971814385	609769	+
chr10	YJR113C	0.969228269	0.954716018	638969	−
chr10	YJR115W	0.952485858	0.960980837	639936	+
chr10	YJR141W	0.957349189	0.956946497	695900	+
chr10	YJR160C	0.978268521	0.997490962	739810	−
chr10	YJR161C	0.965251269	0.976575544	743993	−
chr11	YKL191W	0.950933251	0.978993764	81040	+
chr11	YKL179C	0.975707433	0.967918905	112508	−
chr11	YKL167C	0.956085616	0.955561088	134139	−
chr11	YKL165C	0.960609423	0.966621061	140696	−
chr11	YKL157W	0.95635277	0.9813363	154996	+
chr11	YKL127W	0.955985661	0.982208081	203185	+
chr11	YKL125W	0.976449752	0.960074911	207891	+
chr11	YKL113C	0.971544612	0.954413703	225519	−
chr11	YKL065C	0.96005541	0.969818928	316701	−
chr11	YKL064W	0.974644257	0.972632809	317408	+
chr11	YKL059C	0.957153957	0.961241915	329087	−
chr11	YKL020C	0.955860253	0.977812511	401723	−
chr11	YKL013C	0.972958978	0.953444609	417666	−
chr11	YKR007W	0.981387104	0.972277081	451077	+
chr11	YKR024C	0.971944038	0.981215773	487015	−
chr11	YKR031C	0.952123552	0.969614994	506037	−
chr11	YKR036C	0.959340192	0.988104612	510275	−
chr11	YKR041W	0.982786597	0.957524998	517840	+
chr11	YKR052C	0.954392373	0.96688713	533106	−
chr11	YKR082W	0.962931616	0.991018501	592467	+
chr11	YKR084C	0.953228662	0.976903834	598532	−
chr11	YKR086W	0.952006608	0.976533657	599499	+
chr12	YLL050C	0.976179673	0.990695339	40413	−
chr12	YLL002W	0.972688874	0.981122544	146290	+
chr12	YLR001C	0.966206724	0.98012374	153976	−

Table 3: Continued.

Chromosome	Gene	Correlation coefficient between the control and the *ELP3* deletion	Correlation coefficient between the control and the *HOS2* deletion	Translational start site	Transcription direction
chr12	YLR012C	0.956516633	0.962463803	170280	−
chr12	YLR013W	0.978873849	0.985557204	171338	+
chr12	YLR024C	0.953442868	0.956160837	193282	−
chr12	YLR025W	0.971468039	0.96202948	194453	+
chr12	YLR029C	0.961637928	0.991619508	202591	−
chr12	YLR059C	0.963450626	0.974770886	260548	−
chr12	YLR085C	0.9560163	0.984640913	301990	−
chr12	YLR087C	0.953704478	0.990085627	315732	−
chr12	YLR096W	0.977966477	0.979318401	332591	+
chr12	YLR104W	0.966174553	0.97921157	346586	+
chr12	YLR133W	0.961307333	0.951823465	408446	+
chr12	YLR135W	0.977560494	0.977991425	413282	+
chr12	YLR137W	0.967174193	0.983596278	417007	+
chr12	YLR162W-A	0.968123219	0.962766706	490407	+
chr12	YLR208W	0.96802684	0.969296393	559553	+
chr12	YLR223C	0.966783974	0.974989936	585492	−
chr12	YLR224W	0.974582795	0.978003757	586466	+
chr12	YLR286C	0.972231318	0.978358686	710138	−
chr12	YLR299W	0.962625418	0.970226227	726071	+
chr12	YLR307W	0.962710223	0.990591743	745622	+
chr12	YLR323C	0.953205594	0.975247431	778952	−
chr12	YLR326W	0.979769576	0.991583861	782174	+
chr12	YLR355C	0.967858474	0.973834311	839252	−
chr12	YLR356W	0.968277882	0.974767083	840320	+
chr12	YLR378C	0.974122358	0.977246394	877177	−
chr12	YLR380W	0.973258508	0.97821865	878282	+
chr12	YLR410W-A	0.951545011	0.9529714	941481	+
chr12	YLR410W-B	0.951545011	0.952971407	941481	+
chr12	YLR426W	0.9668981	0.990317101	987059	+
chr12	YLR427W	0.968631588	0.972538491	988425	+
chr12	YLR429W	0.950464544	0.951528134	990774	+
chr12	YLR443W	0.964525656	0.991440098	1022622	+
chr13	YML121W	0.974659011	0.985969584	26930	+
chr13	YML115C	0.97337994	0.996063492	41794	−
chr13	YML080W	0.989302036	0.986381735	108806	+
chr13	YML078W	0.975873108	0.974136336	111002	+
chr13	YML045W	0.993016416	0.993250991	184461	+
chr13	YML045W-A	0.993016416	0.993251	184461	+
chr13	YML041C	0.969801821	0.979224272	195755	−
chr13	YML020W	0.975260683	0.95439232	231149	+
chr13	YML004C	0.966616377	0.98388072	262685	−
chr13	YML003W	0.968338912	0.981828638	263483	+
chr13	YMR010W	0.985774326	0.979191015	285099	+
chr13	YMR011W	0.963732981	0.981283867	288078	+
chr13	YMR027W	0.961041993	0.957831173	325876	+
chr13	YMR036C	0.969260192	0.963784402	343519	−
chr13	YMR060C	0.990018805	0.990517987	392514	−
chr13	YMR066W	0.98321165	0.95995151	401540	+

Table 3: Continued.

Chromosome	Gene	Correlation coefficient between the control and the ELP3 deletion	Correlation coefficient between the control and the HOS2 deletion	Translational start site	Transcription direction
chr13	YMR078C	0.969392941	0.976546196	424727	−
chr13	YMR081C	0.981254312	0.956294421	431094	−
chr13	YMR110C	0.961424621	0.983151943	491991	−
chr13	YMR116C	0.977129849	0.989254306	500687	−
chr13	YMR137C	0.971971845	0.978526214	544962	−
chr13	YMR138W	0.978634859	0.963362392	545154	+
chr13	YMR152W	0.958913422	0.961601086	563095	+
chr13	YMR197C	0.960464224	0.981756931	659197	−
chr13	YMR210W	0.974579381	0.983178959	687515	+
chr13	YMR214W	0.977906077	0.988639434	695349	+
chr13	YMR219W	0.964555974	0.962486321	707132	+
chr13	YMR224C	0.975063127	0.950620617	720652	−
chr13	YMR229C	0.960514554	0.957025636	731122	−
chr13	YMR241W	0.956225901	0.967912184	751960	+
chr13	YMR319C	0.950340769	0.968043162	914536	−
chr14	YNL339C	0.975221324	0.997373072	6098	−
chr14	YNL334C	0.968865734	0.991228175	12876	−
chr14	YNL322C	0.954706246	0.977799203	34234	−
chr14	YNL311C	0.986814003	0.993782307	51687	−
chr14	YNL310C	0.956852702	0.980516471	52430	−
chr14	YNL309W	0.988062215	0.993825499	52661	+
chr14	YNL301C	0.968107997	0.966680876	64562	−
chr14	YNL295W	0.952849161	0.964836983	76946	+
chr14	YNL261W	0.950477134	0.976226949	155101	+
chr14	YNL260C	0.968893453	0.965562883	157456	−
chr14	YNL255C	0.959418743	0.961868386	167791	−
chr14	YNL248C	0.959305496	0.962132738	182609	−
chr14	YNL241C	0.952591706	0.968306617	197944	−
chr14	YNL234W	0.975005906	0.964281927	210234	+
chr14	YNL224C	0.973438019	0.985601741	227100	−
chr14	YNL212W	0.967303893	0.964441355	247462	+
chr14	YNL166C	0.973079673	0.979903307	323567	−
chr14	YNL156C	0.961076874	0.96913283	341970	−
chr14	YNL112W	0.953769466	0.987266001	413641	+
chr14	YNL099C	0.959678116	0.95727388	439285	−
chr14	YNL097C	0.978605808	0.986440434	442360	−
chr14	YNL082W	0.955790378	0.980457797	473392	+
chr14	YNL055C	0.982685169	0.981068004	518846	−
chr14	YNL042W-B	0.956836355	0.979900708	547114	+
chr14	YNL029C	0.976101029	0.988675718	578774	−
chr14	YNL027W	0.96462873	0.978413199	579581	+
chr14	YNR012W	0.958400399	0.977490746	647434	+
chr14	YNR015W	0.967843094	0.979265652	653389	+
chr14	YNR023W	0.975180387	0.972628935	670420	+
chr14	YNR026C	0.966267575	0.984095961	674691	−
chr14	YNR036C	0.955455936	0.956914254	694824	−
chr14	YNR039C	0.97056523	0.9708253	699433	−
chr14	YNR075W	0.950188074	0.953766447	779916	+

TABLE 3: Continued.

Chromosome	Gene	Correlation coefficient between the control and the *ELP3* deletion	Correlation coefficient between the control and the *HOS2* deletion	Translational start site	Transcription direction
chr14	*YNR075C-A*	0.986389631	0.991000914	781603	−
chr15	*YOL166W-A*	0.969218719	0.964760148	585	+
chr15	*YOL157C*	0.977029388	0.98945557	24293	−
chr15	*YOL156W*	0.979765672	0.992685071	25272	+
chr15	*YOL148C*	0.957761501	0.971157209	47573	−
chr15	*YOL104C*	0.976917372	0.981386975	117454	−
chr15	*YOL100W*	0.961328138	0.959394148	129237	+
chr15	*YOL089C*	0.971692903	0.984860678	153490	−
chr15	*YOL086C*	0.976082094	0.966862518	160594	−
chr15	*YOL077C*	0.956963602	0.983941333	186723	−
chr15	*YOL068C*	0.959180595	0.953847182	201879	−
chr15	*YOL062C*	0.954102242	0.967526899	211995	−
chr15	*YOL058W*	0.969401099	0.978364009	219210	+
chr15	*YOL045W*	0.966108727	0.955785397	243496	+
chr15	*YOL031C*	0.971563515	0.959211572	267530	−
chr15	*YOL026C*	0.954115513	0.976281121	274354	−
chr15	*YOL023W*	0.960879668	0.960550998	278057	+
chr15	*YOL006C*	0.971429281	0.978082289	315388	−
chr15	*YOR043W*	0.970409112	0.977199399	410870	+
chr15	*YOR048C*	0.95240535	0.972598244	421651	−
chr15	*YOR056C*	0.95552512	0.980531823	431628	−
chr15	*YOR058C*	0.977288174	0.978577416	436347	−
chr15	*YOR071C*	0.980341941	0.989920087	461278	−
chr15	*YOR075W*	0.983591979	0.981471486	468214	+
chr15	*YOR089C*	0.978934603	0.984066065	490830	−
chr15	*YOR104W*	0.965107091	0.973591736	517643	+
chr15	*YOR124C*	0.97179062	0.983946384	558643	−
chr15	*YOR129C*	0.979558976	0.983142651	569559	−
chr15	*YOR132W*	0.955282168	0.965555781	573176	+
chr15	*YOR148C*	0.987044445	0.98569999	609198	−
chr15	*YOR192C-C*	0.980704362	0.966062169	704225	−
chr15	*YOR193W*	0.973179583	0.992657727	710447	+
chr15	*YOR204W*	0.985619071	0.960639074	722912	+
chr15	*YOR216C*	0.971045967	0.970515948	748980	−
chr15	*YOR247W*	0.959336754	0.957210052	797677	+
chr15	*YOR294W*	0.972999786	0.97236413	868339	+
chr15	*YOR336W*	0.973501735	0.969442343	949770	+
chr15	*YOR365C*	0.981766719	0.98630012	1025570	−
chr15	*YOR372C*	0.955936893	0.971693401	1036469	−
chr15	*YOR389W*	0.964833697	0.976852438	1074211	+
chr15	*YOR390W*	0.98673199	0.997343789	1076782	+
chr15	*YOR391C*	0.952090431	0.990889663	1079256	−
chr15	*YOR393W*	0.952090431	0.990889663	1080274	+
chr16	*YPL283C*	0.990485958	0.9953985	6007	−
chr16	*YPL281C*	0.961921468	0.977623026	10870	−
chr16	*YPL280W*	0.961921468	0.977623026	11887	+
chr16	*YPL278C*	0.965633274	0.959753752	15355	−
chr16	*YPL257W-B*	0.976897418	0.976806903	56748	+

Table 3: Continued.

Chromosome	Gene	Correlation coefficient between the control and the *ELP3* deletion	Correlation coefficient between the control and the *HOS2* deletion	Translational start site	Transcription direction
chr16	YPL257W-A	0.976897418	0.976807	56748	+
chr16	YPL255W	0.96664841	0.970266728	67725	+
chr16	YPL254W	0.95123182	0.962107221	69485	+
chr16	YPL206C	0.975807996	0.972912432	163596	−
chr16	YPL196W	0.978338347	0.984467392	175042	+
chr16	YPL180W	0.976152618	0.987674239	205247	+
chr16	YPL171C	0.96524809	0.961521166	227370	−
chr16	YPL158C	0.980225654	0.985491741	254309	−
chr16	YPL157W	0.98638216	0.988124386	254813	+
chr16	YPL146C	0.98509812	0.956540524	277528	−
chr16	YPL139C	0.981115598	0.987215069	291050	−
chr16	YPL126W	0.988937478	0.982334739	310209	+
chr16	YPL082C	0.980458525	0.992300343	404080	−
chr16	YPL078C	0.957476311	0.95576197	408741	−
chr16	YPL038W	0.953416137	0.966495031	480532	+
chr16	YPL030W	0.973101156	0.975002429	493541	+
chr16	YPL007C	0.954338314	0.983931954	543845	−
chr16	YPR010C	0.983770387	0.983449754	581193	−
chr16	YPR017C	0.986325514	0.989002388	593914	−
chr16	YPR018W	0.981145288	0.96510122	594473	+
chr16	YPR020W	0.950323294	0.983815562	599867	+
chr16	YPR048W	0.980910651	0.976100682	659179	+
chr16	YPR060C	0.960902075	0.969265445	675628	−
chr16	YPR062W	0.96911875	0.967204709	677162	+
chr16	YPR088C	0.957617837	0.976321629	713026	−
chr16	YPR141C	0.978516693	0.973377917	817919	−
chr16	YPR145C-A	0.961127516	0.965220022	824922	−
chr16	YPR156C	0.966866802	0.956230015	839773	−
chr16	YPR158C-D	0.981640164	0.971311933	856253	−
chr16	YPR158C-C	0.981640164	0.971311933	856253	−
chr16	YPR161C	0.969622155	0.984283791	866418	−
chr16	YPR165W	0.959239007	0.967322323	875364	+
chr16	YPR176C	0.986542093	0.977514778	892074	−
chr16	YPR187W	0.964480751	0.98804279	911253	+
chr16	YPR203W	0.968955474	0.985644074	943876	+

on genome-wide nucleosome mapping data for *A. fumigatus* [9] and *S. cerevisiae* [10]. In this analysis, a 1-kb region upstream of the translational start site was defined as a gene promoter. When the length of the gene body region is more than 1 kb, a 1-kb region downstream of the translational start site was defined as the gene body. When the length of the gene body is less than 1 kb, the region between the translational start and end sites was defined as the gene body. Analyses of nucleosome position data including calculation of Spearman's rank correlation coefficient were performed using the statistics software R (http://www.r-project.org/).

3. Results and Discussion

3.1. Nucleosome Position Profiles of Duplicated Genes in Aspergillus fumigatus. I compared nucleosome position profiles in each of the 63 duplicated gene pairs. Nucleosome positioning was conserved more in gene promoters than in gene bodies (Figure 1), as observed in the comparison of nucleosome positioning between trichostatin A-treated and -untreated *A. fumigatus* [4]. This result suggests that nucleosome positioning in the gene promoter plays an important role in transcriptional regulation [14].

TABLE 4: Spearman's rank correlation coefficients between nucleosome position profiles in the promoters of 347 orthologous gene pairs between *Aspergillus fumigatus* and *Saccharomyces cerevisiae*.

S. cerevisiae gene	*A. fumigatus* gene	Correlation coefficient
YCR053W	AFUA_3G08980	0.851750303
YOL006C	AFUA_1G03500	0.814728478
YLR208W	AFUA_4G06090	0.810778356
YJR160C	AFUA_8G07240	0.807607644
YKL064W	AFUA_2G08070	0.771984405
YDL208W	AFUA_1G13570	0.763687527
YDL247W	AFUA_8G07240	0.751799318
YJL034W	AFUA_2G04620	0.751475734
YJL174W	AFUA_2G07590	0.731359519
YGL006W	AFUA_1G10880	0.706118907
YLR307W	AFUA_6G10430	0.701428537
YOR056C	AFUA_5G04000	0.693624039
YJR049C	AFUA_5G12870	0.6873278
YOR132W	AFUA_5G07150	0.684358232
YDR028C	AFUA_8G02720	0.681519128
YKL059C	AFUA_2G06220	0.677236775
YDR019C	AFUA_1G10780	0.673590566
YDR262W	AFUA_8G05360	0.668577442
YJR160C	AFUA_3G01700	0.655149877
YBL087C	AFUA_2G03380	0.654275917
YDR120C	AFUA_3G04200	0.653965099
YMR110C	AFUA_4G13500	0.649573592
YDR479C	AFUA_2G01510	0.649340576
YLR307W	AFUA_4G09940	0.641152119
YLL050C	AFUA_5G10570	0.636927762
YMR011W	AFUA_2G11520	0.631877767
YGR149W	AFUA_3G08240	0.627253467
YKL125W	AFUA_1G02590	0.626251676
YDL247W	AFUA_3G01700	0.621782331
YHR165C	AFUA_2G03030	0.616638486
YLR355C	AFUA_3G14490	0.615946987
YLR307W	AFUA_3G07210	0.615345397
YNL156C	AFUA_4G07680	0.605449075
YOL157C	AFUA_8G07070	0.600254438
YPL196W	AFUA_3G08740	0.596649007
YOL156W	AFUA_2G11520	0.589475104
YGL055W	AFUA_7G05920	0.583471232
YER094C	AFUA_4G07420	0.580111415
YER026C	AFUA_4G13680	0.579632053
YKR031C	AFUA_3G05630	0.578947614
YJR041C	AFUA_1G13200	0.577042969
YCL057C-A	AFUA_1G13195	0.56703028
YKL167C	AFUA_6G12620	0.566497151
YDR334W	AFUA_7G02370	0.556810391
YDL085W	AFUA_1G11960	0.555058025
YIR038C	AFUA_8G02500	0.553657653
YGL056C	AFUA_1G06660	0.550891198
YLR133W	AFUA_1G15930	0.547961885
YLR137W	AFUA_2G09930	0.542448721

TABLE 4: Continued.

S. cerevisiae gene	A. fumigatus gene	Correlation coefficient
YER095W	AFUA_1G10410	0.540102076
YCR090C	AFUA_2G15510	0.540015424
YIR022W	AFUA_3G12840	0.539096148
YJR048W	AFUA_2G13110	0.537701308
YDL247W	AFUA_8G01340	0.536753536
YGL119W	AFUA_6G04380	0.528264148
YDR054C	AFUA_5G09200	0.520282879
YNR036C	AFUA_5G10750	0.51817226
YJL050W	AFUA_4G07160	0.517374235
YKR036C	AFUA_5G13140	0.515007541
YML041C	AFUA_2G05030	0.514819597
YML080W	AFUA_1G16550	0.512494533
YJL093C	AFUA_1G14250	0.5107227
YDR238C	AFUA_1G10970	0.509749906
YJL197W	AFUA_2G14130	0.501902821
YER173W	AFUA_8G02820	0.501581347
YJR160C	AFUA_8G01340	0.499250408
YOL058W	AFUA_2G04310	0.497155736
YGR006W	AFUA_1G16990	0.496383573
YPR161C	AFUA_5G05510	0.495551014
YBR136W	AFUA_4G04760	0.49434371
YJR049C	AFUA_2G01350	0.493221085
YPR060C	AFUA_5G13130	0.490832735
YDR055W	AFUA_4G06820	0.482973721
YLR104W	AFUA_5G04040	0.47610876
YIL129C	AFUA_6G11010	0.473123981
YDR300C	AFUA_2G07570	0.469287567
YOL156W	AFUA_7G00950	0.46726022
YGL201C	AFUA_5G10890	0.462753539
YKR086W	AFUA_1G03820	0.461061937
YPL157W	AFUA_6G08610	0.460257721
YIL046W	AFUA_2G14110	0.45798838
YGL048C	AFUA_4G04660	0.457664607
YHR216W	AFUA_2G03610	0.456902272
YDR424C	AFUA_1G04850	0.455530713
YPR141C	AFUA_2G14280	0.452133975
YLR427W	AFUA_1G07150	0.444534443
YHL004W	AFUA_1G06570	0.439699159
YLL002W	AFUA_5G09540	0.431520497
YKR052C	AFUA_6G12550	0.429429023
YNL260C	AFUA_1G09000	0.425937756
YNL112W	AFUA_2G10750	0.423331447
YFL026W	AFUA_3G14330	0.422540136
YCL038C	AFUA_2G15370	0.420548714
YPR156C	AFUA_4G01140	0.419838437
YHL024W	AFUA_3G06230	0.417835282
YBR173C	AFUA_5G10740	0.414366599
YGR178C	AFUA_1G09630	0.410618083
YMR036C	AFUA_6G08200	0.408252022

TABLE 4: Continued.

S. cerevisiae gene	A. fumigatus gene	Correlation coefficient
YDL147W	AFUA_3G06610	0.40670988
YJR095W	AFUA_2G16930	0.405617695
YPL171C	AFUA_2G17960	0.40043227
YBL061C	AFUA_8G05620	0.398124519
YHL001W	AFUA_6G03830	0.397471165
YNL255C	AFUA_1G07630	0.390527675
YGL006W	AFUA_3G10690	0.389206341
YIL125W	AFUA_4G11650	0.387564721
YMR011W	AFUA_5G01160	0.381368332
YMR010W	AFUA_5G03510	0.378223656
YMR319C	AFUA_4G14640	0.376638524
YDL110C	AFUA_1G09160	0.376540694
YOL086C	AFUA_5G06240	0.376281416
YGR076C	AFUA_4G09000	0.373900773
YOR204W	AFUA_4G07660	0.368301524
YMR229C	AFUA_2G16040	0.367246087
YLR025W	AFUA_1G06420	0.363738278
YPL281C	AFUA_6G06770	0.363265962
YOR294W	AFUA_7G04430	0.363077313
YLR299W	AFUA_7G04760	0.36169498
YCL038C	AFUA_2G06170	0.359699115
YKL113C	AFUA_3G06060	0.351986208
YHL020C	AFUA_5G09420	0.350312962
YMR224C	AFUA_6G11410	0.348946198
YDR444W	AFUA_3G04240	0.345942811
YER107C	AFUA_1G09020	0.34238252
YDR162C	AFUA_2G03680	0.342381673
YNL212W	AFUA_3G08750	0.341561675
YNL097C	AFUA_3G11940	0.339603963
YKL191W	AFUA_6G07100	0.337319743
YPL030W	AFUA_6G02570	0.336821966
YNL224C	AFUA_3G05330	0.336636223
YKR024C	AFUA_5G11050	0.328747023
YPR020W	AFUA_1G16280	0.323479927
YPL171C	AFUA_2G04060	0.322952534
YOR393W	AFUA_6G06770	0.321210691
YPR088C	AFUA_5G03880	0.31284654
YGL058W	AFUA_6G14210	0.312000165
YFL059W	AFUA_5G08090	0.309971957
YBR243C	AFUA_2G11240	0.307271969
YJL048C	AFUA_3G06360	0.306215007
YNL055C	AFUA_4G06910	0.305279025
YOR058C	AFUA_2G16260	0.305221634
YGR240C	AFUA_4G00960	0.304340045
YBR279W	AFUA_2G11000	0.303224
YPL078C	AFUA_8G05440	0.300712891
YLR059C	AFUA_3G11820	0.296403753
YLR380W	AFUA_6G12690	0.295562758
YDL174C	AFUA_1G17520	0.29517154

TABLE 4: Continued.

S. cerevisiae gene	A. fumigatus gene	Correlation coefficient
YCL057W	AFUA_7G05930	0.293762978
YDR234W	AFUA_5G08890	0.293096856
YDR438W	AFUA_5G12140	0.287954196
YGL028C	AFUA_8G05610	0.287899385
YML004C	AFUA_6G07940	0.285261452
YOL045W	AFUA_2G02850	0.284622252
YLR429W	AFUA_2G14270	0.283845667
YDR397C	AFUA_3G02340	0.283499213
YBR260C	AFUA_3G06280	0.282196097
YCR087C-A	AFUA_7G04700	0.281322207
YMR011W	AFUA_7G00950	0.280854319
YOL148C	AFUA_1G16580	0.274850272
YDR379W	AFUA_1G12680	0.273041289
YJL151C	AFUA_5G10590	0.272979399
YML020W	AFUA_5G12090	0.266526854
YGR082W	AFUA_6G11380	0.265129767
YJR113C	AFUA_1G04280	0.262789584
YIL031W	AFUA_5G03200	0.256820972
YDR432W	AFUA_3G10100	0.255963453
YDR359C	AFUA_4G07560	0.255317292
YDR233C	AFUA_6G13670	0.253861938
YGR255C	AFUA_4G12930	0.252312468
YFR009W	AFUA_4G06070	0.245534257
YPL280W	AFUA_3G08490	0.243781447
YOR075W	AFUA_2G09670	0.239487119
YJL219W	AFUA_7G00950	0.239124986
YCR026C	AFUA_2G14770	0.234276623
YOR391C	AFUA_3G08490	0.233098887
YGL180W	AFUA_4G09050	0.231299038
YLR323C	AFUA_5G07720	0.231114462
YGL006W	AFUA_7G01030	0.222088168
YBR179C	AFUA_5G13392	0.219453048
YKR082W	AFUA_4G05840	0.214144582
YLR087C	AFUA_2G13520	0.209434557
YCL064C	AFUA_4G07810	0.204494803
YOL156W	AFUA_5G01160	0.203744073
YBR249C	AFUA_7G04070	0.202306902
YJL093C	AFUA_3G07540	0.201465401
YOL077C	AFUA_1G02210	0.196710807
YLR426W	AFUA_1G06280	0.196041865
YMR060C	AFUA_2G03840	0.194299595
YOR048C	AFUA_1G13730	0.193994391
YDL174C	AFUA_7G02560	0.193117882
YJL221C	AFUA_8G07070	0.187015236
YGR054W	AFUA_3G05970	0.181434929
YPR176C	AFUA_7G04460	0.180215936
YML003W	AFUA_1G09870	0.173517378
YKR084C	AFUA_2G04630	0.171646807
YDR270W	AFUA_4G12620	0.171601357

TABLE 4: Continued.

S. cerevisiae gene	A. fumigatus gene	Correlation coefficient
YOR336W	AFUA_2G02360	0.164107564
YFL058W	AFUA_5G02470	0.160695353
YNL301C	AFUA_2G07380	0.151927636
YJL221C	AFUA_7G06380	0.150927464
YGR280C	AFUA_7G03690	0.150422383
YOL089C	AFUA_6G01960	0.149731857
YGR193C	AFUA_3G08270	0.14875632
YDR328C	AFUA_5G06060	0.143711765
YPL254W	AFUA_2G06060	0.143220591
YPL206C	AFUA_2G00990	0.138505935
YEL021W	AFUA_2G08360	0.13720587
YOL068C	AFUA_4G12120	0.134069861
YJR160C	AFUA_7G06390	0.131179873
YNL310C	AFUA_6G08230	0.130305791
YIL061C	AFUA_5G13480	0.126828721
YGL163C	AFUA_6G12910	0.12597657
YGR173W	AFUA_5G06770	0.123606414
YJL092W	AFUA_2G03910	0.122708164
YJR065C	AFUA_5G11560	0.121854722
YBR023C	AFUA_8G05630	0.120791924
YBR084W	AFUA_3G08650	0.118154138
YMR197C	AFUA_4G10710	0.114428905
YMR241W	AFUA_5G04220	0.114113659
YNL097C	AFUA_7G01870	0.109316946
YGL171W	AFUA_1G16290	0.103112508
YOR365C	AFUA_4G13340	0.094260129
YHR215W	AFUA_6G11330	0.092128581
YKL013C	AFUA_6G02370	0.088958318
YOR043W	AFUA_4G06130	0.087990985
YPR062W	AFUA_1G05050	0.087018417
YDR497C	AFUA_2G07910	0.078863767
YDR477W	AFUA_2G01700	0.074801159
YLR378C	AFUA_5G08130	0.074624328
YBR060C	AFUA_5G08110	0.073589072
YGL120C	AFUA_5G11620	0.069009268
YIL063C	AFUA_2G10810	0.06328373
YOL157C	AFUA_7G06380	0.06047767
YGL043W	AFUA_3G07670	0.059398891
YMR027W	AFUA_5G06710	0.058232831
YHR215W	AFUA_8G01910	0.05787337
YDL247W	AFUA_7G05190	0.05184022
YPR048W	AFUA_5G07290	0.049809159
YBL051C	AFUA_5G01940	0.048514144
YHR148W	AFUA_2G08320	0.047441878
YJR160C	AFUA_7G05190	0.045923723
YJL039C	AFUA_5G12670	0.042943521
YNL029C	AFUA_5G12160	0.031203315
YGR267C	AFUA_5G03140	0.029526621
YNL027W	AFUA_1G06900	0.02747405

TABLE 4: Continued.

S. cerevisiae gene	A. fumigatus gene	Correlation coefficient
YDL247W	AFUA_7G06390	0.023887472
YHR107C	AFUA_5G03080	0.020068991
YKL165C	AFUA_4G03970	0.01938034
YNR012W	AFUA_2G05430	0.019075112
YLR085C	AFUA_4G04420	0.018617278
YPL082C	AFUA_1G05830	0.017285628
YCL064C	AFUA_1G06150	0.014476469
YNR039C	AFUA_1G12090	0.011820922
YOR216C	AFUA_1G08830	0.008439713
YOR148C	AFUA_4G07550	0.006702846
YDR420W	AFUA_4G00500	0.006566287
YPR018W	AFUA_5G03720	−0.000683184
YDR322W	AFUA_5G12810	−0.002517201
YNR015W	AFUA_3G08390	−0.013154917
YBR251W	AFUA_5G11540	−0.01637247
YHL028W	AFUA_5G09020	−0.020657813
YPL146C	AFUA_2G05550	−0.023787247
YKL127W	AFUA_3G11830	−0.026619404
YJR160C	AFUA_2G10910	−0.029508087
YIR038C	AFUA_1G17010	−0.030714781
YDL247W	AFUA_2G10910	−0.033905332
YBR204C	AFUA_1G03540	−0.038361835
YJR010W	AFUA_3G06530	−0.038908707
YMR210W	AFUA_6G04640	−0.046477426
YPR187W	AFUA_1G05160	−0.048502572
YDR037W	AFUA_6G07640	−0.052027721
YIL134W	AFUA_6G05170	−0.053118026
YLR286C	AFUA_5G03760	−0.062116003
YKL179C	AFUA_1G14240	−0.062203351
YDR109C	AFUA_4G04680	−0.063717983
YOL157C	AFUA_3G07380	−0.064624439
YIR038C	AFUA_2G17300	−0.068756542
YDL102W	AFUA_2G16600	−0.088010728
YOR389W	AFUA_2G01940	−0.09375584
YBR244W	AFUA_3G12270	−0.097477745
YJL219W	AFUA_2G11520	−0.098366126
YPL126W	AFUA_7G02610	−0.101632712
YDL189W	AFUA_1G09400	−0.104331651
YHL016C	AFUA_1G04870	−0.106608293
YLR380W	AFUA_4G13930	−0.111437921
YPR165W	AFUA_6G06900	−0.117573203
YOL062C	AFUA_5G07930	−0.122120003
YGL248W	AFUA_1G14890	−0.124318908
YLR029C	AFUA_1G04660	−0.129375946
YMR214W	AFUA_2G08300	−0.130867913
YOR124C	AFUA_6G12270	−0.136050424
YOL023W	AFUA_1G06520	−0.136914472
YPL280W	AFUA_3G01210	−0.13706189
YLR380W	AFUA_7G06760	−0.141209958

TABLE 4: Continued.

S. cerevisiae gene	A. fumigatus gene	Correlation coefficient
YNL241C	AFUA_3G08470	−0.153108798
YOR391C	AFUA_3G01210	−0.154014157
YOL089C	AFUA_6G02330	−0.161608796
YGL215W	AFUA_3G10040	−0.170171082
YMR137C	AFUA_2G15220	−0.174259735
YER004W	AFUA_6G08930	−0.179730339
YDR365C	AFUA_2G05420	−0.181946191
YDR529C	AFUA_4G06790	−0.185927497
YDR062W	AFUA_6G00300	−0.189190157
YOL086C	AFUA_7G01010	−0.190813468
YOL026C	AFUA_5G03630	−0.19138562
YIL033C	AFUA_3G10000	−0.195925157
YBR290W	AFUA_4G13740	−0.19955841
YAR019C	AFUA_4G06750	−0.204744755
YHL007C	AFUA_2G04680	−0.20737573
YMR078C	AFUA_7G05480	−0.222707833
YDR311W	AFUA_4G11690	−0.22686955
YOL100W	AFUA_3G12670	−0.230714484
YGR162W	AFUA_2G09490	−0.232235851
YGL255W	AFUA_1G01550	−0.248116851
YPL038W	AFUA_6G01910	−0.252388553
YCL035C	AFUA_1G06100	−0.255759517
YNL082W	AFUA_2G13410	−0.259394819
YDR062W	AFUA_1G11890	−0.263523207
YOR390W	AFUA_2G16210	−0.269616243
YJR141W	AFUA_3G07970	−0.2738444
YOR365C	AFUA_2G17650	−0.274036262
YDL116W	AFUA_1G10860	−0.278315121
YKL020C	AFUA_1G12550	−0.278379939
YDL174C	AFUA_3G06820	−0.286419799
YBR029C	AFUA_1G07010	−0.296926336
YLR096W	AFUA_1G11080	−0.297540349
YFL028C	AFUA_6G05080	−0.303502181
YOR089C	AFUA_3G10740	−0.30892361
YPR010C	AFUA_2G13000	−0.309890067
YJL219W	AFUA_5G01160	−0.321415181
YNL097C	AFUA_4G11660	−0.329160424
YMR116C	AFUA_4G13170	−0.329415498
YPL280W	AFUA_5G01430	−0.331554435
YHR091C	AFUA_2G14030	−0.337931682
YIL030C	AFUA_2G08650	−0.350942099
YFR037C	AFUA_7G05510	−0.356342064
YOR391C	AFUA_5G01430	−0.356843862
YDL174C	AFUA_1G00510	−0.402217656
YML121W	AFUA_5G09650	−0.410825283
YFL060C	AFUA_2G08580	−0.411505108
YDR301W	AFUA_8G04040	−0.413112754
YJL221C	AFUA_3G07380	−0.413321446
YLR326W	AFUA_5G12410	−0.458648708

TABLE 4: Continued.

S. cerevisiae gene	A. fumigatus gene	Correlation coefficient
YBL108C-A	AFUA_4G03360	−0.487139446
YGR165W	AFUA_5G08380	−0.519339895
YKL157W	AFUA_4G09030	−0.604021491
YNL334C	AFUA_2G08580	−0.720610634

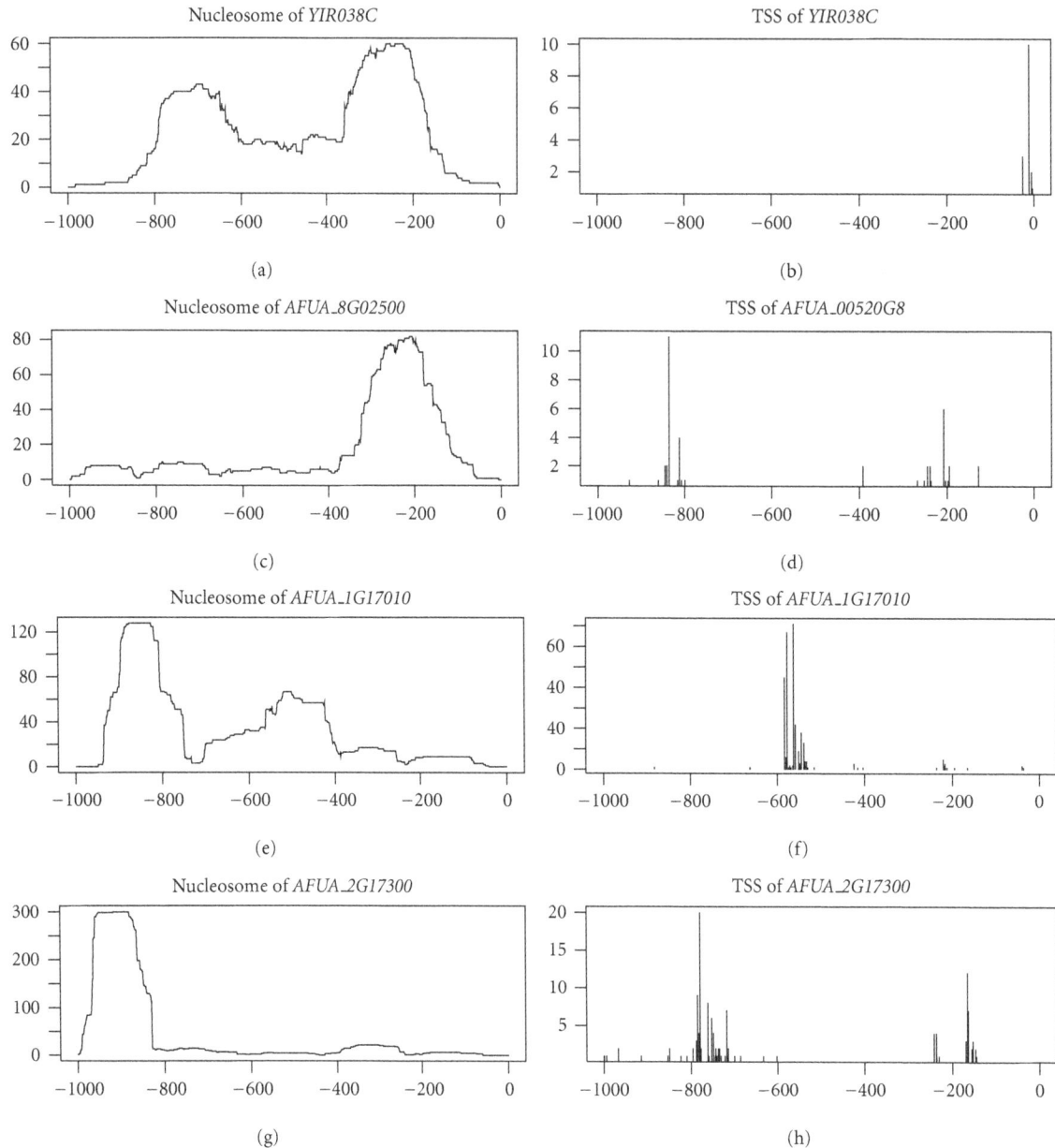

FIGURE 3: Mapping numbers of nucleosomes and transcription start sites in the promoter regions of YIR038C, AFUA_8G02500, AFUA_1G17010, and AFUA_2G17300. Position 0 indicates the translational start site.

Single-gene duplications and gene cluster duplications consisting of multiple genes were identified. One cluster of 4 genes (AFUA_1G00420 to AFUA_1G00470) is a duplication of another 4-gene cluster (AFUA_8G04120 to AFUA_8G04080) (Table 2). Among these gene pairs, the nucleosome position profile was poorly conserved in the gene promoter between AFUA_1G00470 and AFUA_8G04080 and in the gene body between AFUA_1G00440 and AFUA_8G04110 (Spearman's rank correlation coefficients were 0.43 and 0.23, resp.) (Table 2). With the exception of these 2 cases, the nucleosome position profile was highly conserved (correlation coefficients were higher than 0.7) (Table 2).

We analyzed another pair of duplicated clusters (9 genes) (*AFUA_1G16030* to *AFUA_1G16120* and *AFUA_5G14930* to *AFUA_5G15030*). The genes in each cluster have evolved for the same period after the duplication (Table 2). At present, conservation of the nucleosome position profiles varies among the 9 genes (Table 2). For example, the nucleosome position profile is poorly conserved in the gene promoters of 3 gene pairs (*AFUA_1G16050* and *AFUA_5G14950*, *AFUA_16110* and *AFUA_15020*, *AFUA_1G16120* and *AFUA_15030*) (Spearman's rank correlation coefficients are −0.35, −0.26, and −0.14, resp.). On the other hand, the nucleosome position profile is highly conserved in the promoters of *AFUA_16070* and *AFUA_5G14980* and was strongly correlated (correlation coefficient = 0.93). These results suggest that transcriptional regulation of duplicated genes is associated with nucleosome positions in the gene promoters.

3.2. Nucleosome Position Profiles of Orthologous Gene Promoters in Aspergillus fumigatus and Saccharomyces cerevisiae. I compared the nucleosome position profiles in the promoters of 347 orthologous pairs of yeast genes that showed notably high conservation in the control and mutant strains. In the 63 duplicated *A. fumigatus* gene pairs, 13 (20.6%) gene promoter profiles and 11 (17.5%) gene body profiles were highly correlated (Spearman's rank correlation coefficient > 0.7) (Table 1, Figure 1). On the other hand, of the 347 orthologous gene pairs, only 11 (3.2%) nucleosome position profiles were highly correlated (Spearman's rank correlation coefficient > 0.7) (Table 4, Figure 2). The distribution of correlation coefficients of the 347 orthologous gene promoters did not significantly differ from that of the control (gene pairs chosen at random) (P-value = 0.28 Kolmogorov-Smirnov test). One potential cause of this low conservation is the large evolutionary distance between the 2 fungi. *A. fumigatus* and *S. cerevisiae* belong to the subphyla Pezizomycotina and Saccharomycotina, respectively. Alternatively, this low conservation may represent a difference in mechanisms regulating the nucleosome arrangement, since the nucleosomal (nucleosome-bound) DNA lengths differ between the 2 fungi [9, 10].

Nucleosome position profiles in gene promoters are thought to be related to gene function. For example, *YIR038C* of *S. cerevisiae* encodes an amino acid sequence protein (glutathione S-transferase) similar to 3 genes (*AFUA_1G17010*, *AFUA_2G17300*, and *AFUA_8G02500*) in *A. fumigatus* (Table 4). Although the nucleosome position profiles show some conservation between *YIR038C* and *AFUA_8G02500* (Spearman's rank correlation coefficient = 0.55, except for one nucleosome position loss in *A. fumigatus*), they are poorly conserved between *YIR038C* and *AFUA_1G17010* (Spearman's rank correlation coefficient = −0.03) and between *YIR038C* and *AFUA_2G17300* (Spearman's rank correlation coefficient = −0.07) (Table 4, Figure 3).

Interestingly, although the nucleosome position profile of *AFUA_8G02500* is completely different from that of *AFUA_2G17300*, the transcription start site patterns are very similar between these genes (Figure 3), suggesting that the relationship between transcription start site and nucleosome position in the gene promoter varies.

Acknowledgments

The author thanks Dr. Shinji Kondo for helpful comments and critical review of the paper. This study was supported in part by a grant from the Institute for Fermentation, Osaka (IFO), Japan.

References

[1] K. Luger, A. W. Mäder, R. K. Richmond, D. F. Sargent, and T. J. Richmond, "Crystal structure of the nucleosome core particle at 2.8 Å resolution," *Nature*, vol. 389, no. 6648, pp. 251–260, 1997.

[2] C. K. Lee, Y. Shibata, B. Rao, B. D. Strahl, and J. D. Lieb, "Evidence for nucleosome depletion at active regulatory regions genome-wide," *Nature Genetics*, vol. 36, no. 8, pp. 900–905, 2004.

[3] G. C. Yuan, Y. J. Liu, M. F. Dion et al., "Molecular biology: genome-scale identification of nucleosome positions in *S. cerevisiae*," *Science*, vol. 309, no. 5734, pp. 626–630, 2005.

[4] H. Nishida, T. Motoyama, Y. Suzuki, S. Yamamoto, H. Aburatani, and H. Osada, "Genome-wide maps of mononucleosomes and dinucleosomes containing hyperacetylated histOnes of *Aspergillus fumigatus*," *PLoS One*, vol. 5, no. 3, article e9916, 2010.

[5] A. M. Tsankov, D. A. Thompson, A. Socha, A. Regev, and O. J. Rando, "The role of nucleosome positioning in the evolution of gene regulation," *PLoS Biology*, vol. 8, no. 7, article e1000414, 2010.

[6] E. Segal, Y. Fondufe-Mittendorf, L. Chen et al., "A genomic code for nucleosome positioning," *Nature*, vol. 442, no. 7104, pp. 772–778, 2006.

[7] A. B. Lantermann, T. Straub, A. Strålfors, G. C. Yuan, K. Ekwall, and P. Korber, "*Schizosaccharomyces pombe* genome-wide nucleosome mapping reveals positioning mechanisms distinct from those of *Saccharomyces cerevisiae*," *Nature Structural and Molecular Biology*, vol. 17, no. 2, pp. 251–257, 2010.

[8] S. A. Teichmann and M. M. Babu, "Gene regulatory network growth by duplication," *Nature Genetics*, vol. 36, no. 5, pp. 492–496, 2004.

[9] H. Nishida, T. Motoyama, S. Yamamoto, H. Aburatani, and H. Osada, "Genome-wide maps of mono- and di-nucleosomes of *Aspergillus fumigatus*," *Bioinformatics*, vol. 25, no. 18, pp. 2295–2297, 2009.

[10] T. Matsumoto, C.-S. Yun, H. Yoshikawa, and H. Nishida, "Comparative studies of genome-wide maps of nucleosomes between deletion mutants of *elp3* and *hos2* genes of Saccharomyces cerevisiae," *PLoS One*, vol. 6, no. 1, article e16372, 2011.

[11] H. Nishida, "Evolutionary conservation levels of subunits of histOne-modifying protein complexes in fungi," *Comparative and Functional Genomics*, vol. 2009, Article ID 379317, 6 pages, 2009.

[12] I. Uchiyama, T. Higuchi, and M. Kawai, "MBGD update 2010: toward a comprehensive resource for exploring microbial genome diversity," *Nucleic Acids Research*, vol. 38, no. 1, pp. D361–D365, 2009.

[13] H. Nishida, "Calculation of the ratio of the mononucleosome mapping number to the dinucleosome mapping number for

each nucleotide position in the *Aspergillus fumigatus* genome," *Open Access Bioinformatics*, vol. 1, pp. 1–6, 2009.

[14] D. E. SchOnes, K. Cui, S. Cuddapah et al., "Dynamic regulation of nucleosome positioning in the human genome," *Cell*, vol. 132, no. 5, pp. 887–898, 2008.

3

Mixed Sequence Reader: A Program for Analyzing DNA Sequences with Heterozygous Base Calling

Chun-Tien Chang,[1] Chi-Neu Tsai,[2] Chuan Yi Tang,[1] Chun-Houh Chen,[3] Jang-Hau Lian,[4] Chi-Yu Hu,[4] Chia-Lung Tsai,[5,6] Angel Chao,[5] Chyong-Huey Lai,[5] Tzu-Hao Wang,[5,6] and Yun-Shien Lee[4,6]

[1] Department of Computer Science, National Tsing Hua University, Hsin-Chu, Taiwan
[2] Graduate Institutes of Clinical Medical Sciences, Chang Gung University, No. 259 Wen-Hwa, 1st Road, Kwei-Shan, Tao-Yuan 333, Taiwan
[3] Institute of Statistical Science, Academia Sinica, Taipei, Taiwan
[4] Department of Biotechnology, Ming Chuan University, Tao-Yuan, Taiwan
[5] Department of Obstetrics and Gynecology, Lin-Kou Medical Center, Chang Gung Memorial Hospital, Chang Gung University, Fu-Hsing Street, Kwei-Shan, Tao-Yuan 333, Taiwan
[6] Genomic Medicine Research Core Laboratory, Chang Gung Memorial Hospital, No. 5, Fu-Hsing Street, Kwei-Shan, Tao-Yuan 333, Taiwan

Correspondence should be addressed to Tzu-Hao Wang, knoxtn@cgmh.org.tw and Yun-Shien Lee, bojack@mail.mcu.edu.tw

Academic Editors: A. Amorim, P.-O. Angrand, and H. Yang

The direct sequencing of PCR products generates heterozygous base-calling fluorescence chromatograms that are useful for identifying single-nucleotide polymorphisms (SNPs), insertion-deletions (indels), short tandem repeats (STRs), and paralogous genes. Indels and STRs can be easily detected using the currently available Indelligent or ShiftDetector programs, which do not search reference sequences. However, the detection of other genomic variants remains a challenge due to the lack of appropriate tools for heterozygous base-calling fluorescence chromatogram data analysis. In this study, we developed a free web-based program, Mixed Sequence Reader (MSR), which can directly analyze heterozygous base-calling fluorescence chromatogram data in .abi file format using comparisons with reference sequences. The heterozygous sequences are identified as two distinct sequences and aligned with reference sequences. Our results showed that MSR may be used to (i) physically locate indel and STR sequences and determine STR copy number by searching NCBI reference sequences; (ii) predict combinations of microsatellite patterns using the Federal Bureau of Investigation Combined DNA Index System (CODIS); (iii) determine human papilloma virus (HPV) genotypes by searching current viral databases in cases of double infections; (iv) estimate the copy number of paralogous genes, such as β-defensin 4 (DEFB4) and its paralog HSPDP3.

1. Introduction

The detection of genomic variations is important in studying the relationships between causative genes and diseases and the relationships between predisposing genes and complex trait diseases, such as type 2 diabetes, coronary heart disease, and cancers [1–4]. Structural genomic variations also provide important information about both genetic diversity and human evolution [5]. Human genomic variations include single-nucleotide polymorphisms (SNPs), variable number of tandem repeats (VNTRs), short tandem repeats (STRs, or microsatellites), and copy number variations (CNVs) [6]. Among these genomic variants, there are currently 51,810,853 reference SNPs for the human genome, which include 6,516,668 indel sequences and 5,214 microsatellite markers, according to dbSNP Build 135. For instance, a recent genome sequence of a human individual revealed 292,102 heterozygous indel events and 559,473 homozygous indels [7].

Genomic variants are frequently identified with heterozygous base-calling fluorescence chromatogram data generated from the direct sequencing of genomic PCR products using the dye-terminator method with Applied Biosystems (ABIs) autosequencers, such as models 3700 or 3730. Several groups have developed programs to analyze heterozygous base-calling fluorescence chromatogram data. For instance, Shift-Detector is a program for detecting shift mutations and calculating a probability score for sequences reconstructed from .abi sequence files [8]. The Indelligent program uses dynamic programming optimization to decode the heterozygous indels with International Union of Pure and Applied Chemistry DNA code system (IUPAC code) [9, 10]. The CHILD (CHromatogram In/Del Location and Detection) program was specifically designed to detect indels in DNA mixtures where one variant is rare, and it can also estimate the ratio of two variants [11]. Generally speaking, all of the currently available programs can be applied for the analysis of indel genomic variants even without reference sequences. However, the alignment of heterozygous base-calling fluorescent data with a reference database can be used to detect the physical position of indel within the genome. Even so, some heterozygous indels may not be easily visualized (Figures 1(b-4) and 2(a)).

Short tandem repeats (STRs) are the most important markers in forensic genetic analysis, and several commercial kits for STR analysis are available [12, 13]. Two-nucleotide repeats are the most prevalent STRs, while repeats with more nucleotides ($n = 3\sim6$) are less common in the human genome. Among the more than 5000 human STR markers, the Federal Bureau of Investigation (FBI) has included thirteen loci in its Combined DNA Index System (CODIS) database, which contains information from more than 6,384,379 individuals [14–16]. Current STR genotyping uses multiplex PCR with fluorescent STR primers to amplify genomic regions containing VNTRs or STRs, and the PCR products are then separated with capillary electrophoresis, which distinguishes fragments by length but does not display the actual sequence [17]. Nevertheless, sequencing methods remain very useful for analysis of STRs because they reveal the actual sequences of STRs [18], although the number may be too ambiguous to interpret (Figure 3(c)). To resolve this ambiguity, DNA cloning is often required to identify the different sequences of the two alleles. Currently, no program is able to analyze microsatellite repeat units (or CODIS) directly using the chromatogram trace data, even though the chromatogram trace profiles of heterozygous microsatellites are similar to those of heterozygous indels with 2 or more nucleotide deletions (Supplementary Figure 1 in Supplementary Materials available online at doi:10.1100/2012/365104).

Some genotypes of HPV are oncogenic. Therefore, routine screening and genotyping of HPV in women are crucial for prevention of cervical cancer. Current HPV detection methods are either using PCR techniques or HPV genotyping array system [19]. About 90% of cervical cancer tissues are infected by a single HPV genotype, while about 10% of specimens by two or more types of HPV. In the single HPV-infected samples, the viral genotype can be easily identified by genotyping array or PCR. For those with mixed infection by double types of HPV, mixed chromatogram traces are observed, in a similar way to those of SNP and indel sequences (Figure 4(b)).

CNVs, another type of genomic variation, are segments of DNA with variable copy numbers; the length of a single CNV may range from one kilobase to several megabases [20]. One well-known CNV occurs in the β-defensin 4 locus (*DEFB4*), which is known to have copy numbers that range from two to seven. The CNV of the defensin genes is associated with increased susceptibility to infectious diseases, autoimmune diseases, inflammatory disorders, Crohn's disease, and certain cancers [21, 22]. Array-based comparative genomic hybridization (aCGH) is used to detect genomic CNVs. Other methods for CNV validation include multiplex amplifiable probe hybridization combined with restriction enzyme digest variant ratios (MAPH/REDVRs) [23, 24], multiplex ligation-dependent probe amplification (MLPA) [25, 26], paralog ratio test (PRT) [24, 27, 28], and real-time polymerase chain reaction [29]. The PRT method uses a pair of primers to amplify two paralogous sequences, which are then separated with capillary electrophoresis. By detecting the ratio between the chromatographic intensities of the two PCR products, PRT can estimate the copy number of the CNV [24, 27, 28]. However, the chromatogram traces of the PCR products derived from paralogous genes are often too heterozygous to be analyzed with any currently available programs (Figure 5(b)).

To address the above issues, we have developed a program, Mixed Sequencer Reader (MSR), which can be used to identify indels, microsatellite copy numbers, and CODIS combinations. On the basis of the Indelligent method, the heterozygous sequences are identified as two distinct sequences, which are further aligned with reference genomic sequences to provide information about the physical position of indels, copy number or types of STR, and paralogous genes. We also applied this program to identify double HPV infections in cervical cancer tissues and estimate the copy number of paralogous human genes (e.g., the β-defensin 4 gene *DEFB4* and its paralog *HSPDP3*). The software is freely available at http://MSR.cs.nthu.edu.tw/.

2. Materials and Methods

2.1. Mixed Sequence Reader (MSR). The MSR program was developed to detect heterogeneity in chromatographic traces, determine the physical positions of the detected variants in the human genome, and identify the type of genomic variation present. The algorithm used in MSR was modified from that of Indelligent, but MSR is designed to use reference database alignment. The analytic steps used in MSR are described below and shown in Figure 1(a).

2.1.1. Importing the DNA Sequence Chromatographs and Selecting the Reference Database. The imported files are chromatography traces in the .abi format. The base peak

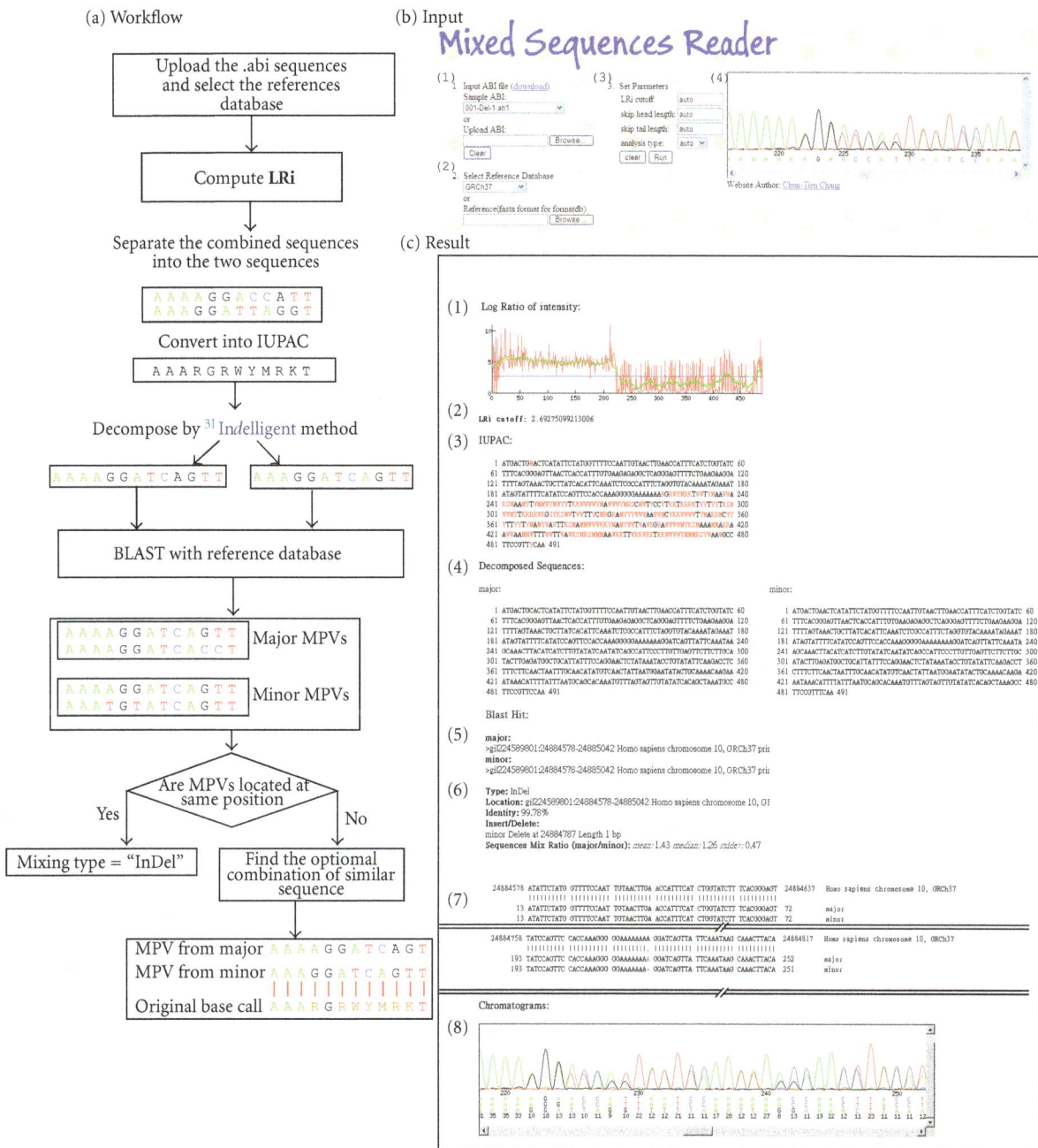

FIGURE 1: Workflow and user interfaces (input and result output) of the Mixed Sequences Reader (MSR). A flowchart describing the MSR workflow is shown in the left-side panel (a); the input interface of MSR is shown in the right-side panel (b); the result output is shown in panel (c). To input data, users can upload .abi sequences (b-1) and select the desired reference sequences (b-2). MSR defines an LRi (log ratio of intensity) value for each sequence as the log ratio of the two intensities of the combined signal peaks. On the basis of the smooth LRi curve (c-1), the LRi can be calculated (c-2); this value is used to separate the combined sequences into two sequences and then convert them into IUPAC code (c-3). The IUPAC codes are decomposed by Indelligent method into major and minor sequences (c-4). The major and minor sequences are BLASTed against a set of reference sequences to obtain the major and minor most possible variances (MPV) (c-5). If the MPVs are located at same position, the variant type is defined as an "indel"; if they are at different positions, the variant is defined as a "mixed" type (c-6). The variant type is then reported (c-7). The chromatographs of the analyzed sequences are shown in (b-4). The combination of major and minor sequences is shown in (c-8). Users have the option to define the LRi cut-off value, sequence type, and the ignored head and tail sequence lengths (b-3).

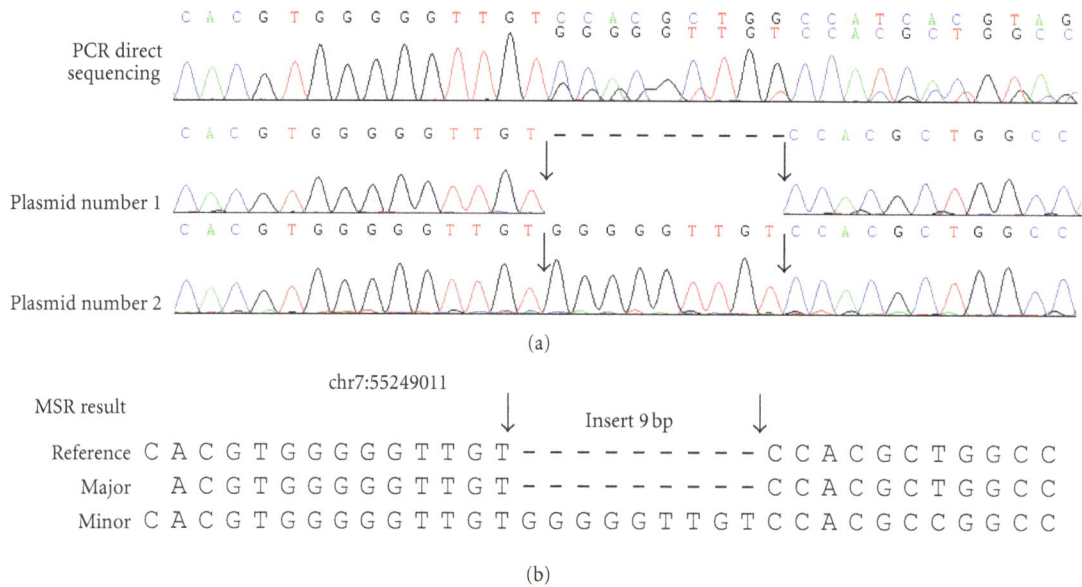

FIGURE 2: Experimental confirmation of the Indel identified by Mixed Sequence Reader. A 9-bp insertion at chromosome 7:55249011 was detected by Mixed Sequence Reader from a directly sequenced PCR product. (a) PCR direct sequencing chromatography trace, (b) MSR results. The PCR products were cloned, and at least ten single colonies were analyzed by DNA sequencing. One plasmid contained the wild type sequence (plasmid 1); whereas the other plasmids contained an insertion of "GGGGGTTGT," as shown in plasmid 2 (a). The insertion sequence is shown between two arrows.

positions, quality values, and four channel signals (A, C, G, T) recorded in the .abi file are extracted and analyzed to identify the major and the minor signals at each base location (Figures 1(b-1) and 1(b-4)). Users can select the desired reference database (Figure 1(b-2)).

2.1.2. Defining the Log Ratio of Intensity (LRi). The LRi value for each base position is defined as the log ratio of the chromatographic intensities of two combined sequences. The formula for LRi is:

$$LRi = \log_2\left(\frac{\text{major fluorescence intensity}}{\text{minor fluorescence intensity}}\right). \quad (1)$$

If the sequence position contained only one major band (no heterogeneity in the chromatographic trace), the value of LRi should be infinite ($\log_2(1/0)$). If the DNA sequences contain two heterogeneous chromatographic traces with equal intensities, the LRi should be 0 ($\log_2(1/1)$). After the .abi file is imported, users can define an LRi cut-off value (Figure 1(b-3)), or the MSR program can automatically set the cut-off value using smooth LRi (Figure 1(c-1), green line). For example, in the results shown in Figure 1(c-1), the shift of signal intensity at 212 bp was detected by the MSR program and considered an "indel"-type heterozygosity. The LRi values of "indel" sequences are higher than the cutoff value for sequences without heterogeneous fluorescence chromatography traces, so the LRi line drops when heterogeneous fluorescence traces are identified. When there is no obvious shift of the LRi line, the sequences might be a "mixed" type, and the LRi cutoff value is automatically set at 2.0 by the MSR program. The sequences of either "indel" or

"mixed" heterozygosities are then displayed in IUPAC code (Figure 1(c-3)).

2.1.3. Decomposing the IUPAC Code Using the Indelligent Algorithm. In the Indelligent algorithm, dynamic programming is used to convert the IUPAC code into two nucleotides (i.e., M is converted into A/C, W into A/T, Y into C/T, K into G/T, and S into G/C) [9, 10]. For ambiguous bases that cannot be decomposed with Indelligent, major and minor sequences are assigned according to the intensities of the corresponding fluorescence signals (Figure 1(c-4)).

2.1.4. Finding the DNA Sequence of the Most Possible Variances (MPVs) by BLAST. The major and minor DNA sequences derived from Section 2.1.3 were further analyzed by BLASTN against the databases that were built into the MSR program (GRCh37 primary reference assembly or CODIS, HPV, *DEFB4/HSPDP3* reference sequences) or user-defined reference sequences (Figure 1(b-2)) to detect the most possible variances (Figure 1(c-5)). The major and minor sequences were BLASTed against reference sequences to obtain the major and minor MPVs. Once the physical positions of the major and minor MPVs were identified, the sequences were categorized as indels (Figure 1(c-6)) or considered as a "mixed" type. For "mixed" type heterogeneities, MSR continues with the following procedures.

2.1.5. Deriving the Optimal Combinations. Each top M (major) MPV and top N (minor) MPV are combined pairwise into IUPAC code, resulting in (M × N) combinations. Each IUPAC code combination is aligned to the original signal

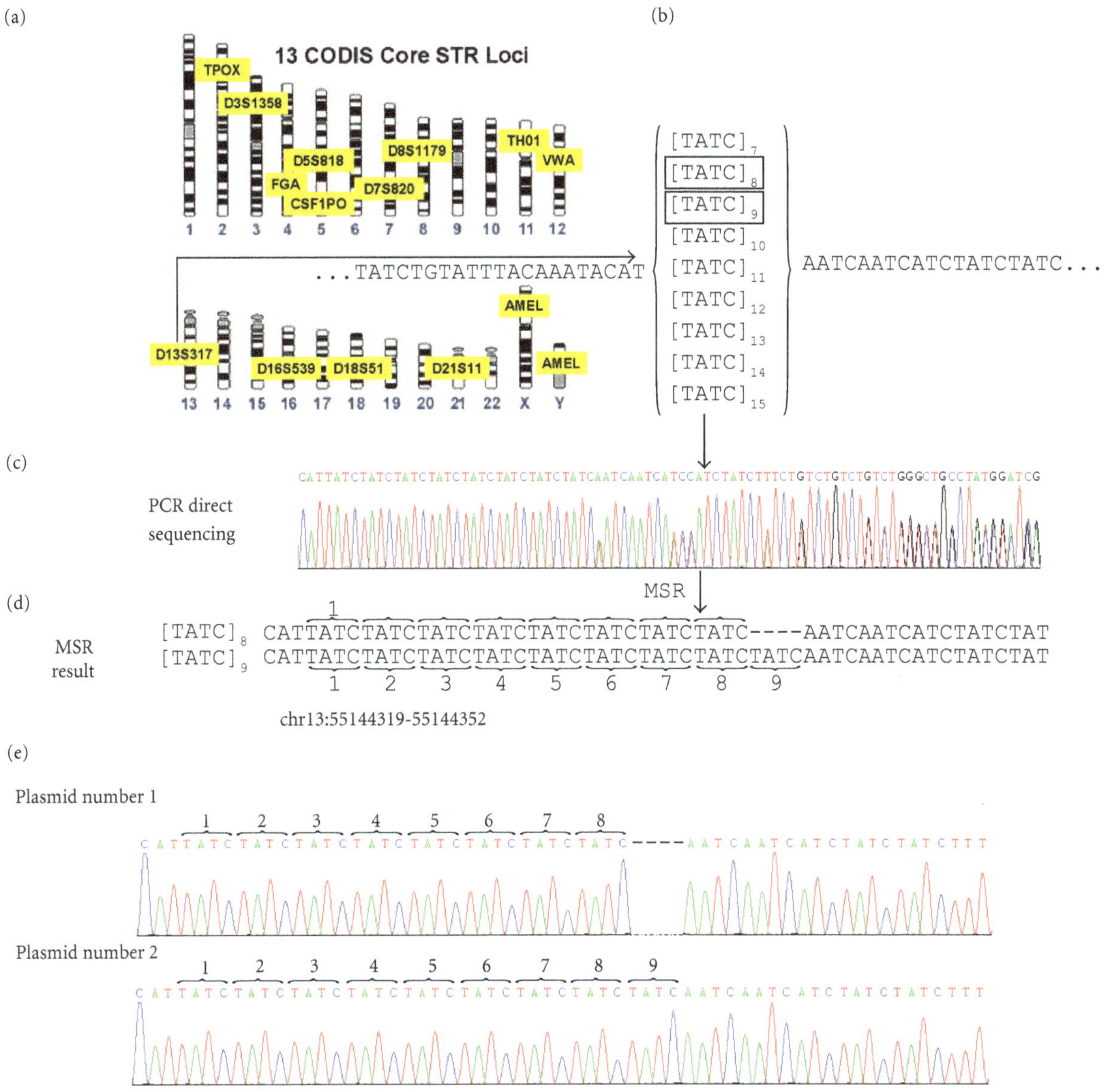

FIGURE 3: Experimental confirmation of one CODIS-STR locus identified by the Mixed Sequencer Reader. (a) The FBI CODIS Core STR Loci map (adapted from http://www.cstl.nist.gov/strbase/fbicore.htm). The D13S317 locus (chromosome 13:55144219-55144352 region) was randomly selected for analysis in this study. (b) The common repeat structure in the D13S317 locus in each person represents the combination of 2 structures in the CODIS database. (c) The direct sequencing chromatography trace of a PCR product containing the D13S317 locus at chromosome 13:55144219-55144352. (d) The MSR results indicated the presence of 8 and 9 copies of (TATC) within the individual. (e) The sequences of cloned PCR products were consistent with the MSR prediction. One of plasmid contained 8 copies of (TATC) (e.g., plasmid 1); whereas the other plasmids contained 9 copies of [TATC] (e.g., plasmid 2).

IUPAC code (Figure 1(c-3)), and the (M × N) IUPAC code combinations with the highest scores are identified as the optimal MPV combinations by the MSR program (Figure 1(c-7)). Using the selected reference database, such as the default MSR human genome sequence database or a user-imported database, the MSR program identifies the genotypes of the mixed sequences.

2.1.6. Calculating the Ratio between the Major and Minor Sequences. For all heterogeneous bases, the LRi values were calculated as described in Section 2.1.2. MSR calculated the

medium LRi value of heterogeneous nucleotides as the "Sequence Mix Ratio" (example shown in Figure 4(d)). The Sequence Mix Ratio is proportional to the signal ratio between the major and minor sequences.

2.2. DNA Cloning. To validate DNA sequences, PCR products amplified from the experimental samples were selected for cloning experiments to confirm the variations. The PCR products were cloned into the pCRII-TOPO cloning system (Invitrogen) according to the manufacturer's recommendations. The cloned DNA sequences were then analyzed with an ABI 3770 autosequencer.

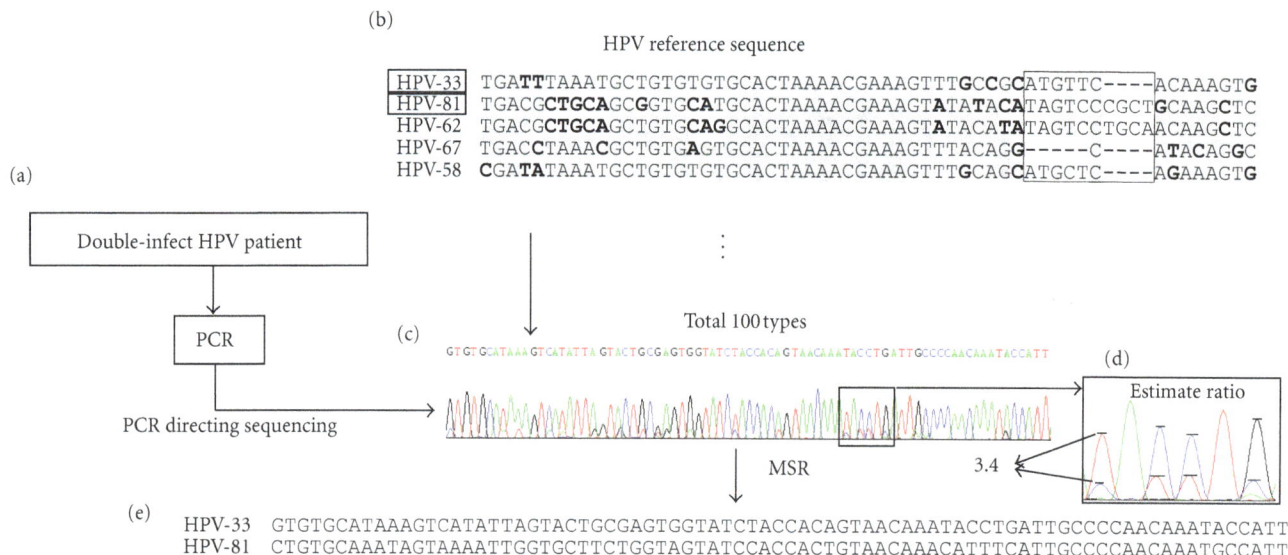

FIGURE 4: Application of MSR in identifying double-infection of HPV in a cervical cancer sample. (a) A sample infected with two strains of HPV was selected according to HPV array results. The HPV sequences in the specimen were amplified with a pair of PCR primers specific to the L1 region. (b) Alignment of the L1 sequences of several HPV genotypes. (c) PCR direct sequencing results showed a mixed chromatography trace representing two HPV genotypes. (d) MSR estimated the LRi ratio between the major/minor sequences. (e) The MSR output showed that the sample was infected with HPV genotypes 33 and 81.

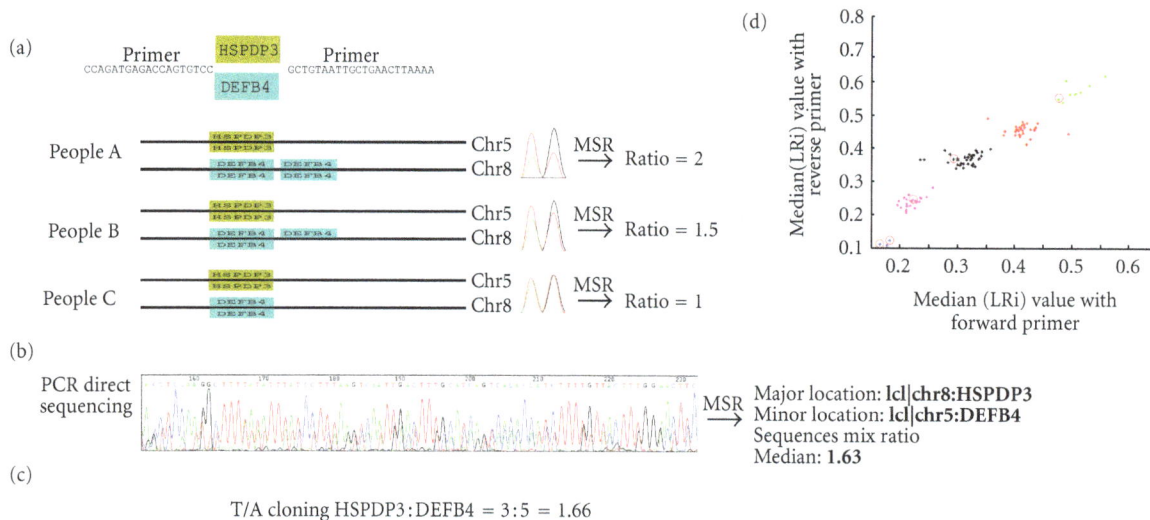

FIGURE 5: Experimental confirmation of *HSPDP3* and *DEFB4* copy numbers identified by Mixed Sequence Reader. (a) The primers used to amplify the *DEF4B* and *HSPDP3* genes in the human genome. The LRi ratio between *DEFB4* and *HSPDP3* was calculated for each chromatography trace. The copy number of *HSPDP3* is constant in the human genome ($n = 2$). (b) The heterozygous chromatography trace of PCR products comprising both *HSPDP3* and *DEFB4*. The median *DEFB4/HSPDP3* ratio was 1.63, which was estimated by MSR program. (c) The PCR product from (b) was cloned, and at least twenty colonies were picked for DNA sequencing to calculate the ratio between *DEFB4* and *HSPDP3*. The observed ratio was 1.66, compatible with that derived from MSR. (d) Comparison of median LRi values from the forward and reverse primers. Using the k-means algorithm, pairs of median LRi values were clustered into 5 groups corresponding to different copy numbers (2, 3, 4, 5, and 6). Five samples (marked with circles) were cloned into the T and A vectors to validate the *DEFB4* copy number.

2.3. Determination of the β-Defensin 4 (DEFB4) Copy Number. All DNA samples used in this study were unlinked from clinical information, and the DNA collection was approved by the Institutional Review Board (IRB) of Chang Gung Memorial Hospital (CGMH) (#99-0229B, IRB#100-2900A3). Genomic DNA samples from 100 normal individuals were tested for CNVs of the paralogous *DEFB4/HSPDP3* genes. *DEFB4* and its paralog, the *HSPDP3* pseudogene, were amplified from 50 genomic DNA specimens. PCR using a previously reported pair of primers [27] amplifies two products of similar size, one from the *DEFB4* gene on chromosome 8 and the other from the *HSPDP3* pseudogene on chromosome 5 (the copy number of *HSPDP3* is always 2). These PCR products were directly sequenced (see Figure 5(b) for an example). For each peak in the chromatogram trace, we calculated the LRi value, which indicated the *DEFB4/HSPDP3* ratio. For each sample, the median values of all heterogeneous sequences were used to determine the copy number of the *DEFB4* gene. The k-means method was used to partition the different copy numbers.

3. Results

3.1. The Mixed Sequences Reader Interface. The data import interface of the Mixed Sequence Reader is divided into three parts (Figure 1(b)): (1) data import, (2) reference database selection, and (3) MSR parameter settings. At the data import step, users can import an .abi file into MSR (Figure 1(b-1)). Users can also test the performance of MSR using 260 sample .abi files (Figure 1(b-1)), some of which were experimentally validated in this study. After the .abi files are imported, users can preview the chromatography data by moving the cursor in the right side panel (Figure 1(b-4)). Users can then either select the reference database to use (GRCh37, CODIS, HPV, *DEFB4/HSPDP3*) or import their own reference sequences in FASTA format (Figure 1(b-2)). Then, users can define the LRi cutoff value, sequence type and specify the ignored head and tail sequence lengths (Figure 1(b-3)). Then, users click the "Run" button to execute the MSR program. The results of a sample analysis are shown in Figure 1(c).

3.2. Indel and Short Tandem Repeat Sequences. We first analyzed indel sequences with MSR (Figure 2(b)). From the NCBI dbSNP database, we selected 6 indel sites of high heterozygosity, PCR amplified the corresponding genomic DNA, and directly sequenced the PCR products to obtain .abi files. Fourteen .abi files were successfully analyzed by MSR (Supplementary Table 1). To validate the MSR results, the predicted indel sequences were confirmed by cloning the PCR products, followed by sequencing of the plasmid DNAs. For example, one heterozygous chromatography trace was predicted to contain a nine nucleotide insertion at human chr7:55249011 (Figure 2(b)). Although similar results were also obtained when the sequences were analyzed with In*d*elligent or other programs, the MSR was able to identify the physical position of the indel by BLASTing the sequences against the GRCh37 database (Figure 2(b)). The experimental results confirmed the results predicted by MSR (Figure 2(a)).

We also selected several simple short tandem repeats (STRs) from the NCBI dbSNP database. Three STR-containing regions of the genome were amplified by PCR, sequenced, and analyzed with MSR (13 .abi files of STR in Supplementary Table 2). Among the 13 .abi files analyzed, the PCR products corresponding to 2 .abi files were further

cloned and validated by single-plasmid DNA sequencing. The number of each repeats were also defined by MSR (Supplementary Figure 1).

3.3. The Repeat Structure of a Short Tandem Repeat in the CODIS Database. Because CODIS is the largest STR database currently available, the repeat patterns of the 13 loci in CODIS were documented (Figure 3(a)) [14, 15]. The D13-S317 locus ($[TATC]_{7-15}$) was amplified by PCR using specific primers (Supplementary Table 3), and the resulting PCR products were directly sequenced (Figures 3(b) and 3(c)). According to the MSR results, the D13S317 locus genotype in the tested individual should be $[TATC]_8/[TATC]_9$ (Figure 3(d)). The PCR products were further cloned and sequenced, and the results confirmed the MSR prediction (Figure 3(e)). Similar results were obtained when the same .abi file was analyzed using the In*d*elligent program.

Some STRs in the CODIS database contain complex repeat structures, such as the D21S11 locus with the 65-repeat structure $[TCTA]_{4-11}[TCTG]_{3-14}[TCTA]_{0-3}[TA]_{0-1}$ $[TCTA]_3TCA[TCTA]_2TCCATA [TCTA]_{6-15}$. The alignment of the two sequences in the sample 211-CODIS-D21S11-3 .ab1 file detected two 4-bp gap structures (Supplementary Figure 2). Because of the complexity of the CODIS STR repeat structure, its pattern was not easily solved using the In*d*elligent program without a reference database. However, the optimal STR repeat structure was successfully identified by the MSR program using the reference databases. A total of 29 sequences for 10 CODIS STR sites were analyzed with MSR, and the results are shown in Supplementary Table 4.

3.4. Infection with Two Genotypes of HPV in Cervical Cancer Specimens. Another type of "mixed" sequence is represented by viral coinfection in the same specimen. For example, double HPV genotypes within one sample could be analyzed by EasyChip using probes designed to amplify the variable regions that are unique to each HPV genotype [19]. To validate the microarray data, dually infected specimens were amplified at the L1 region of HPV and the resulting PCR products were sequenced directly (Figure 4(c)). The alignment of sequences representing the different genotypes of HPV revealed both indels and SNPs (Figures 4(b) and 4(c)). One of the samples was identified by MSR as coinfected with HPV-33 and HPV-81, compatible with the EasyChip results (Figure 4(e)). In addition, the ratio between the major (HPV-33) and minor (HPV-81) sequences calculated by MSR could be used to estimate the relative ratio of the two genotypes of HPV. The ratio between HPV-33 and HPV-81 was 3.4 : 1. Genotype-specific PCR confirmed the results of MSR prediction for all 7 specimens that were defined by EasyChip as infected with multiple HPV strains (Supplementary Table 5).

3.5. Copy Number Variations of Paralogous Genes. We also applied MSR to detect the copy number of the *DEFB4* gene, which is well known for its multiple copy number variations. The PCR primers used amplified *DEFB4* and its paralog,

TABLE 1: Feature comparison between MSR and other currently available DNA sequencing software.

Feature/software	MSR	Indelligent	ShiftDetector	CHILD	PolyScan
Directly reads trace files (abi format)	Yes	No	Yes	Yes	No
Maps sequence to reference database	Yes	No	No	No	No
Detects indels	Yes	Yes	Yes	Yes	Yes
Decomposes 2 mixed sequences from a single trace	Yes	Yes	No	Yes	No
Estimates the ratio of 2 mixed sequences in one trace	Yes	No	No	Yes	No
Accepts NGS data	No	No	No	No	Yes

the pseudogene *HSPDP3* (Figure 5(a)) [27]; thus, the chromatography traces generated by PCR direct sequencing were very heterozygous (Figure 5(b)) because the chromatographs contained sequences from both chromosome 5 (*HSPDP3* as reference, copy number $n = 2$) and chromosome 8 (*DEFB4*, often variable copy numbers). The LRi value of each chromatography peak was used to estimate the ratio between *DEFB4*/*HSPDP3* and calculate the *DEFB4* copy number (Figure 5(a), right-side panel). The med(LRi) was defined as the median LRi value in all heterogeneous sequences. The scatter plot of the med(LRi) values of the DNA sequences amplified by the forward and reverse primers for 98 individual DNA samples is shown in Figure 5(d). The five groups shown in different colors were partitioned with the K-means clustering algorithm. To confirm these results, 5 specimens were randomly selected and the PCR products were cloned into the pCRII-TOPO cloning system for single-plasmid sequencing (Figure 5(d), circles; Supplementary Table 6). For each specimen, at least 20 clones were sequenced to calculate the *HSPDP3* and *DEFB4* ratio. All of the cloning-sequencing results were identical to the results predicted by MSR.

4. Discussion

Structural variations in the human genome are clinically important [6]. For instance, CNVs in *DEFB4* are associated with susceptibility to infectious disease, autoimmune, inflammatory disorders, and even cancers [5, 20]. The copy numbers of the genes *CCL3 L1*, *CCL4 L1*, and *TBC1D3* vary between 0 and 10 copies, and such variations have been found to be associated with susceptibility to HIV-1 [30, 31]. In addition, microsatellite markers are used as indicators for global genome stability and are especially useful in genomic research of cancer [32]. In short, structural variations shape the genome and determine disease susceptibility at the individual level. Therefore, analytical methods that can detect structural genomic variations are acutely needed to study the relationship between genomic variations and disease.

The MSR program introduced in this study can be used to directly analyze heterozygous base-calling chromatographs to detect multiple structural variations, including SNPs, microsatellites, and CNVs, in the human genome. The fluorescence intensity of chromatographs has already been used to detect SNPs [33], but we have extended this analysis of heterozygous base-calling chromatography to explore more structural variations in the human genome. The

accuracy of the MSR predictions was validated by other methods. Our analyses show that MSR can also be used to identify double infections by different genotypes of human papilloma virus (HPV) in cervical cancer tissues [34, 35]. It is worth testing whether MSR can be used to determine the presence of multiple viral infections in other cancer tissues [34, 35]. The ability to identify dual viral infections is limited because the reference sequences for these viruses often mutate rapidly, especially for RNA viruses [36]. Therefore, to improve this program for the aforementioned applications, we would need more clinical specimens to challenge this software.

The MSR program does have some limitations, however. First, the majority of structural variations in the human genome, which can be readily identified with MSR, are small indels. On the other hand, deletions of large fragments are likely better analyzed by array-based comparative genomic hybridization methods (aCGH) [37]. Second, some of the heterozygosity identified by base-calling chromatography may be caused by the formation of DNA secondary structures (likely in the GC-rich or AT-rich regions) that result in band compression; these sequences are difficult to analyze with MSR. Third, the STRs in the human genome are not yet fully characterized. Therefore, the MSR may not be able to predict all STRs. In this study, we used some types of STRs as examples of the potential of MSR, but more samples are needed to analyze all 13 core loci in the CODIS database and other STRs. Fourth, MSR and all other web sources only provide tools to analyze or predict structural variations in the human genome. We should rely on experimental data to confirm these results. Fifth, the MSR program is designed to read only fluorescence chromatography tracings derived from the ABI 3730 autosequencer, but not for those generated by the recently developed ultrahigh-throughput sequencers, such as Roche 454, ABI SoLid, and Illumina/Solexa.

A comparison of the features of the MSR program and other currently available DNA sequence analysis software is summarized in Table 1. All of the available programs can detect indels. Most of the programs can read files in .abi format, with the exception of Indelligent, which only processes IUPAC code data. Only MSR can map sequence data to a reference database and report the most possible mixed sequence. MSR and CHILD can estimate the ratio of sequences in a mixture. Only PolyScan can process next-generation sequencing data [38].

In conclusion, we have developed a user-friendly web-based program, Mixed Sequences Reader (MSR), to analyze heterozygous fluorescent chromatographs derived from an autosequencer. Using this program, several types of human genomic variations, including SNPs, indels, and CNVs of microsatellites or genes can be detected from a single DNA sequence read. Furthermore, MSR is useful for detecting viral infection with double genotypes in clinical specimens.

Authors' Contribution

Chun-Tien Chang and Chi-Neu Tsai contributed equally to this paper.

Conflict of Interests

The authors declare that they have no conflict of interests.

Acknowledgments

This work was supported by the National Research Program for Genomic Medicine (NSC-97-3112-B-001-020 to Y.-S. Lee and C.-H. Chen); The National Science Council, Taiwan (NSC-98-2320-B-182-034-MY3 to C.-N. Tsai and NSC-100-2221-E-126-011-MY3 to C. Y. Tang); Chang-Gung Memorial Hospital (CMRPD190571 to C.-N. Tsai); and the Department of Health, Taiwan (DOH99-TD-I-111-TM013, DOH-99-TD-C-111-006 to T.-H. Wang).

References

[1] A. C. Janssens and C. M. van Duijn, "Genome-based prediction of common diseases: advances and prospects," *Human Molecular Genetics*, vol. 17, no. 2, pp. R166–R173, 2008.

[2] T. A. Manolio, "Genomewide association studies and assessment of the risk of disease," *The New England Journal of Medicine*, vol. 363, no. 2, pp. 166–176, 2010.

[3] I. Menashe, D. Maeder, M. Garcia-Closas et al., "Pathway analysis of breast cancer genome-wide association study highlights three pathways and one canonical signaling cascade," *Cancer Research*, vol. 70, no. 11, pp. 4453–4459, 2010.

[4] S. Wacholder et al., "Performance of common genetic variants in breast-cancer risk models," *The New England Journal of Medicine*, vol. 362, no. 11, pp. 986–993, 2010.

[5] D. F. Conrad, D. Pinto, R. Redon et al., "Origins and functional impact of copy number variation in the human genome," *Nature*, vol. 464, no. 7289, pp. 704–712, 2010.

[6] Y. Nakamura, "DNA variations in human and medical genetics: 25 years of my experience," *Journal of Human Genetics*, vol. 54, no. 1, pp. 1–8, 2009.

[7] S. Levy et al., "The diploid genome sequence of an individual human," *PLoS Biology*, vol. 5, no. 10, article e254, 2007.

[8] E. Seroussi, M. Ron, and D. Kedra, "ShiftDetector: detection of shift mutations," *Bioinformatics*, vol. 18, no. 8, pp. 1137–1138, 2002.

[9] D. A. Dmitriev and R. A. Rakitov, "Decoding of superimposed traces produced by direct sequencing of heterozygous indels," *PLoS Computational Biology*, vol. 4, no. 7, Article ID e1000113, 2008.

[10] D. A. Dmitriev and R. A. Rakitov, Indelligent v.1.2, 2008, http://ctap.inhs.uiuc.edu/dmitriev/indel.asp.

[11] I. Zhidkov, R. Cohen, N. Geifman, D. Mishmar, and E. Rubin, "CHILD: a new tool for detecting low-abundance insertions and deletions in standard sequence traces," *Nucleic Acids Research*, vol. 39, no. 7, article e47, 2011.

[12] B. Budowle, A. Masibay, S. J. Anderson et al., "STR primer concordance study," *Forensic Science International*, vol. 124, no. 1, pp. 47–54, 2001.

[13] E. A. Cotton, R. F. Allsop, J. L. Guest et al., "Validation of the AMPFlSTR SGM Plus system for use in forensic casework," *Forensic Science International*, vol. 112, no. 2-3, pp. 151–161, 2000.

[14] J. M. Butler, "Genetics and genomics of core short tandem repeat loci used in human identity testing," *Journal of Forensic Sciences*, vol. 51, no. 2, pp. 253–265, 2006.

[15] J. M. Butler, "Short tandem repeat typing technologies used in human identity testing," *BioTechniques*, vol. 43, no. 4, pp. 2–5, 2007.

[16] The Federal Bureau of Investigation, "Combined DNA Index System (CODIS)," http://www.fbi.gov/about-us/lab/codis/codis.

[17] J. M. Butler, E. Buel, F. Crivellente, and B. R. McCord, "Forensic DNA typing by capillary electrophoresis using the ABI Prism 310 and 3100 genetic analyzers for STR analysis," *Electrophoresis*, vol. 25, no. 10-11, pp. 1397–1412, 2004.

[18] A. M. Divne, H. Edlund, and M. Allen, "Forensic analysis of autosomal STR markers using Pyrosequencing," *Forensic Science International*, vol. 4, no. 2, pp. 122–129, 2010.

[19] C. Y. Lin, A. Chao, Y. C. Yang et al., "Human papillomavirus typing with a polymerase chain reaction-based genotyping array compared with type-specific PCR," *Journal of Clinical Virology*, vol. 42, no. 4, pp. 361–367, 2008.

[20] E. H. Cook Jr. and S. W. Scherer, "Copy-number variations associated with neuropsychiatric conditions," *Nature*, vol. 455, no. 7215, pp. 919–923, 2008.

[21] M. Gersemann, J. Wehkamp, K. Fellermann, and E. F. Stange, "Crohn's disease-defect in innate defence," *World Journal of Gastroenterology*, vol. 14, no. 36, pp. 5499–5503, 2008.

[22] M. Groth, C. Wiegand, K. Szafranski et al., "Both copy number and sequence variations affect expression of human DEFB4," *Genes and Immunity*, vol. 11, no. 6, pp. 458–466, 2010.

[23] P. M. R. Aldred, E. J. Hollox, and J. A. L. Armour, "Copy number polymorphism and expression level variation of the human α-defensin genes DEFA1 and DEFA3," *Human Molecular Genetics*, vol. 14, no. 14, pp. 2045–2052, 2005.

[24] E. J. Hollox, U. Huffmeier, P. L. J. M. Zeeuwen et al., "Psoriasis is associated with increased β-defensin genomic copy number," *Nature Genetics*, vol. 40, no. 1, pp. 23–25, 2008.

[25] M. Groth, K. Szafranski, S. Taudien et al., "High-resolution mapping of the 8p23.1 beta-defensin cluster reveals strictly concordant copy number variation of all genes," *Human Mutation*, vol. 29, no. 10, pp. 1247–1254, 2008.

[26] J. P. Schouten, C. J. McElgunn, R. Waaijer, D. Zwijnenburg, F. Diepvens, and G. Pals, "Relative quantification of 40 nucleic acid sequences by multiplex ligation-dependent probe amplification," *Nucleic acids research*, vol. 30, no. 12, p. e57, 2002.

[27] J. A. L. Armour, R. Palla, P. L. J. M. Zeeuwen, M. D. Heijer, J. Schalkwijk, and E. J. Hollox, "Accurate, high-throughput typing of copy number variation using paralogue ratios from dispersed repeats," *Nucleic Acids Research*, vol. 35, no. 3, article e19, 2007.

[28] S. Deutsch, U. Choudhury, G. Merla, C. Howald, A. Sylvan, and S. E. Antonarakis, "Detection of aneuploidies by paralogous sequence quantification," *Journal of Medical Genetics*, vol. 41, no. 12, pp. 908–915, 2004.

[29] R. M. Linzmeier and T. Ganz, "Copy number polymorphisms are not a common feature of innate immune genes," *Genomics*, vol. 88, no. 1, pp. 122–126, 2006.

[30] J.R. Townson, Barcellos L.F., and R.J. Nibbs and, "Gene copy number regulates the production of the human chemokine CCL3-L1," *European Journal of Immunology*, vol. 32, no. 10, pp. 3016–3026, 2002.

[31] E. Gonzalez, H. Kulkarni, H. Bolivar et al., "The influence of CCL3L1 gene-containing segmental duplications on HIV-1/AIDS susceptibility," *Science*, vol. 307, no. 5714, pp. 1434–1440, 2005.

[32] G. H. Perry, "The evolutionary significance of copy number variation in the human genome," *Cytogenetic and Genome Research*, vol. 123, no. 1–4, pp. 283–287, 2009.

[33] C. Ngamphiw, S. Kulawonganunchai, A. Assawamakin, E. Jenwitheesuk, and S. Tongsima, "VarDetect: a nucleotide sequence variation exploratory tool," *BMC Bioinformatics*, vol. 9, no. 12, article S9, 2008.

[34] C. H. Lai, C. J. Chang, H. J. Huang et al., "Role of human papillomavirus genotype in prognosis of early-stage cervical cancer undergoing primary surgery," *Journal of Clinical Oncology*, vol. 25, no. 24, pp. 3628–3634, 2007.

[35] S. W. Yang, Y. S. Lee, T. A. Chen, C. J. Wu, and C. N. Tsai, "Human papillomavirus in oral leukoplakia is no prognostic indicator of malignant transformation," *Cancer Epidemiology*, vol. 33, no. 2, pp. 118–122, 2009.

[36] S. Duffy, L. A. Shackelton, and E. C. Holmes, "Rates of evolutionary change in viruses: patterns and determinants," *Nature Reviews Genetics*, vol. 9, no. 4, pp. 267–276, 2008.

[37] Y. S. Lee, A. Chao, A. S. Chao et al., "CGcgh: a tool for molecular karyotyping using DNA microarray-based comparative genomic hybridization (array-CGH)," *Journal of Biomedical Science*, vol. 15, no. 6, pp. 687–696, 2008.

[38] K. Chen, M. D. McLellan, L. Ding et al., "PolyScan: an automatic indel and SNP detection approach to the analysis of human resequencing data," *Genome Research*, vol. 17, no. 5, pp. 659–666, 2007.

4

The Evolutionary Relationship between Microbial Rhodopsins and Metazoan Rhodopsins

Libing Shen, Chao Chen, Hongxiang Zheng, and Li Jin

State Key Laboratory of Genetic Engineering and Key Laboratory of Contemporary Anthropology of Ministry of Education, School of Life Sciences, Fudan University, Shanghai 200433, China

Correspondence should be addressed to Li Jin; ljin007@gmail.com

Academic Editors: L. Han, X. Li, Z. Su, and X. Xu

Rhodopsins are photoreceptive proteins with seven-transmembrane alpha-helices and a covalently bound retinal. Based on their protein sequences, rhodopsins can be classified into microbial rhodopsins and metazoan rhodopsins. Because there is no clearly detectable sequence identity between these two groups, their evolutionary relationship was difficult to decide. Through ancestral state inference, we found that microbial rhodopsins and metazoan rhodopsins are divergently related in their seven-transmembrane domains. Our result proposes that they are homologous proteins and metazoan rhodopsins originated from microbial rhodopsins. Structure alignment shows that microbial rhodopsins and metazoan rhodopsins share a remarkable structural homology while the position of retinal-binding lysine is different between them. It suggests that the function of photoreception was once lost during the evolution of rhodopsin genes. This result explains why there is no clearly detectable sequence similarity between the two rhodopsin groups: after losing the photoreception function, rhodopsin gene was freed from the functional constraint and the process of divergence could quickly change its original sequence beyond recognition.

1. Introduction

Rhodopsin is a class of proteins whose common features are a seven-transmembrane alpha-helix apoprotein and a cofactor of retinal [1, 2]. Retinal works as a rhodopsin's chromophore which is responsible for light absorption. It reversibly and covalently binds to a lysine in the seventh helix of apoprotein. So to speak, the protein part of rhodopsin is its structural foundation while the retinal is rhodopsin's functional backbone. Rhodopsins are ubiquitously found in three domains of life—archaea, eubacteria, and eukaryotes [3–7]. According to their protein sequences, rhodopsins can be classified into two groups—Type 1 rhodopsins and Type 2 rhodopsins [2]. Type 1 rhodopsins exist in single-celled organisms while Type 2 rhodopsins only appear in multicellular animals. For convenience, we call Type 1 rhodopsins microbial rhodopsins and Type 2 rhodopsins metazoan rhodopsins in this study. Microbial rhodopsins function as phototaxis receptors (sensory rhodopsin), light-driven proton or chloride ion transporters (bacteriorhodopsin and halorhodopsin) [2, 3, 5, 6, 8]. Metazoan rhodopsins mainly function as visual receptors in animal's eyes such as rod or cone opsins [9–11]. Like microbial rhodopsins, metazoan rhodopsins also perform nonsensory functions. Melanopsin, expressed in brain and eyes, may be involved in circadian rhythms and papillary reflex [12]. Neuropsin (Opn5) is expressed in predominantly neural tissues [13]. Encephalopsin is expressed in brain and visceral organs [14]. RGR opsin, expressed in the retinal pigment epithelium (RPE) and Müller cells, functions as the photoisomerase [15, 16]. Peropsin is expressed in the retinal pigment epithelium (RPE) cells [17]. So far researchers have identified nine subgroups of nonvisual opsins in Metazoa [18–21].

The evolutionary relationship between microbial rhodopsins and metazoan rhodopsins is difficult to decide, because they show no clearly detectable identity at sequence level. Although lacking in sequence identity cannot be used to prove that they are not homologous proteins, sequence identity is the cornerstone for conventional knowledge of protein homology [22]. Due to evolutionary divergence, the sequence identity in different homologous proteins decreases with time. Our ability to detect sequence homology in related proteins depends on their divergence rate and evolutionary distance [23]. Using PAM matrix, Dayhoff et al. show that the limitation of sequence identity for deducing protein homology is around 20% identity [23]. If two proteins share less than 20% sequence identity, it means either they are not homologous proteins or their common origin is obliterated in evolution.

There are two possible evolutionary scenarios for microbial rhodopsins and metazoan rhodopsins: (1) using retinal as chromophore, binding retinal with a lysine and similar seven-transmembrane domain are the result of convergent evolution; (2) their common features are the legacy of a common ancestor, yet their sequence identity is hardly detectable because of the quick and/or longtime divergence.

To investigate the evolutionary relationship between microbial rhodopsins and metazoan rhodopsins, we have to bypass the problem of lacking sequence similarity. Fitch developed a statistical method to distinguish homologous proteins from nonhomologous ones [24]. His method compares the ancestral state from one protein group with the ancestral state from another. It circumvents the need of sequence identity to decide the evolutionary relationship between two groups of proteins. In this study, we used his method to test whether microbial rhodopsins and metazoan rhodopsins are homologous proteins or not.

2. Materials and Methods

2.1. Structure Data. A direct search in PDB database came back with two metazoan rhodospins and five microbial rhodopsins with structure data (Table 1).

2.2. Sequence Data. The whole genome protein sequences and corresponding cDNA sequences for twenty-seven metazoan species were downloaded from Ensembl database, NCBI database, and VectorBase [25]. These species cover seven phyla—Porifera, Cnidaria, Nematoda, Arthropoda, Chordata, Hemichordata, and Echinodermata. The species in Chordata also represented major classes in this phylum. We used a Perl script to extract the longest transcripts for each genome in this study.

2.3. BLAST and FASTA Search for Rhodopsin Genes in Genome Data. We used BLAST to search for rhodopsin genes in microbial genomes [26]. Using five microbial rhodopsins with structure data as queries, we searched the complete microbe genome database, fungi genome database, and green algae genome database on NCBI website. The BLAST parameters were set as follows: max target sequences were 500, expect threshold was 0.001, and the others were default.

We used FASTA 3.5 to search for rhodopsin genes in each metazoan genome [27]. Two metazoan rhodopsins with structure data served as queries. The E-value for FASTA search was set as 0.001.

Hits in BLAST or FASTA search result were aligned back to query sequences using MUSCLE with default parameters [28]. The hits were identified as candidate rhodopsins only when they share a conserved retinal-binding lysine in the

seventh helix as the same position as queries. We removed redundant candidate hits and any sequence shorter than 200 amino acids or longer than 1000 amino acids.

2.4. Structure Alignment. Using their PBD files, two metazoan rhodospin protein structures and five microbial rhodopsin protein structures were aligned with CE-MC multiple protein structure alignment server with default parameters [29].

2.5. Sequence Alignment. Microbial or metazoan rhodopsin protein sequences were aligned using MUSCLE with default parameters [28]. All nucleotide sequences in this study were aligned according to their protein sequence alignment result.

2.6. Test Region Selection. Although there is no clearly detectable sequence identity, protein structure is something comparable between microbial and metazoan rhodopsins. The selection of test region between microbial and metazoan rhodopsins was based on their structure alignment. The problem we encountered here is that structure data are far scarcer than sequence data in both groups of rhodopsins. Only two metazoan rhodospins and five microbial rhodopsins have structure data. So we have to use their structure alignment as a guide to infer seven-transmembrane domain in their sequence alignment.

All microbial rhodopsins share a clearly detectable sequence homology as well as all metazoan rhodopsins, so sequence alignment result is reliable within microbial or metazoan group. However, structure alignment result does not always coincide with sequence alignment result; that is, the positional homology proposed by microbial structure alignment may not be the same one proposed by microbial sequence alignment. Our solution is that we first aligned all microbial rhodospin sequences using MUSCLE. Then we picked out five microbial rhodopsin sequences with structure data in MUSCLE alignment result and compared their sequence alignment with their structure alignment. By doing so, we could identify the positional homology agreed by both alignment methods. We repeated this practice in metazoan rhodopsins using squid and bovine rhodopsins' structure alignment as a guide. The final test region is the alignment result agreed by both structure and sequence alignments.

2.7. Phylogenetic Analysis and Ancestral State Inference. Neighbor-joining, Bayesian, and maximum-likelihood methods were used to construct phylogenetic tree for microbial or metazoan rhodopsins. ProtTest was used to select evolution models for our phylogenetic analyses [30]. MEGA 5 was used to construct NJ tree with "pairwise deletion" option and "JTT" model [31]. Rates and patterns were set as "Gamma Distributed", and Gamma parameter was set as "4". Bootstrap method was used to test phylogeny, and number of bootstrap replications was set as "500". PhyML 3.0 was used to construct ML tree with "WAG" model [32]. Proportion of invariable sites and gamma shape parameter were estimated from alignment result. Approximate likelihood-ratio test was used to test for branch reliability [33]. MrBayes 3.1.1 was used to

TABLE 1: PDB accession numbers for two metazoan rhodopsins and five microbial rhodopsins.

PDB number	Protein name	Species	Classification
1U19	Rhodopsin	*Bos taurus* (bovine)	Eukaryota (Animalia)
2Z73	Rhodopsin	*Todarodes pacificus* (Japanese flying squid)	Eukaryota (Animalia)
1GU8	Sensory rhodopsin II	*Natronobacterium pharaonis*	Archaea (Halobacteria)
1JV6	Bacteriorhodopsin	Halobacterium salinarum	Archaea (Halobacteria)
1XIO	*Anabaena* sensory rhodopsin	*Nostoc* sp. pcc 7120	Bacteria (Cyanobacteria)
3A7K	Halorhodopsin	*Natronomonas pharaonis* dsm 2160	Archaea (Halobacteria)
3DDL	Xanthorhodopsin	Salinibacter ruber	Bacteria (Sphingobacteria)

construct Bayesian tree with "WAG" model [34]. We ran for 500,000 generations and sampled posterior probability trees every 1000 generations. We summarized 25% of both parameter values and trees to get the consensus tree.

PHYLIP package was used to construct Fitch-Margoliash tree for rhodopsin genes within each metazoan species [35]. Within-species rhodopsin tree was built with "JTT" model and tested with 100 bootstrap replicates.

Phylogenetic trees served as the evolutionary history for our ancestral state inference. Parsimony method was used to infer ancestral states [24]. We wrote a Perl script to implement this method.

2.8. Test for Relatedness in Ancestral States. The test for relatedness in two ancestral states is a statistic method Fitch devised in his 1970 paper [24]. The basic idea behind this test is that the probability of relatedness can be calculated by comparing the observed mutation distance between two ancestral states with the expected mutation distance between them. The observed mutation distance is the actual nucleotide differences between two ancestral states. The expected mutation distance between two ancestral states is the probability of randomly chosen disjoint nucleotide sets between them multiplied by the length of their sequence. The standard deviation between two distances is the square root of expected distance multiplied by the probability of randomly chosen intersectant nucleotide sets between them. The number of standard deviations between the observed mutation distance and the expected mutation distance follows normal distribution. The probability of its value could be found in the table of normal probability and it is used as the probability of significance.

3. Results

3.1. Structural Homology between Microbial Rhodopsin and Metazoan Rhodopsin. The structure alignment of five microbial rhodopsins and two metazoan rhodospins shows that all rhodopsins share a remarkable structural homology (Figure 1). Seven-transmembrane helices are conserved within microbial or metazoan rhodopsins and between them.

Although there is no clearly detectable sequence homology between these two groups of rhodopsins, the structure alignment reveals that they share a conserved WXXY sequence motif in the sixth helix. Interestingly, the lysine that binds retinal in the seventh is not structurally conserved and locates in different position between them. There is also an/a insertion/deletion in the seventh helix between these two groups of rhodopsins, which is just one amino acid before the crucial lysine in microbial rhodopsins (insertion) or one amino acid after the crucial lysine in metazoan rhodopsins (deletion). So the position of retinal-binding lysine shifts three amino acids forward in metazoan rhodopsins.

3.2. Rhodopsin Genes in Microbial and Metazoan Genomes. BLAST search for microbial rhodopsins came back with 62 microbial rhodopsins (See Table S1 in Supplementary Material available online at http://dx.doi.org/10.1155/2013/435651). FASTA search for metazoan rhodopsins came back with 227 metazoan rhodopsins from 25 species (Table 2).

In 62 microbial rhodopsins, thirty-five of them are from bacteria, twenty-four are from archaea, and three are from eukaryotes. Bacterium *Salinibacter ruber M8* and archaea *Haloarcula marismortui ATCC 43049* have four different copies of rhodopsin gene. One bacterium species and four archaea species have three different rhodopsin genes. Eleven microbial species have two different rhodopsin genes. Among three eukaryotic microbial rhodopsins, two of them are from single-celled green alga *Chlamydomonas reinhardtii* and one is from encapsulated yeast *Cryptococcus neoformans* var. *neoformans*.

Table 2 shows the number of rhodopsin genes in each metazoan species. We named rhodospin genes in numeric order within each metazoan species. The number of rhodopsin genes varies drastically in each metazoan species. In insects, malaria mosquito has nine rhodopsin genes while body blouse only has three. There is no rhodopsin gene found in sponge *Amphimedon queenslandica* and nematode *Caenorhabditis elegans*, although they do have rhodopsin-related genes.

Helix A   Helix B

```
2Z73:A|PDB ----MGRDLR DNETWWYNPS IVVHPHWREF DQVPDAVYYS LGIFIGICGI IGCGGNGIVI YLFTKTKSLQ TPANMFIINL AFSDFTFSLV NGFPLMTISC FLKKWIF--- ----------
1U19:A|PDB XMNGTEGPNF YVPFSNKTGV VRSPFEAPQY YLAEPWQFSM LAAYMFLLIM LGFPINFLTL YVTVQH-KKL RTPLNYILLN LAVADLFMVF GGFTTTLYTS LHGYFVF--- ----------
1XIO:A|PDB ---------- ---------- ---------- ----MNLESL LHWIYVAGMT IGALHFWSLS R-------NP RGVPQYEYLV AMFIPIWSGL AY-MAMAIDQ GKVEAAGQ-- ----------
1GU8:A|PDB ---------- ---------- ---------- ----MVGLTT LFWLGAIGML VGTLAFAWAG R-------DA GSGERRYYVT LVGISGIAAV AY-VVMALGV GWVPVA---- ----------
1JV6:A|PDB ---------- ---------- --------Q AQITGRPEWI WLALGTALMG LGTLYFLVKG M------GVS DPDAKKFYAI TTLVPAIAFT MY-LSMLLGY GLTMVPFG-- ----------
3A7K:A|PDB ----MTETLP PVTESAVALQ AEVTQRELFE FVLNDPLLAS SLYINIALAG LSILLFFVMT R------GLD DPRAKLIAVS TILVPVVSIA SYTGLASGLT ISVLEMPAGH FAEGSSVMLG
3DDL:A|PDB ---------- ---------- --MLQELPTL TPGQYSLVFN MFSFTVATMT ASFVFFVLAR N-------NV APKYRISMMV SALVVFIAGY HY-FRITSSW EAAYALQNGM YQPTGELFN-
```

Helix C   Helix D   Helix E

```
-----GFAAC KVYGFIGGIF GFMSIMTMAM ISIDRYNVIG RPMAASKKMS HRRAFIMIIF VWLWSVLWAI GPIFGWGAYT LEGVLCNCSF DYIS--RDST TRSNILCMFI LGFFGPILII
-----GPTGC NLEGFFATLG GEIALWSLVV LAIERYVVVC KP-MSNFRFG ENHAIMGVAF TWVMALACAA PPLVGWSRYI PEGMQCSCGI DYYTPHEETN NESFVIYMFV VHFIIPLIVI
-------IAH YARYIDWMVT TPLLLLSLSW TAMQFIKK-- ---------- ---DWTLIGF LMSTQIVVIT SGL--IADLS ---------- -----ERDWV RYLWYICGVC A-FLIILWGI
-----ERTVF APRYIDWILT TPLIVYFLGL LAGL------ ---------- ---DSREFGI VITLNTVVML AGF--AGAMV ---------- ------PGIE RYALFGMGAV A-FLGLVYYL
---GEQNPIY WARYASWLFT TPLLLLDLAL LVDA------ ---------- ---DQGTILA LVGADGIMIG TGL--VGALT ---------- -----KVYSY RFVWWAISTA A-MLYILYVL
GEEVDGVVTM WGRYLTWALS TPMILLALGL LAGS------ ---------- ---NATKLFT AITFDIAMCV TGL--AAALT ---------- ----TSSHLM RWFWYAISCA C-FIVVLYIL
--------D AYRYVDWLLT VPLLTVELVL VMGLPKNE-- ---------- ---RGPLAAK LGFLAALMIV LGY--PGEVS ---------- --ENAALFGT RGLWGFLSTI P-FVWILYIL
```

Helix E   Helix F   Helix G

```
FFCYFNIVMS VSNHEKEMAA MAKRLNAKEL RKAQAGANAE MRLAKISIVI VSQFLLSWSP YAVVALL-AQ FG----PLEW VTPYAAQLPV MFA KA-SAIH NPMIYSVSHP KFREAISQTF
FFCYGQLVFT VK-------- ----EAAAQQ QESATTQKAE KEVTRMVIIM VIAFLICWLP YAGVAFYIFT HQ----GSDF GP-IFMTIPA FFA KT-SAVY NPVIYIMMNK QFRNCMVTTL
WNPLRAKTRT Q--------- ---------- --------S SELANLYDKL VTYFTVLWIG YPIVWIIGPS G------FGW INQTIDTFLF CLLPFFS KVG FSFLDLHGLR NLNDSRQTTG
VGPMTESASQ R--------- ---------- --------S SGIKSLYVRL RNLTVILWAI YPFIWLLGPP G------VAL LTPTVDVALI VYLDLVIKVG FGFIALDAAA TLRAEHGESL
FFGFTSKAES M--------- ---------- --------R PEVASTFKVL RNVTVVLWSA YPVVWLIGSE G------AGI VPLNIETLLF MVLDVSAKVG LGLILLRSRA IFGEAEAPEP
LVEWAQDAKA A--------- ---------- --------G --TADIFSTL KLLTVVMWLG YPIVWALGVE G------VAV LPVGYTSWAY SALDIVAKYI FAFLLLNYLT SNEGVVSGSI
FTQLGDTIQR Q--------- ---------- --------S SRVSTLLGNA RLLLLATWGF YPIAYMIPMA FPEAFPSNTP GTIVALQVGY TIADVLAKAG YGVLIYNIAK AKSEEEGFNV
                                                            *  *
```

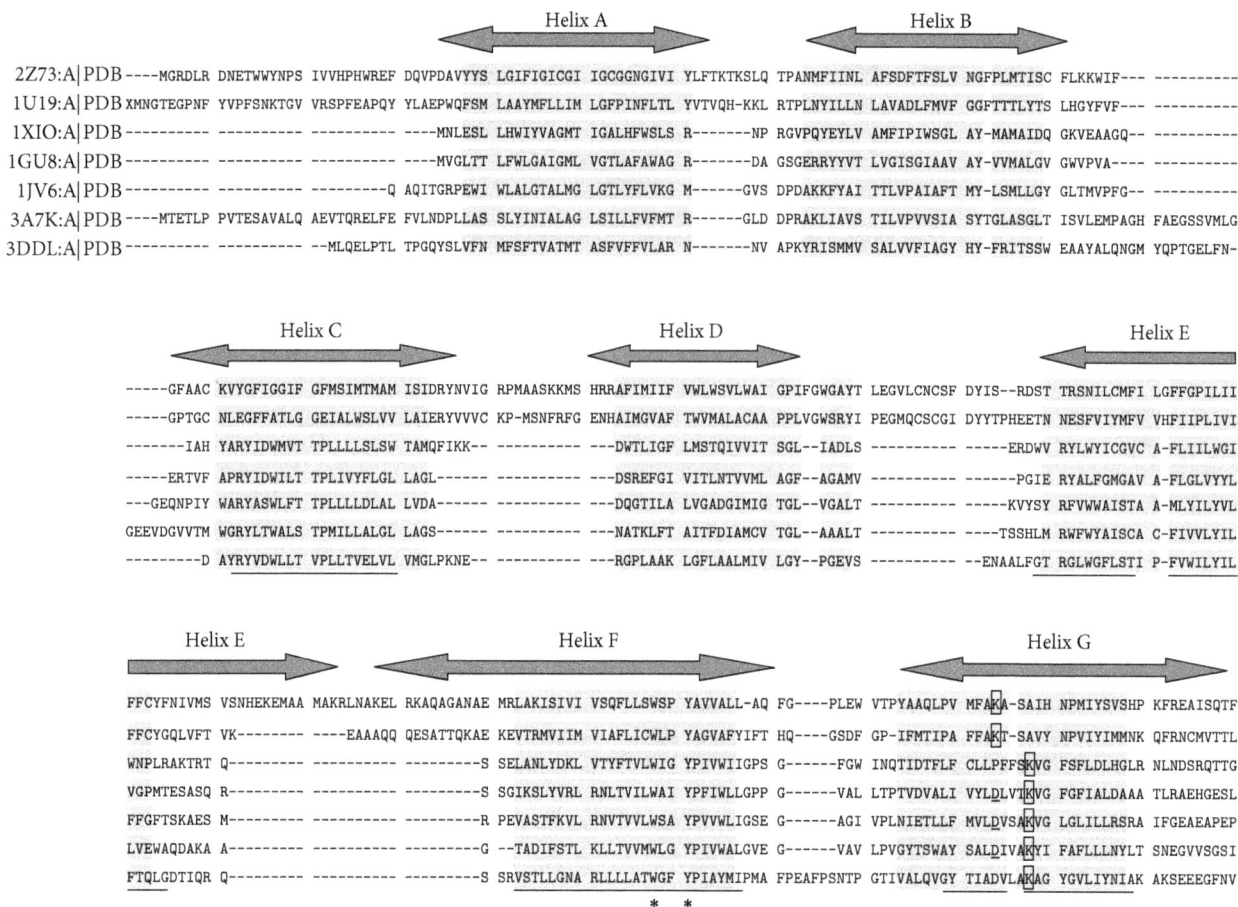

FIGURE 1: Structure alignment of squid rhodopsin (2Z73:A|PDB, metazoan rhodopsin), bovine rhodopsin (1U19:A|PDB, metazoan rhodopsin), *Anabaena* sensory rhodopsin (1XIO:A|PDB, microbial rhodopsin), *Natronomonas* sensory rhodopsin II (1GU8:A|PDB, microbial rhodopsin), *Halobacterium salinarum* bacteriorhodopsin (1JV6:A|PDB, microbial rhodopsin), *Natronomonas* halorhodopsin (3A7 K:A|PDB, microbial rhodopsin), and *Salinibacter ruber* xanthorhodopsin (3DDL:A|PDB, microbial rhodopsin). Squid rhodopsin is used as the template for delineating seven-transmembrane helices. Shaded residues are structural homologues. Conserved tryptophan and tyrosine in WXXY motif are marked with black asterisks. The retinal-binding lysine is in bold style and boxed. The aspartic acid in microbial rhodopsin corresponding to the retinal-binding lysine in metazoan rhodopsin is in bold style and underlined. The test region is marked with thin lines.

3.3. Final Test Region.

The final test region we selected is the consistent alignment result between structure and sequence alignments. There is no consistent region found in helices A, B, or D. In helix C, there is an 18-amino acid consistent region. In helix E, there are two consistent regions: one is 11 amino acid long and the other is 14 amino acid long. In helix F, there is a 25-amino acid consistent region. In helix G, there is an 18-amino acid consistent region. The total test region is 86 amino acid long and equals 258 nucleotides.

3.4. The Evolutionary History and Ancestral State Inference in Metazoan Rhodopsins.

We used three different methods to construct phylogenetic trees for all metazoan rhodopsins in this study. *Hydra* rhodopsins serve as an outgroup to root metazoan trees. In our study, *Hydra* is the only animal from Cnidaria. It is the basal phylum to Arthropoda, Chordata, Hemichordata, and Echinodermata. Rooted with *Hydra* rhodopsins, three trees show three different overall topologies. Neighbor-joining tree shows all rhodopsin genes

divided into three major clades except *Hydra* rhodopsins (Supplemental Figure 1). One clade mainly consists of chordate rhodopsins and no arthropod rhodopsins. The other two clades contain both chordate and arthropod rhodopsins. Maximum-likelihood tree shows a different evolutionary history from NJ tree (Supplemental Figure 2). ML tree has four major clades instead of three. Bayesian tree shows a more complicated evolutionary history (Supplemental Figure 3). Three separate clades in NJ tree are mixed in Bayesian tree. We did not know which tree is the most reliable one in all three trees. Three trees produced three different ancestral states. Only one state is true, because all metazoan rhodopsins share only one evolutionary history.

In order to get reliable ancestral state, we constructed the phylogenetic tree for rhodopsins within each metazoan species instead of for all metazoan rhodopsins (Figures 2(a) and 2(b)). By reducing the number of taxa in tree construction, we could get more reliable trees for ancestral state inference. Nevertheless, by doing so, we had to infer one ancestral

TABLE 2: The number of rhodopsin genes in each metazoan species.

Common name	Phylum	Scientific name	Number of rhodopsins
Sponge	Porifera	*Amphimedon queenslandica*	0
Hydra	Cnidaria	*Hydra magnipapillata*	4
Nematode	Nematoda	*Caenorhabditis elegans*	0
Carolina anole	Chordata	*Anolis carolinensis*	15
Malaria mosquito	Arthropoda	*Anopheles gambiae*	9
Honey bee	Arthropoda	*Apis mellifera*	5
Bovine	Chordata	*Bos taurus*	6
Amphioxus	Chordata	*Branchiostoma floridae*	20
Dog	Chordata	*Canis lupus familiaris*	6
Sea squirt	Chordata	*Ciona intestinalis*	5
Zebra fish	Chordata	*Danio rerio*	35
Armadillo	Chordata	*Dasypus novemcinctus*	2
Fruit fly	Arthropoda	*Drosophila melanogaster*	6
Atlantic cod	Chordata	*Gadus morhua*	25
Chicken	Chordata	*Gallus gallus*	12
Human	Chordata	*Homo sapiens*	8
Coelacanth	Chordata	*Latimeria chalumnae*	11
Opossum	Chordata	*Monodelphis domestica*	8
Mouse	Chordata	*Mus musculus*	8
Brown bat	Chordata	*Myotis lucifugus*	6
Platypus	Chordata	*Ornithorhynchus anatinus*	4
Body louse	Arthropoda	*Pediculus humanus*	3
Lamprey	Chordata	*Petromyzon marinus*	3
Acorn worm	Hemichordata	*Saccoglossus kowalevskii*	1
Sea urchin	Echinodermata	*Strongylocentrotus purpuratus*	2
Dolphin	Chordata	*Tursiops truncatus*	5
Clawed frog	Chordata	*Xenopus tropicalis*	17

state for each metazoan species. Using one *Hydra* rhodopsin as an outgroup, we constructed 24 metazoan rhodopsin trees and inferred 24 ancestral states based on these trees. Eighteen of them are possible metazoan rhodopsin's ancestral states in Chordata. Four of them are possible ancestral states in Arthropoda. Two of them are possible ancestral states in Hemichordata and Echinodermata.

3.5. The Evolutionary History and Ancestral State Inference in Microbial Rhodopsins. We also used three different methods to construct phylogenetic trees for all microbial rhodopsins. Three microbial trees are consistent in overall topologies, although they differ in the position of one branch which contains six bacteria rhodopsins (Supplemental Figures 4, 5, and 6). The problem is that bacteria and archaea are sister clades in biological systematics. It means that we are unable to root microbial trees. If we could not decide an outgroup for microbial trees, we would not be capable of inferring any ancestral state with them.

To overcome this problem, we first tried to find which microbial subtree is the most possible candidate tree for ancestral state inference. Using Fitch's method, we tested each extant microbial rhodospin gene with 24 metazoan ancestral states. We found that three microbial rhodopsins are distantly related to metazoan ancestral states with statistical

significance (Supplemental Table 2). These three rhodopsins are all located in one single subtree which contains 13 microbial rhodopsins (Figure 3). Then we inferred all possible microbial rhodopsin's ancestral states on this subtree.

3.6. The Relatedness between Microbial Rhodopsins' Ancestral States and Metazoan Rhodopsins' Ancestral States. We tested 24 metazoan rhodopsin's ancestral states with all possible microbial rhodopsin's ancestral states on the candidate subtree. Among all inferred microbial rhodopsin's ancestral states, one microbial rhodopsin's ancestral state has the smallest mutation distance with metazoan rhodopsin's ancestral states (Figure 3). This microbial ancestral state is reconstructed upon one fungi rhodopsin, one bacteria rhodopsin, and eight archaea rhodopsins. Test result shows that 13 metazoan rhodopsin's ancestral states are divergently related to it with statistical significance (Table 3). These ancestral states cover Arthropoda, Chordata, Hemichordata, Echinodermata, and two subphyla in Chordata—Tunicata (sea squirt) and Cephalochordata (amphioxus).

4. Discussion

4.1. Structural Homology versus Common Origin. Microbial rhodopsins and metazoan rhodopsins share a remarkable

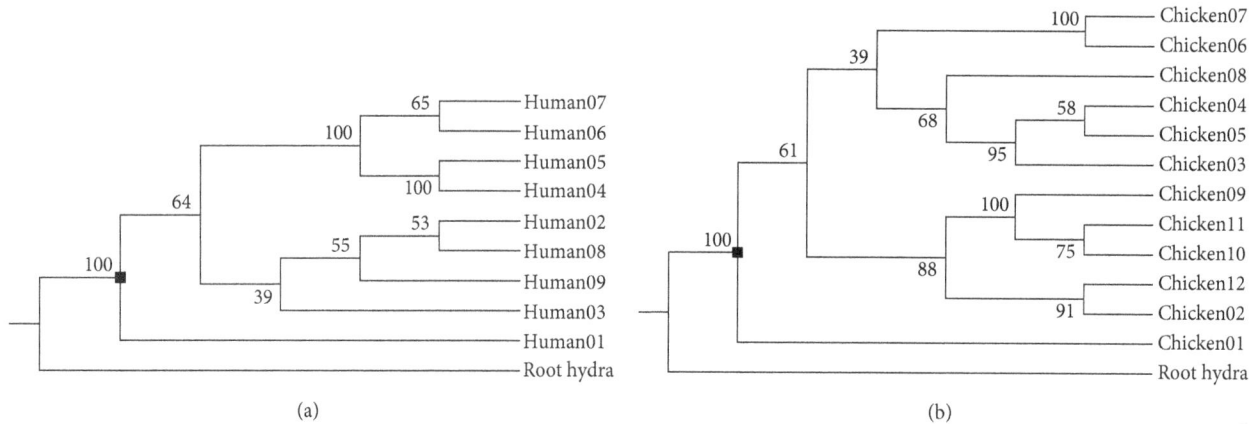

FIGURE 2: (a) Fitch-Margoliash tree for all rhodopsin genes in human. (b) Fitch- Margoliash tree for all rhodopsin genes in chicken. *Hydra* rhodopsin gene serves as outgroup. The numbers adjacent to tree nodes are bootstrap values. The tree node where ancestral state is built on is marked with a filled black square ■.

structural homology in their seven helices (Figure 1). However, the structural homology does not necessarily indicate the common origin. The empirical view of common origin is based on sequence homology. Convergent evolution is also a probable cause for structural homology [36]. Through both structure alignment and sequence alignment, we found that the vast majority of microbial and metazoan rhodopsins share a conserved WXXY sequence motif in the sixth helix. The tryptophan and tyrosine in this motif are crucial amino acids which form retinal-binding pocket in both groups of rhodopsins [37–41]. The conservation of WXXY motif in both groups of rhodopsin can be explained by either convergent evolution or common origin. In this case, common origin seems to be more plausible than convergent evolution. According to PAM matrix, tryptophan is the least mutable amino acid and tyrosine is the fifth-least mutable amino acid [23].

4.2. The Convoluted Evolutionary History of Metazoan Rhodopsins.

There is only one rhodopsin gene found in acorn worm while there are 35 rhodopsin genes found in zebra fish. No rhodopsin gene found in sponge and nematode indicates that rhodopsin is not essential for the survival of metazoa. However, photoreception capability does grant animals a great advantage for their survival. Nonessentiality and advantage for survival render the evolution of metazoans rhodopsins a birth-and-death process, in which gene duplication event creates new genes and some newly-created genes are kept in genome while others vanish from genome by accumulating deleterious mutations [42]. This process led to the various number of rhodopsin genes in different metazoan species; for example, body louse has three different rhodopsin genes while malaria mosquito has nine, and both of them are insects. It also made divergence and subfunctionalization rampant among duplicated rhodopsin genes. There are at least ten different subgroups of metazoan rhodopsins, and only one subgroup directly functions as visual opsins [11, 18–21]. The birth-and-death process produced a very complicated evolutionary history for metazoan rhodopsins. Due to

their convoluted evolutionary history and the large number of sequences used in phylogenetic analysis, we could not acquire an accurate phylogenetic tree for all metazoan rhodopsins. So in ancestral state inference, we used each species' rhodopsin genes to perform phylogenetic analysis in order to build a reliable tree within each metazoan species.

4.3. Gene Duplication and Horizontal Gene Transfer in Microbial Rhodopsins.

Gene duplication and horizontal gene transfer are common in microbial rhodopsins. Two microbial species have four rhodopsin genes, five species have three rhodopsin genes, and eleven species have two rhodopsin genes (Supplemental Table 1). Both of the gene duplication and horizontal gene transfers contribute to multiple rhodopsin copies in these species. For example, bacterium *Salinibacter ruber M8* has four rhodopsin genes. Its two sensory rhodopsins (Bac_Sal_s1 and Bac_Sal_s2) were the result of a gene duplication event, but they are clustered with archaea rhodopsins in microbial tree. It means that *Salinibacter ruber M8* got its original sensory rhodopsin from archaea through horizontal gene transfer. Horizontal gene transfer makes the origin of microbial rhodopsins untraceable. The fact that all three domains of life have microbial rhodopsins proposes that microbial rhodopsin is a very ancient gene. It could be as old as life itself.

4.4. Are Metazoan Rhodopsins and Microbial Rhodopsins Homologous Genes?

The main purpose of this study is to answer the question: are metazoan rhodopsins and microbial rhodopsins homologous genes? Due to the lack of direct evidence—sequence homology, we tried to answer this question by comparing their ancestral states. The complicated evolutionary history of metazoan rhodopsins made a reliable overall phylogenetic tree hardly possible. We circumvented this problem by building the phylogenetic tree for metazoan rhodopsins within each species. Then using these reliable trees, we inferred one ancestral state for each metazoan species.

TABLE 3: Mutation distance between 24 metazoan rhodopsin's ancestral states and their evolutionarily closest microbial rhodopsin's ancestral state (Figure 3). Within test region, the average mutation distance between existent microbial and metazoan rhodopsins is 119 ± 5 mutations in the first and second codon positions.

	Observed mutation distance	Expected mutation distance (standard deviation)	The probability of observed mutation distance is caused by chance
Carolina anole	66	79.7 (±6.5)	0.0183 (<0.05)
Malaria mosquito	72	90.1 (±6.5)	0.0029 (<0.05)
Honey bee	79	84.9 (±6.6)	0.1814
Bovine	81	87.1 (±6.6)	0.1736
Amphioxus	75	92.9 (±6.5)	0.0031 (<0.05)
Dog	80	89.3 (±6.6)	0.0778
Sea squirt	68	84.5 (±6.6)	0.006 (<0.05)
Zebra fish	77	92.8 (±6.5)	0.0078 (<0.05)
Armadillo	75	82.4 (±6.6)	0.1292
Fruit fly	74	86.4 (±6.6)	0.0294 (<0.05)
Atlantic cod	86	92.4 (±6.5)	0.1611
Chicken	63	80.1 (±6.5)	0.0045 (<0.05)
Human	69	81.8 (±6.5)	0.0256 (<0.05)
Coelacanth	69	78.7 (±6.5)	0.0681
Opossum	67	82.6 (±6.6)	0.0084 (<0.05)
Mouse	69	73.5 (±6.5)	0.2451
Brown bat	74	74.8 (±6.5)	0.4522
Platypus	80	92.3 (±6.5)	0.0294 (<0.05)
Body louse	70	86.8 (±6.6)	0.0052 (<0.05)
Lamprey	68	78.4 (±6.5)	0.0548
Acorn worm	87	105.8 (±6.4)	0.0016 (<0.05)
Sea urchin	73	84.5 (±6.6)	0.0392 (<0.05)
Dolphin	69	79.6 (±6.5)	0.0516
Clawed frog	83	93 (±6.5)	0.063

In our 24 metazoan rhodopsin's ancestral states, more than half of them are divergently related to the microbial rhodopsin's ancestral state with statistical significance and less than half of them without statistical significance (Table 2). There are two possible explanations for the reason why the other 11 metazoan rhodopsin's ancestral states show no statistical significance: (1) the birth-and-death process eliminated some basal metazoan rhodopsins in these species. Therefore, their phylogenetic trees only allowed us to trace back to a recent ancestral state instead of a much more ancient one; (2) in these species, the existent metazoan rhodopsins diverge from their ancestor so greatly that there is no traceable information left in their sequences. These two explanations are not mutually exclusive.

For thirteen metazoan rhodopsin's ancestral states divergently related to the microbial rhodopsin's ancestral states with statistical significance, does it mean that metazoan rhodopsin and microbial rhodopsin are homologous genes? By the definition of Fitch's test, the answer is yes. The test region we selected is total 86 amino acids. Within test region, the average mutation distance between existent metazoan and microbial rhodopsin is 119 ± 5 mutations in the first and second codon positions. Assuming one mutation in the first or second codon position would change its coding amino acid, each paired codon in the test region averagely shares about 1.38 mutations between two rhodopsin groups. It explains why we cannot find clearly detectable sequence homology between microbial and metazoan rhodopsins. After ancestral state reconstruction, the shortest mutation distance between microbial and metazoan ancestral states was 63 mutations. It is found between chicken and one microbial ancestral state inferred on nine microbial rhodopsins, with a P value of 0.0045. There are total 86 amino acids in the test region. If mutations were evenly distributed in each codon, there would be 63 amino acid differences between microbial and metazoan ancestral states. In another words, the sequence identity between microbial and metazoan ancestral states would be 23 amino acids. 23 divided by 86, it is about 26.7% sequence identity.

In pairwise sequence alignment, over 30% sequence identity is the safe standard for homologous proteins. Proteins sharing from 15% to 30% sequence identity are in the twilight zone, which means their homologous status is still in ambiguity [22]. Even when tracing back in time by reconstructing ancestral states, our result shows that only 26.7% sequence identity might exist in four helices

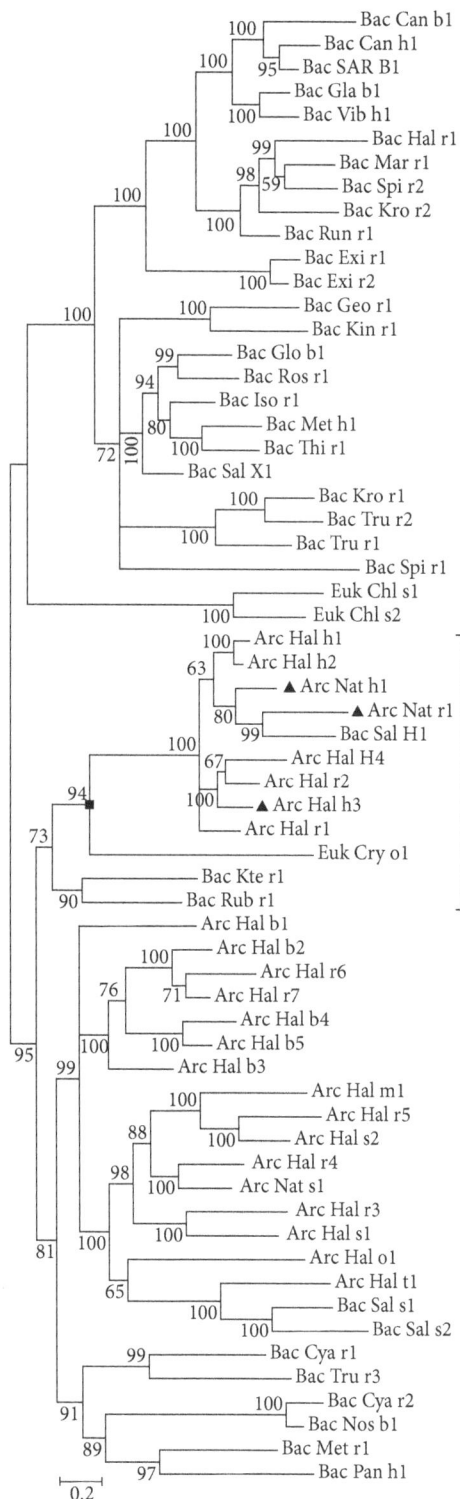

FIGURE 3: Unrooted Bayesian tree for all microbial rhodopsin genes. The numbers adjacent to the nodes are posterior probability values. The length of branch reflects evolutionary divergence. Microbial rhodopsin genes distantly related to metazoan rhodopsin's ancestral states (>95% quantile) are marked with a filled black triangle ▲. The tree node where microbial rhodopsin's ancestral state is built on is marked with a filled black square ■. In all possible ancestral states on that branch (marked with vertical line), this microbial rhodopsin's ancestral state has the smallest mutation distance with metazoan rhodopsin's ancestral states.

between ancestral microbial and metazoan rhodopsins. In conventional viewpoint, such result still cannot prove that metazoan and microbial rhodopsins are homologous proteins. Using *Hydra* rhodopsin as an outgroup, we can only infer metazoan ancestral rhodopsin states as early as in bilaterian ancestors. Fossil records show that the earliest bilaterian animal appeared about 580 million years ago [43]. However, based on the estimation of nuclear genes, early metazoan divergence can be traced back to 830 million years ago [44]. There is no rhodopsin gene found in sponge, and the closest microbe species related to Metazoa in this study is fungus *Cryptococcus neoformans* var. *neoformans*. So we have at least 250-million-year divergence time between microbial and metazoan ancestral states. Such longtime divergence could explain the low sequence identity between microbial and metazoan ancestral states. Certainly, the low sequence identity could also be seemingly explained by convergent evolution, which means rhodopsin gene appeared independently in microbes and Metazoa. But our result shows that ancestral microbial rhodopsins and ancestral metazoan rhodopsins shared about 26.7% sequence identity in four helices. It is implausible to believe that random mutations would create an almost identical structure by generating long strings of amino acids with similar sequences.

4.5. The Position of Retinal-Binding Lysine in the Seventh Helix. The structure alignment of microbial and metazoan rhodopsins shows an intriguing phenomenon: although both groups of rhodopsins have a retinal-binding lysine in the seventh helix, the position of this lysine is not structurally conserved between them (Figure 1). Its position shifts three amino acids forward in metazoan rhodopsins. Once again the different position of retinal-binding lysine could be simply explained by convergent evolution. However, most microbial rhodopsins have an aspartic acid in the position where metazoan rhodopsins have a retinal-binding lysine. In microbial rhodopsins, this aspartic acid functions as a part of counterion which balances the positive charge of retinal-binding lysine [45, 46]. Since structure alignment and ancestral state tests suggest that microbial and metazoan rhodopsins are homologous proteins, it means that this negatively charged aspartic acid in microbial rhodopsin mutated to the positively charged retinal-binding lysine in metazoan rhodopsin. The genetic code for aspartic acid is GAC or GAT while the genetic code for lysine is AAG or AAA. These two amino acids share the same adenine at the second codon position. The second codon position tends to have the slowest mutation rate among three codon positions [47]. It is probable that Asp (GAC or GAT coding) first mutated to Asn (AAC or AAT coding) and then Asn mutated to Lys (AAG or AAA coding) during the evolution of rhodopsin gene.

The retinal-binding lysine in the seventh helix is the most crucial amino acid for rhodopsin's photoreception function. It binds the chromophore retinal which is responsible for light absorption [1, 2]. If microbial and metazoan rhodopsins are homologous proteins, their retinal-binding lysine at different positions means that the function of photoreception was once lost during the evolution of rhodopsin gene. In metazoan rhodopsin, rescue mutation of this lysine salvaged the function of photoreception in metazoan rhodopsin. The once-lost lysine explains why there is no clearly detectable sequence homology between microbial and metazoan rhodopsins. During the evolution from single-celled organisms to multicellular animals, the rhodopsin gene in early metazoan ancestor lost retinal-binding lysine and therefore lost its function of photoreception. Loss of function freed the rhodopsin gene from functional constraint, and the process of divergence quickly changed its original sequence beyond recognition. Inexplicably in the later metazoan evolution, one of those loss-function rhodopsin genes managed to retrieve a lysine in its seventh helix through random mutation and therefore rescued its function of photoreception.

5. Conclusion

Based on our analysis, we propose that microbial and metazoan rhodopsins are homologous proteins and the function of photoreception was once lost during the evolution of rhodopsin gene. This conclusion may be controversial under the conventional view for homologous proteins. Logically, the view that microbial and metazoan rhodopsins are homologous proteins is the most parsimonious one. It does not require another protein to be the precursor of metazoan rhodopsins. Nature just recycled seven-transmembrane-helix protein for photoreception. However, the alternative view that the nearly identical structure between microbial and metazoan rhodopsins is the result of convergent evolution requires random mutations to create seven-transmembrane-helix domain twice through generating long strings of amino acids with similar sequences. Seven-transmembrane-helix domain does perform other functions than photoreception in Metazoa [48]. They form a large protein family of G-protein-coupled receptors which include metazoan rhodopsin and olfactory receptor. Research shows that most of these seven-transmembrane receptors share a common origin [49]. It is natural for someone to wonder what was the origin of all these seven-transmembrane receptors. There is no ancient seven-transmembrane receptor other than microbial rhodopsins which could be as old as life itself. For those who believe that the identical structure between microbial and metazoan rhodopsins is a result of convergent evolution, they will have to answer such two questions: (1) what was the precursor for all seven-transmembrane receptors in Metazoa; (2) if such a precursor existed, how could random mutations shape it into seven-transmembrane helices through generating long strings of amino acids which are also similar to a subset of microbial rhodopsins? On the other hand, our ancestral state inference failed to provide a decisive sequence identity between microbial and metazoan ancestral rhodopsins. The ambiguous sequence identity could be explained by once-relieved functional constraint and the long divergence time between microbes and metazoa. The divergence-time gap might be filled by using rhodopsin-related genes from basal animals for ancestral state inference. The future genome projects for basal animals could hold the ultimate answer to the question of the evolutionary relationship between microbial rhodopsin and metazoan rhodopsin.

Acknowledgments

The authors thank Dr. Xun Gu (Fudan University and Iowa State University) for his research suggestions in this research. This research was supported by grants from the National Basic Research Program (2012CB944600), Ministry of Science and Technology (2011BAI09B00), and National Science Foundation of China (30890034).

References

[1] K. Nakanishi, "11-cis-retinal, a molecule uniquely suited for vision," *Pure and Applied Chemistry*, vol. 63, pp. 161–170, 1991.

[2] J. L. Spudich, C. S. Yang, K. H. Jung, and E. N. Spudich, "Retinylidene proteins: structures and functions from archaea to humans," *Annual Review of Cell and Developmental Biology*, vol. 16, pp. 365–392, 2000.

[3] B. Schobert and J. K. Lanyi, "Halorhodopsin is a light-driven chloride pump," *Journal of Biological Chemistry*, vol. 257, no. 17, pp. 10306–10313, 1982.

[4] O. Beja, L. Aravind, E. V. Koonin et al., "Bacterial rhodopsin: evidence for a new type of phototrophy in the sea," *Science*, vol. 289, no. 5486, pp. 1902–1906, 2000.

[5] O. A. Sineshchekov, K. H. Jung, and J. L. Spudich, "Two rhodopsins mediate phototaxis to low- and high-intensity light in Chlamydomonas reinhardtii," *Proceedings of the National Academy of Sciences of the United States of America*, vol. 99, no. 13, pp. 8689–8694, 2002.

[6] K. H. Jung, V. D. Trivedi, and J. L. Spudich, "Demonstration of a sensory rhodopsin in eubacteria," *Molecular Microbiology*, vol. 47, no. 6, pp. 1513–1522, 2003.

[7] A. Terakita, "The opsins," *Genome Biology*, vol. 6, no. 3, article 213, 2005.

[8] S. A. Waschuk, A. G. Bezerra, L. Shi, and L. S. Brown, "Leptosphaeria rhodopsin: bacteriorhodopsin-like proton pump from a eukaryote," *Proceedings of the National Academy of Sciences of the United States of America*, vol. 102, no. 19, pp. 6879–6883, 2005.

[9] D. Arendt, K. Tessmar-Raible, H. Snyman, A. W. Dorresteijn, and J. Wittbrodt, "Ciliary photoreceptors with a vertebrate-type opsin in an invertebrate brain," *Science*, vol. 306, no. 5697, pp. 869–871, 2004.

[10] M. Koyanagi, K. Kubokawa, H. Tsukamoto, Y. Shichida, and A. Terakita, "Cephalochordate melanopsin: evolutionary linkage between invertebrate visual cells and vertebrate photosensitive retinal ganglion cells," *Current Biology*, vol. 15, no. 11, pp. 1065–1069, 2005.

[11] M. Koyanagi and A. Terakita, "Gq-coupled rhodopsin subfamily composed of invertebrate visual pigment and melanopsin," *Photochemistry and Photobiology*, vol. 84, no. 4, pp. 1024–1030, 2008.

[12] I. Provencio, G. Jiang, W. J. De Grip, W. Pär Hayes, and M. D. Rollag, "Melanopsin: an opsin in melanophores, brain, and eye," *Proceedings of the National Academy of Sciences of the United States of America*, vol. 95, no. 1, pp. 340–345, 1998.

[13] E. E. Tarttelin, J. Bellingham, M. W. Hankins, R. G. Foster, and R. J. Lucas, "Neuropsin (Opn5): a novel opsin identified in mammalian neural tissue," *FEBS Letters*, vol. 554, no. 3, pp. 410–416, 2003.

[14] S. Blackshaw and S. H. Snyder, "Encephalopsin: a novel mammalian extraretinal opsin discretely localized in the brain," *Journal of Neuroscience*, vol. 19, no. 10, pp. 3681–3690, 1999.

[15] M. Jiang, S. Pandey, and H. K. W. Fong, "An opsin homologue in the retina and pigment epithelium," *Investigative Ophthalmology and Visual Science*, vol. 34, no. 13, pp. 3669–3678, 1993.

[16] D. Shen, M. Jiang, W. Hao, L. Tao, M. Salazar, and H. K. W. Fong, "A human opsin-related gene that encodes a retinaldehyde-binding protein," *Biochemistry*, vol. 33, no. 44, pp. 13117–13125, 1994.

[17] H. Sun, D. J. Gilbert, N. G. Copeland, N. A. Jenkins, and J. Nathans, "Peropsin, a novel visual pigment-like protein located in the apical microvilli of the retinal pigment epithelium," *Proceedings of the National Academy of Sciences of the United States of America*, vol. 94, no. 18, pp. 9893–9898, 1997.

[18] M. Max, P. J. McKinnon, K. J. Seidenman et al., "Pineal opsin: a nonvisual opsin expressed in chick pineal," *Science*, vol. 267, no. 5203, pp. 1502–1506, 1995.

[19] S. Blackshaw and S. H. Snyder, "Parapinopsin, a novel catfish opsin localized to the parapineal organ, defines a new gene family," *Journal of Neuroscience*, vol. 17, no. 21, pp. 8083–8092, 1997.

[20] A. R. Philp, J. M. Garcia-Fernandez, B. G. Soni, R. J. Lucas, J. Bellingham, and R. G. Foster, "Vertebrate ancient (VA) opsin and extraretinal photoreception in the Atlantic salmon (Salmo salar)," *Journal of Experimental Biology*, vol. 203, no. 12, pp. 1925–1936, 2000.

[21] P. Moutsaki, D. Whitmore, J. Bellingham, K. Sakamoto, Z. K. David-Gray, and R. G. Foster, "Teleost multiple tissue (tmt) opsin: a candidate photopigment regulating the peripheral clocks of zebrafish?" *Molecular Brain Research*, vol. 112, no. 1-2, pp. 135–145, 2003.

[22] W. R. Pearson, "Protein sequence comparison and Protein evolution," Tutorial, ISMB2000, 2001.

[23] M. O. Dayhoff, R. M. Schwartz, and B. C. Orcutt, "A model of evolutionary change in proteins," *Atlas of Protein Sequence and Structure*, vol. 5, no. 3, pp. 345–352, 1978.

[24] W. M. Fitch, "Distinguishing homologous from analogous proteins," *Systematic zoology*, vol. 19, no. 2, pp. 99–113, 1970.

[25] D. Lawson, P. Arensburger, P. Atkinson et al., "VectorBase: a data resource for invertebrate vector genomics," *Nucleic Acids Research*, vol. 37, no. 1, pp. D583–D587, 2009.

[26] S. F. Altschul, W. Gish, W. Miller, E. W. Myers, and D. J. Lipman, "Basic local alignment search tool," *Journal of Molecular Biology*, vol. 215, no. 3, pp. 403–410, 1990.

[27] W. R. Pearson and D. J. Lipman, "Improved tools for biological sequence comparison," *Proceedings of the National Academy of Sciences of the United States of America*, vol. 85, no. 8, pp. 2444–2448, 1988.

[28] R. C. Edgar, "MUSCLE: multiple sequence alignment with high accuracy and high throughput," *Nucleic Acids Research*, vol. 32, no. 5, pp. 1792–1797, 2004.

[29] C. Guda, S. Lu, E. D. Scheeff, P. E. Bourne, and I. N. Shindyalov, "CE-MC: a multiple protein structure alignment server," *Nucleic Acids Research*, vol. 32, pp. W100–W103, 2004.

[30] F. Abascal, R. Zardoya, and D. Posada, "ProtTest: selection of best-fit models of protein evolution," *Bioinformatics*, vol. 21, no. 9, pp. 2104–2105, 2005.

[31] K. Tamura, D. Peterson, N. Peterson, G. Stecher, M. Nei, and S. Kumar, "MEGA5: molecular evolutionary genetics analysis using maximum likelihood, evolutionary distance, and maximum parsimony methods," *Molecular Biology and Evolution*, vol. 28, no. 10, pp. 2731–2739, 2011.

[32] S. Guindon, J. F. Dufayard, V. Lefort, M. Anisimova, W. Hordijk, and O. Gascuel, "New algorithms and methods to estimate maximum-likelihood phylogenies: assessing the performance of PhyML 3.0," *Systematic Biology*, vol. 59, no. 3, pp. 307–321, 2010.

[33] M. Anisimova and O. Gascuel, "Approximate likelihood-ratio test for branches: a fast, accurate, and powerful alternative," *Systematic Biology*, vol. 55, no. 4, pp. 539–552, 2006.

[34] J. P. Huelsenbeck and F. Ronquist, "MRBAYES: bayesian inference of phylogenetic trees," *Bioinformatics*, vol. 17, no. 8, pp. 754–755, 2001.

[35] J. Felsenstein, "Phylogeny Inference Package (PHYLIP)," Version 3.5., University of Washington, Seattle, Wash, USA, 1993.

[36] R. F. Doolittle, "Convergent evolution: the need to be explicit," *Trends in Biochemical Sciences*, vol. 19, no. 1, pp. 15–18, 1994.

[37] H. Luecke, B. Schobert, H. T. Richter, J. P. Cartailler, and J. K. Lanyi, "Structure of bacteriorhodopsin at 1.55 A resolution," *Journal of Molecular Biology*, vol. 291, no. 4, pp. 899–911, 1999.

[38] K. Edman, A. Royant, P. Nollert et al., "Early structural rearrangements in the photocycle of an integral membrane sensory receptor," *Structure*, vol. 10, no. 4, pp. 473–482, 2002.

[39] T. Okada, M. Sugihara, A. N. Bondar, M. Elstner, P. Entel, and V. Buss, "The retinal conformation and its environment in rhodopsin in light of a new 2.2 A crystal structure," *Journal of Molecular Biology*, vol. 342, no. 2, pp. 571–583, 2004.

[40] L. Vogeley, O. A. Sineshchekov, V. D. Trivedi, J. Sasaki, J. L. Spudich, and H. Luecke, "Anabaena sensory rhodopsin: a photochromic color sensor at 2.0 A," *Science*, vol. 306, no. 5700, pp. 1390–1393, 2004.

[41] M. Murakami and T. Kouyama, "Crystal structure of squid rhodopsin," *Nature*, vol. 453, no. 7193, pp. 363–367, 2008.

[42] M. Nei, X. Gu, and T. Sitnikova, "Evolution by the birth-and-death process in multigene families of the vertebrate immune system," *Proceedings of the National Academy of Sciences of the United States of America*, vol. 94, pp. 7799–7806, 1997.

[43] A. H. Knoll and S. B. Carroll, "Early animal evolution: emerging views from comparative biology and geology," *Science*, vol. 284, no. 5423, pp. 2129–2137, 1999.

[44] X. Gu, "Early metazoan divergence was about 830 million years ago," *Journal of Molecular Evolution*, vol. 47, no. 3, pp. 369–371, 1998.

[45] T. Marti, S. J. Rosselet, H. Otto, M. P. Heyn, and H. G. Khorana, "The retinylidene Schiff base counterion in bacteriorhodopsin," *Journal of Biological Chemistry*, vol. 266, no. 28, pp. 18674–18683, 1991.

[46] H. Luecke, B. Schobert, J. Stagno et al., "Crystallographic structure of xanthorhodopsin, the light-driven proton pump with a dual chromophore," *Proceedings of the National Academy of Sciences of the United States of America*, vol. 105, no. 43, pp. 16561–16565, 2008.

[47] L. Bofkin and N. Goldman, "Variation in evolutionary processes at different codon positions," *Molecular Biology and Evolution*, vol. 24, no. 2, pp. 513–521, 2007.

[48] V. Katritch, V. Cherezov, and R. C. Stevens, "Diversity and modularity of G protein-coupled receptor structures," *Trends in Pharmacological Sciences*, vol. 33, no. 1, pp. 17–27, 2012.

[49] K. J. Nordström, M. Sällman Almén, M. M. Edstam, R. Fredriksson, and H. B. Schiöth, "Independent HHsearch, Needleman—Wunsch-based, and motif analyses reveal the overall hierarchy for most of the G protein-coupled receptor families," *Molecular Biology and Evolution*, vol. 28, no. 9, pp. 2471–2480, 2011.

5

Genome-Wide Analysis of DNA Methylation in Human Amnion

Jinsil Kim,[1] Mitchell M. Pitlick,[2] Paul J. Christine,[2] Amanda R. Schaefer,[2] Cesar Saleme,[3] Belén Comas,[4] Viviana Cosentino,[4] Enrique Gadow,[4] and Jeffrey C. Murray[1,2]

[1] Department of Anatomy and Cell Biology, University of Iowa, 500 Newton Road, 2182 ML, Iowa City, IA 52242, USA
[2] Department of Pediatrics, University of Iowa, 500 Newton Road, 2182 ML, Iowa City, IA 52242, USA
[3] Departamento de Neonatología, Instituto de Maternidad y Ginecología Nuestra Señora de las Mercedes, 4000 San Miguel de Tucumán, Argentina
[4] Dirección de Investigación, Centro de Educación Médica e Investigaciones Clínicas (CEMIC), 1431 Buenos Aires, Argentina

Correspondence should be addressed to Jeffrey C. Murray; jeff-murray@uiowa.edu

Academic Editors: L. Han, X. Li, Z. Su, and X. Xu

The amnion is a specialized tissue in contact with the amniotic fluid, which is in a constantly changing state. To investigate the importance of epigenetic events in this tissue in the physiology and pathophysiology of pregnancy, we performed genome-wide DNA methylation profiling of human amnion from term (with and without labor) and preterm deliveries. Using the Illumina Infinium HumanMethylation27 BeadChip, we identified genes exhibiting differential methylation associated with normal labor and preterm birth. Functional analysis of the differentially methylated genes revealed biologically relevant enriched gene sets. Bisulfite sequencing analysis of the promoter region of the oxytocin receptor (OXTR) gene detected two CpG dinucleotides showing significant methylation differences among the three groups of samples. Hypermethylation of the CpG island of the solute carrier family 30 member 3 (SLC30A3) gene in preterm amnion was confirmed by methylation-specific PCR. This work provides preliminary evidence that DNA methylation changes in the amnion may be at least partially involved in the physiological process of labor and the etiology of preterm birth and suggests that DNA methylation profiles, in combination with other biological data, may provide valuable insight into the mechanisms underlying normal and pathological pregnancies.

1. Introduction

The human amnion is the inner layer of the fetal membranes composed of a monolayer of epithelial cells attached to a basement membrane overlying a collagen-rich stroma [1, 2]. This tissue, which encloses the amniotic fluid, protects the fetus from external mechanical forces and provides an environment that supports fetal movement and growth [3, 4]. The amnion is also a metabolically active tissue involved in the synthesis of various substances with important functions during pregnancy, including prostaglandins and cytokines [1, 5, 6]. It is particularly well known as a major source of prostaglandin E2, a potent molecule mediating cervical ripening and myometrial contraction [7–10], whose levels dramatically increase before and during labor [11, 12].

The amniotic membrane provides most of the tensile strength of the fetal membranes, and alterations in its integrity can lead to undesirable pregnancy outcomes such as preterm premature rupture of membranes (PPROMs) [1, 13], which complicates 3% of all pregnancies and is responsible for approximately one-third of all preterm births (PTBs) [14]. Given the important role of the amnion in the maintenance of pregnancy and parturition, investigation into molecular events occurring in this tissue may contribute to a better understanding of physiological and pathological processes involved in pregnancy.

Considering that the amniotic fluid is in a constantly changing state, it may be critical that the amnion properly responds to environmental cues from the amniotic fluid to accommodate the dynamic needs of the fetus, which could be mediated through epigenetic processes. A previous study by Wang et al. [15] has shown that matrix metalloproteinase 1 (MMP1), whose genetic variation is associated with susceptibility to PPROM [16], is regulated at the epigenetic level, specifically by DNA methylation, and that *MMP1* promoter methylation status correlates with its expression in the amnion and association with PPROM. This finding suggests that the amnion represents an intriguing source of tissue for studying epigenetic events of potential physiological and pathological relevance.

In this study, we performed genome-wide methylation profiling of human term and preterm amnion in order to explore the possible importance of DNA methylation in physiologic labor as well as the etiology of PTB. In addition, independent of the genome-wide methylation study, we carried out methylation analysis of the promoter region of the oxytocin receptor (*OXTR*) gene whose role in human parturition is well established [17]. Given that *OXTR* expression in the amnion increases in association with the onset of labor [18] and that its aberrant methylation in other tissue types has been implicated in autism [19], a disorder that has been associated with PTB [20, 21], we sought to investigate if DNA methylation could represent one mechanism regulating *OXTR* gene function in the contexts of normal parturition and prematurity.

2. Materials and Methods

2.1. Placental Tissue Collection and Preparation. Fresh human placentas were collected in 2009 and 2010 at the University of Iowa Hospitals and Clinics in IA, USA and Instituto de Maternidad y Ginecología Nuestra Señora de las Mercedes in Tucumán, Argentina with signed informed consent and an institutional review board approval. We examined 121 placentas from three groups of patients undergoing: term cesarean delivery without labor (term no labor (TNL) group, $n = 18$), normal term vaginal delivery (term labor (TL) group, $n = 40$), and spontaneous preterm (<37 weeks of gestation) delivery (preterm labor (PTL) group, $n = 63$). Gestational age (GA) was determined using the first day of the last menstrual period as well as by ultrasound examination and was confirmed by assessment at birth. Each placenta was dissected into fetal (amnion, chorion) and maternal (decidua basalis) components within an hour of delivery. The amnion and chorion obtained from the extraplacental membranes (reflected membranes) were separated by blunt dissection under sterile conditions. Decidual tissue samples were macroscopically isolated from the surface of the basal plate of the placenta. After being cut into small pieces, the dissected tissues were placed in RNA later solution (Applied Biosystems, Carlsbad, CA, USA) and stored per manufacturer's recommendations until used. A subset of these samples was selected for genome-wide methylation analysis on the basis of their informativity in relation to our previous gene expression profiling study (unpublished). Additional samples used for validation experiments were selected primarily based on the quality of DNA or RNA extracted from the tissue samples.

2.2. DNA Preparation and Methylation Standards. Genomic DNA was extracted from placental tissue samples using the DNeasy Blood & Tissue Kit (QIAGEN, Valencia, CA, USA)

following the manufacturer's protocol. The quality of the extracted DNA was evaluated by agarose gel electrophoresis. 500 ng of DNA was bisulfite-converted using the EZ DNA Methylation Kit (Zymo Research, Irvine, CA, USA) according to the manufacturer's instructions, and used in subsequent experiments. Universal Methylated Human DNA Standard (Zymo Research), which is enzymatically methylated *in vitro* at all cytosines in CpG dinucleotides, was used as a positive control in the Illumina Infinium methylation assay. We also used Human Methylated and Non-methylated DNA Standards (Zymo Research) as positive and negative controls for methylation-specific PCR. Both of the standards are purified from DNMT1 and DNMT3b double-knockout HCT116 cells, but the methylated standard is enzymatically methylated at all cytosines in CpG dinucleotides.

2.3. Genome-Wide DNA Methylation Analysis

2.3.1. Illumina Infinium Methylation Assay. DNA methylation profiling was performed by the W.M. Keck Biotechnology Resource Laboratory at Yale University, using the Illumina Infinium HumanMethylation27 BeadChip (Illumina, San Diego, CA, USA). Details of the design and general properties of this platform have been previously described [22]. A total of 24 samples were assayed on two BeadChips (12 samples per chip) following the standard protocol provided by Illumina. The samples examined included 9 individual and 1 pooled amnion samples each from the TNL and TL groups, one pooled amnion sample from the PTL group obtained by combining 6 individual samples, and 3 controls (methylated DNA control treated with M.SssI methyltransferase (New England Biolabs, Ipswich, MA, USA), Universal Methylated Human DNA Standard (Zymo Research), and bisulfite-untreated control). These samples were selected from among patients who had participated in our previous gene expression profiling study (unpublished), performed independently of the current work. Based on this previous study, which showed heterogeneous global gene expression patterns among PTL samples, we only included one pooled PTL sample to assess a group DNA methylation average. The samples were arranged randomly on each chip and were processed in a blinded fashion. Table 1 summarizes the clinical characteristics of the three groups of samples studied.

2.3.2. Quality Control and Statistical Analysis. Data analysis was conducted on a fee-for-service basis by the W.M. Keck Biostatistics Resource at Yale University with GenomeStudio Methylation Module v1.0 (Illumina). We evaluated the quality of the data based on the signals of assay built-in control probes (staining, hybridization, target removal, extension, bisulfite conversion, methylation signal specificity, background determination, and overall assay performance) and three experimental controls (two positive methylated controls and one non-bisulfite-converted control), and confirmed the reliability of our data. Principal component analysis (PCA) demonstrated that there is no significant batch effect among the three groups of samples examined. The methylation status of each interrogated CpG site was

TABLE 1: Clinical characteristics of the three subject groups studied by genome-wide DNA methylation profiling.

Parameter	TNL ($n = 9$)[1]	TL ($n = 9$)[1]	PTL ($n = 6$)[2]
Gestational age (weeks)[3]	39.1 ± 0.8	38.8 ± 0.8	33.5 ± 2.6
Race			
White	4	3	3
Black	0	0	1
Other	5	6	2
Maternal age at delivery (years)[3]	29.7 ± 5.5 (Range 22–38)	27 ± 4.4 (Range 20–33)	28.7 ± 3.5 (Range 25–33)
Antibiotics during pregnancy or labor			
Yes	6	1	5
No	3	7	0
Unknown	0	1	1
Birth weight (grams)[3]	3508.3 ± 267.2	3354.1 ± 348.6	2102.2 ± 724.6
Infant gender			
Female	6	4	2
Male	3	5	4

[1] Examined both individually and as a pooled sample.
[2] Examined as a pooled sample.
[3] Data are presented as mean ± standard deviation (SD).
Abbreviations: TNL: term no labor; TL: term labor; PTL: preterm labor.

determined employing the β-value (defined as the fraction of methylation, calculation details described in a previous study [23]) method. An average β-value (AVG_Beta) for each CpG locus ranging from 0 (unmethylated) to 1 (completely methylated) was extracted utilizing the GenomeStudio software and used in further analyses. For the determination of differential methylation between two given groups, we used the Illumina custom error model. This model assumes a normal distribution of the methylation value (β) among replicates corresponding to a set of biological conditions (TNL, TL, and PTL). We prioritized differentially methylated CpG sites by difference score (DiffScore). DiffScore, which takes into account background noise and sample variability [24], was calculated using the following formula: DiffScore $= 10 \, \text{sgn}(\beta_{\text{condition}} - \beta_{\text{reference}})\log_{10}P$, where $\beta_{\text{condition}} = \beta_{\text{TL}}/\beta_{\text{PTL}}$, $\beta_{\text{reference}} = \beta_{\text{TNL}}$ or $\beta_{\text{condition}} = \beta_{\text{TNL}}/\beta_{\text{PTL}}$, $\beta_{\text{reference}} = \beta_{\text{TL}}$. The resulting differentially methylated CpG sites were annotated with respect to their nearest gene based on the information provided by Illumina. A more detailed description of the Illumina custom error model and the DiffScore has been provided previously [25].

2.3.3. Functional Enrichment Analysis.
Differentially methylated genes (DMGs) with a DiffScore of >20 (equivalent to P-value of <0.01) were evaluated for functional enrichment using predefined gene sets from the Molecular Signatures Database (MSigDB) [26]. We searched for significantly enriched gene sets by computing overlaps between the lists of DMGs and the CP collection (canonical pathways, 880 gene sets) or the C5 collection (GO gene sets, 1454 gene sets) in the MSigDB. Gene sets with a P-value (based on the hypergeometric distribution) less than 0.05 were considered significant.

2.4. Bisulfite Sequencing (BS).
To validate methylation differences revealed by the genome-wide methylation assay, we performed bisulfite sequencing on urocortin (UCN), a gene identified as differentially methylated between the TL and PTL groups, and $OXTR$, a gene whose methylation status has recently been shown to be important in the pathogenesis of autism [19]. We investigated the methylation status of $OXTR$, given its significant role in parturition [17] and its labor-associated expression pattern in the amnion [18], which makes it a potential candidate gene for PTB. There are two $OXTR$ CpG sites targeted by the Illumina Infinium BeadChip assay, both of which were not identified as being differentially methylated. However, because there is currently no evidence supporting the biological importance of the regions containing the two sites, we focused our BS analysis on CpG sites of known biological significance that are located in a different region of the $OXTR$ gene. Primers for UCN were designed to cover the CpG site identified as being differentially methylated by genome-wide methylation profiling, using the default parameters of MethPrimer [27]. PCR amplification using the primer pair results in a 278 bp product that spans part of the promoter, exon 1, and part of intron 1 of UCN (-439 to -162 relative to translation start site (TSS)) containing 16 CpG sites. For $OXTR$, we used the same primers and PCR conditions as those used in the previous study [19]. PCR amplification using the primer set results in a 358 bp product that spans the $OXTR$ promoter (-1195 to -838 relative to TSS) containing 22 CpG sites that has been associated with tissue-specific $OXTR$ expression [28] and the development of autism [19]. The regions examined in both genes were located within CpG islands. We carried out our analysis using the same samples assayed on the BeadChips and eight additional independent PTL samples

(TNL, n = 9; TL, n = 9; PTL, n = 14). Bisulfite-converted DNA was PCR-amplified using ZymoTaq DNA polymerase (Zymo Research). The resulting PCR products were run on an agarose gel and cloned into the pGEM-T Easy vector (Promega, Madison, WI, USA). Individual clones were isolated, amplified following standard protocols, and purified using the PureLink Quick Plasmid Miniprep Kit (Invitrogen, Carlsbad, CA, USA) per manufacturer's instructions. Ten clones per sample, on average, were isolated and sequenced at the University of Iowa DNA facility. Percentage methylation was determined for each CpG site similarly as done in previous work [19]. Statistical analysis was conducted using SigmaPlot 11.0 (Systat Software, San Jose, CA, USA). Significance of differential methylation (DM) was assessed using the t-test (two-tailed), Mann-Whitney (M-W) rank sum test (two-sided), one-way ANOVA, or Kruskal-Wallis (K-W) one-way ANOVA by ranks, as indicated in the text and/or figure legends. Post hoc analysis following ANOVA was performed using either the Holm-Sidak or Dunn's Method. A $P < 0.05$ was considered significant.

2.5. Encyclopedia of DNA Elements (ENCODE) ChIP-Seq Data. We examined the potential functional significance of the region of the *OXTR* gene containing CpG sites with statistically different DNA methylation status (CpGs-959 and -1084) using the ChIP-seq data from the ENCODE project available in the University of California Santa Cruz (UCSC) genome browser [29, 30]. We specifically used the suppressor of zeste 12 homolog (Drosophila) (*SUZ12*) and *Pol2* ChIP-seq data generated by the laboratories of Michael Snyder at Stanford University and Vishy Iyer at the University of Texas Austin. The ChIP-Seq data were obtained using human cells (NT2-D1 for the *SUZ12* data; GM18526, 18951, 19099, 19193, and ProgFib for the *Pol2* data).

2.6. Methylation-Specific PCR (MSP). Validation of DM was additionally carried out using methylation-specific PCR (MSP). Two pairs of primers (unmethylated and methylated) for each of the lysophosphatidic acid receptor 5 (*LPAR5*), paternally expressed 10 (*PEG10*), and solute carrier family 30 member 3 (*SLC30A3*) genes were designed using the MSP-specific default parameters of the MethPrimer program [27]. Bisulfite-converted DNA extracted from amnion tissues (TNL, n = 9; TL, n = 9; PTL, n = 14) was PCR-amplified using Biolase DNA polymerase (Bioline, Taunton, MA, USA). The resulting PCR products were visualized on a 2% agarose gel. Human Methylated and Non-methylated DNA Standards from Zymo Research were used as positive and negative controls.

2.7. RNA Extraction and Real-Time qRT-PCR. Total RNA was extracted from amnion (TNL, n = 14; TL, n = 34; PTL, n = 59) and decidua (TNL, n = 12; TL, n = 16; PTL, n = 31) tissues using TRIzol reagent (Invitrogen) according to the manufacturer's protocol. The quality of extracted RNA was checked using the Agilent 2100 Bioanalyzer (Agilent Technologies, Santa Clara, CA, USA). Reverse transcription was carried out with the High Capacity cDNA Reverse Transcription Kit (Applied Biosystems), using random hexamers

as primers following the manufacturer's instructions. Real-time qRT-PCR was performed using synthesized cDNA as a template, gene-specific primers (*UCN* and *OXTR*) and Power SYBR Green PCR Master Mix (Applied Biosystems). The reactions (including no-template controls) were run in triplicate on the 7900HT Fast Real-Time PCR System (Applied Biosystems) using *ACTB* (beta actin) [31] as an endogenous reference. Data were analyzed with the SDS 2.4 software (Applied Biosystems), employing the comparative CT method [32]. Absence of nonspecific amplification was confirmed by dissociation curve analysis. Samples with a value that falls outside ±2 standard deviations of the group mean were defined as outliers and removed from the study. Statistical analysis was performed similarly as described above in the bisulfite sequencing section. Data were presented as mean ± standard error of the mean (SEM).

3. Results

3.1. Genome-Wide Patterns of DNA Methylation and Differentially Methylated CpG Loci between Term (Non-Labored and Labored) and Preterm Amnion Tissues. To investigate the possible involvement of epigenetic mechanisms in the physiology of normal labor and the pathogenesis of PTB, we examined the genome-wide methylation profiles of the amnion obtained following term (TNL and TL, n = 9 for each) and preterm (PTL, n = 6) deliveries using the Illumina Infinium BeadChip platform. The overall levels of DNA methylation in the experimental samples were low with third quartile AVG_Beta values between 0.4 and 0.55. Principal component analysis (PCA) placed the pooled TL and PTL samples close to each other and very distant from the pooled TNL sample (Figure 1), which indicates that the genome-wide methylation patterns in amnion tissues from the two spontaneous labor groups (regardless of gestational age (GA) at delivery) are more similar to each other than to those observed in non-labor tissues.

We also performed gene/locus level analysis of differential methylation (DM), searching for methylation changes associated with labor and/or PTB at specific CpG sites. Using the Illumina custom error model algorithm, we identified 65 CpG sites in 64 and 61 autosomal genes each that are differentially methylated between the TNL and TL groups and the TL and PTL groups, respectively with a DiffScore of >30 (equivalent to P-value of <0.001). Listed in Table 2 are the 15 most highly differentially methylated genes (DMGs). It was noted that among the genes with differentially methylated sites, although very few, were those belonging to special classes of genes, including noncoding RNAs and imprinted genes (such as Down syndrome critical region gene 10 (*DSCR10*), FBXL19 antisense RNA 1 (*FBXL19-AS1*), and paternally expressed 10 (*PEG10*) as shown in Table 2), many of which have regulatory functions in diverse biological processes.

3.2. Functional Enrichment Analysis. To determine the biological significance of DMGs, functional annotation analysis was performed. Our approach involved examining the extent of overlap between our lists of DMGs and predefined

TABLE 2: List of top 15 differentially methylated autosomal genes in amnion tissues from term (TNL, TL) and preterm (PTL) deliveries ranked by statistical significance[1].

	TNL versus TL[2]			TL versus PTL[3]	
Gene	Locus	CpG island[4]	Gene	Locus	CpG island[4]
IL32	16p13.3	No	TOB1	17q21	No
EDARADD	1q42.3	No	PNPLA3	22q13.31	No
STK19	6p21.3	No	ZNF671	19q13.43	No
EXTL1	1p36.1	No	DAB2IP[7]	9q33.1–q33.3	No
HLA-DQB2	6p21	No	MFNG	22q12	No
MFSD3	8q24.3	Yes	UCN	2p23–p21	Yes
RAB31	18p11.3	Yes	EXOC3L2	19q13.32	Yes
PNPLA3	22q13.31	No	SLC44A2	19p13.1	Yes
GRHPR	9q12	Yes	FBXL19-AS1[6]	16p11.2	Yes
MPHOSPH10	2p13.3	No	DLGAP5	14q22.3	Yes
PEG10[5]	7q21	No	SLC30A3	2p23.3	Yes
DSCR10[6]	21q22.13	No	CHFR	12q24.33	No
SRRD	22q12.1	Yes	C11orf1	11q23.1	No
POLI	18q21.1	Yes	SLC24A4	14q32.12	No
OSTalpha	3q29	No	PI4KB	1q21	No

[1] Statistical significance was determined based on P-values calculated from DiffScores. All genes listed here have a DiffScore >40 (corresponding to P-value of <0.0001).
[2] Genes most highly methylated in the TL group compared to the TNL group.
[3] Genes most highly methylated in the PTL group compared to the TL group.
[4] Defined by the CpG island track in the UCSC Genome Browser.
[5] An imprinted gene.
[6] Non-protein coding genes.
[7] A gene identified as having three non-island CpG sites with a DiffScore >40.
Abbreviations: TNL: term no labor; TL: term labor; PTL: preterm labor.

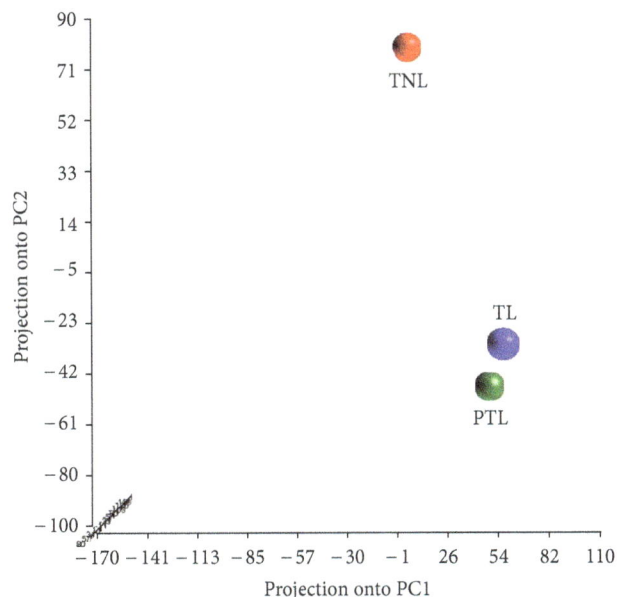

FIGURE 1: Principal component analysis (PCA) plot of DNA methylation profiles in term (non-labored and labored) and preterm amnion. Each colored dot represents a pooled DNA sample from term no labor (TNL), term labor (TL), or preterm labor (PTL) group. Note that the TNL sample is placed distantly from the TL or PTL samples, indicating that the TNL group displays distinctly different methylation patterns compared to the other two groups.

annotated gene sets from the MSigDB [26] (see Section 2 for further details). For this analysis, we used gene lists with a less stringent P-value cutoff of <0.01 (corresponding to a DiffScore of >20), given the small number of DMGs ($n = 65$) with a P-value below 0.001. We found that 7 gene sets were significantly overrepresented ($P < 0.05$) in the list of 110 DMGs between the TNL and TL groups. The seven enriched gene sets included cation transport, ion channel activity, and those shown in Table 3, most of which are highly relevant to molecular processes involved in physiologic labor. Among the 186 DMGs between the TL and PTL groups, 17 gene sets were overrepresented. Many of the enriched gene sets were found to be associated with the regulation of cell behavior and extracellular matrix-cell interactions, including focal adhesion, cell junction, cell-substrate adherens junction, and integrin binding (Table 3).

3.3. Bisulfite Sequencing (BS) Analysis of Differential Methylation. To validate DM detected by genome-wide methylation profiling, we performed BS analysis on *UCN*, a gene identified as being overmethylated in the PTL group compared with the TL group with a DiffScore >50 (Table 2). We performed the same analysis on one additional gene named oxytocin receptor (*OXTR*) whose mRNA and protein expression has been shown to be markedly upregulated in association with labor in primary human amnion epithelial cells [18]. Previous

TABLE 3: Gene sets overrepresented among differentially methylated genes in amnion tissues from term (TNL, TL) and preterm (PTL) deliveries[1].

Gene set[2]	P-value[3]
TNL versus TL	
HEART_DEVELOPMENT	0.012
POSITIVE_REGULATION_OF_CYTOKINE_PRODUCTION	0.017
GATED_CHANNEL_ACTIVITY	0.024
REGULATION_OF_HEART_CONTRACTION	0.041
REGULATION_OF_CYTOKINE_PRODUCTION	0.044
TL versus PTL	
NEGATIVE_REGULATION_OF_TRANSFERASE_ACTIVITY	0.007
ADHERENS_JUNCTION[4]	0.011
HEPARIN_BINDING	0.014
FOCAL_ADHESION_FORMATION	0.02
FOCAL_ADHESION	0.024

[1] Presented are the top 5 most significantly enriched gene sets from C5 collection (GO gene sets).
[2] Defined in the Molecular Signatures Database (MSigDB).
[3] The cutoff for statistical significance was $P = 0.05$.
[4] Also identified as being enriched ($P = 0.049$) in the analysis performed with the CP collection.
Abbreviations: TNL: term no labor; TL: term labor; PTL: preterm labor.

studies have demonstrated that the methylation status of the promoter region of this gene is associated with tissue-specific *OXTR* expression [28] and the development of autism [19], a disorder linked to PTB [20, 21, 33, 34]. These findings intrigued us to investigate whether DNA methylation could represent one mechanism regulating the labor-associated activity of *OXTR* in the amnion. We selected the two genes (*UCN* and *OXTR*), given their crucial role in normal labor and parturition, which makes them potential candidate genes for PTB. Details on the regions amplified, samples used in the BS experiments, and statistical tests performed for the analysis of the sequencing results are given in Section 2 and Figure 2.

All 16 CpG dinucleotides interrogated in the *UCN* gene showed some degree of methylation with the ones at positions -361, -335, and -319, being more highly methylated (22.9–55.7%, Table 4) compared with those at other positions (1.1–17.1%). All except two CpG sites were overmethylated in the PTL samples compared to the TL samples, showing the expected direction of DM. However, the differences were not statistically significant.

For *OXTR*, since we had no *priori* data on the methylation status of the 22 CpG sites in amnion tissue, all three groups of samples (TNL, TL, and PTL) were examined. Consistent with the finding of Gregory et al. [19], 5 CpG sites at positions -959, -934, -924, -901, and -860 showed the highest levels (22.2–68.6%, Table 4) and variation in methylation, whereas very little or no methylation (0–5.7%) was observed at the other sites. We found that one (CpG-959) of the five sites was significantly differentially methylated among the three groups tested (one-way ANOVA, $P = 0.014$, Table 4). Pairwise comparisons (Holm-Sidak test) revealed significant differences between the TNL and TL groups ($P = 0.017$) and the TNL and PTL groups ($P = 0.025$) and borderline significant difference between the TL and PTL

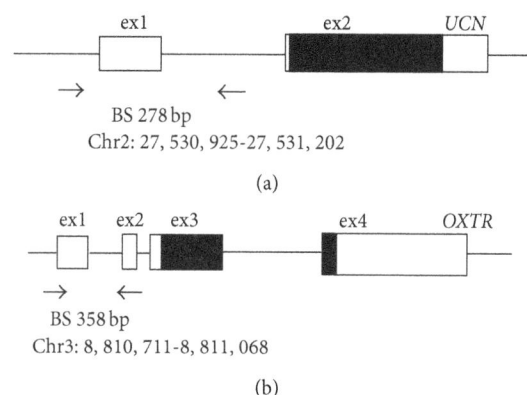

FIGURE 2: Schematic representation of CpG island regions of *UCN* (a) and *OXTR* (b) analyzed by bisulfite sequencing (BS). Black horizontal arrows denote BS PCR primer binding sites. Solid box: coding region; open box: untranslated region. The expected PCR product sizes and positions of the primer binding sites (chromosome and base count, NCBI Build GRCh37/hg19) are indicated. Further details on the PCR-amplified regions are provided in Section 2.

groups ($P = 0.050$), demonstrating more distinct differences in methylation at this site between non-labor and labor tissues than between term and preterm tissues.

To determine if the observed DM also occurs in other parts of the placenta where the genes are known to be expressed [17, 35], we extended our study to decidua tissues from the same groups of individuals. The decidua, which is of maternal origin, unlike the amnion of fetal origin [4], was selected, given that the function of *OXTR* in parturition has been well demonstrated in maternal tissue [17], and therefore, the examination of the decidua, along with the amnion, may allow us to compare the methylation state of

Table 4: *UCN* and *OXTR* promoter methylation status in the amnion and decidua from term (TNL, TL) and preterm (PTL) deliveries[1].

	UCN						
	Amnion			Decidua			
Site[2]	TL	PTL	*P*-value	TL	PTL	*P*-value	
-190[3]	2.2%	8.6%	0.39	6.7%	14.8%	0.07	
-279[3]	2.2%	11.4%	0.07	15.6%	12.5%	0.97	
-319[4]	24.4%	22.9%	0.83	19.9%	30.8%	0.08	
-335[4]	23.3%	27.1%	0.56	23.2%	25.8%	0.75	
-361[4]	48.9%	55.7%	0.33	32.1%	36.4%	0.48	

	OXTR							
	Amnion				Decidua			
Site[2]	TNL	TL	PTL	*P*-value	TNL	TL	PTL	*P*-value
-860	24.4%	22.2%	24.3%	0.96	20%	15.6%	24.6%	0.19
-901	37.8%	30%	45.7%	0.17	45.6%	38.9%	46.9%	0.66
-924	56.7%	62.2%	68.6%	0.29	60%	52.2%	67.2%	0.09
-934[5]	50%	41.1%	59.3%	0.22	46.7%	42.2%	57.1%	0.02
-959[5]	43.3%	24.4%	27.1%	0.014	33.3%	33.3%	30.6%	0.91
-1084[5]	4.4%	4.4%	5.7%	0.97	10%	0%	8.7%	0.008

[1] Presents average % methylation at each CpG site.
[2] Nucleotide positions relative to translation start site.
[3] CpG sites in *UCN* with the lowest *P*-value in each tissue type.
[4] CpG sites methylated at higher levels in both tissues than the average methylation level of all sites examined.
[5] CpG sites with statistically significant (*P* < 0.05) differential methylation in either tissue. Details of statistical tests used are described in Section 2.
Abbreviations: TNL: term no labor; TL: term labor; PTL: preterm labor.

the *OXTR* gene and possibly its importance in both fetal and maternal tissues.

The overall methylation patterns observed in the decidua were similar to those identified in the amnion. However, unlike in the amnion tissues, the methylation levels not at CpG-959, but at different sites (CpGs-934 and -1084), were found to be statistically significantly different (*P* = 0.02, 0.008, resp., K-W one-way ANOVA by ranks) among the three groups of the decidua tissues (Table 4). The CpG-1084 site, interestingly, was completely unmethylated in the TL group, whereas it was methylated to some small degree in the other two groups (TNL, 10%; PTL, 8.7%) (Table 4). Significant differences between the TL and TNL or PTL groups were confirmed by Dunn's post hoc test (*P* < 0.05). In the case of CpG-934, the difference was significant only between the TL and PTL groups. Taken together, it appears that there exist compartment-specific *OXTR* methylation patterns in the placenta.

3.4. Analysis of UCN and OXTR Gene Expression in the Amnion and Decidua. To evaluate the functional significance of the methylation status of the two genes, we performed gene expression analysis using qRT-PCR on an extended set of amnion and decidua tissues (*n* = 107, 59, resp.) from the three groups. Although the DM of *UCN* was not validated by BS, we observed a statistically significant 2.3-fold increase in its transcript levels in the PTL amnion samples compared to the TNL and TL samples (*P* < 0.001, K-W one-way ANOVA by ranks, Figure 3). There was also a statistically significant, but less than twofold increase in *OXTR* mRNA

levels in the PTL amnion samples compared with the TL samples (*P* < 0.05, K-W one-way ANOVA by ranks, Dunn's post hoc test, Figure 3). The results were not replicated in the decidua samples for either gene. These findings suggest that the upregulation of *UCN* is specific to the amnion from spontaneous preterm deliveries, and that the DM observed in *OXTR* may not correlate with *OXTR* expression given that methylation generally plays a role in gene silencing.

3.5. Methylation-Specific PCR (MSP) Analysis of Differential Methylation. As an alternative approach to validate DNA methylation differences captured by our genome-wide methylation study, we carried out methylation-specific PCR (MSP) for selected 3 DMGs between the TNL and TL groups (*PEG10*) and between the TL and PTL groups (*LPAR5* and *SLC30A3*). Analysis of the same set of amnion samples used in BS revealed no intergroup differences in *PEG10* and *LPAR5* methylation (data not shown). However, the methylation status of *SLC30A3* was in good agreement with our genome-wide methylation data with methylated MSP products present in 10 out of 14 (71%) PTL samples and none of the TL samples (Figure 4).

4. Discussion

The present study investigated if there exist unique genome-wide methylation signatures that distinguish among term (non-labored and labored) and preterm amnion tissues. Our methylation profiling revealed a higher degree of similarity between the methylation patterns in the TL and PTL pooled

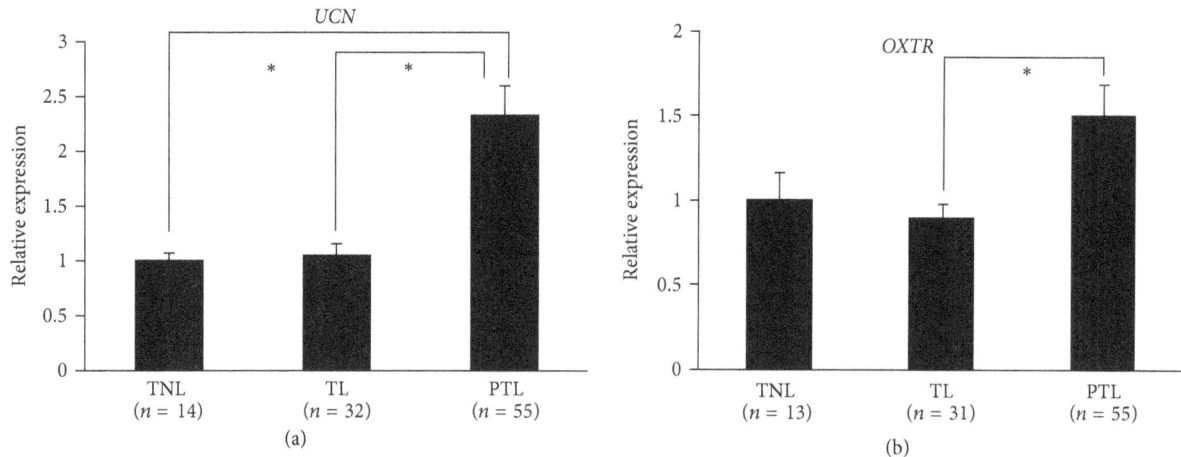

(a)

(b)

FIGURE 3: *UCN* and *OXTR* mRNA expression levels in term (non-labored and labored) and preterm amnion. Expression levels were normalized to that of beta-actin *(ACTB)*. Experiments were performed in triplicate. Data presented are mean ± standard error of the mean (SEM). Asterisks represent statistically significant differences ($P < 0.05$, K-W one-way ANOVA by ranks followed by Dunn's post hoc test) between specified groups.

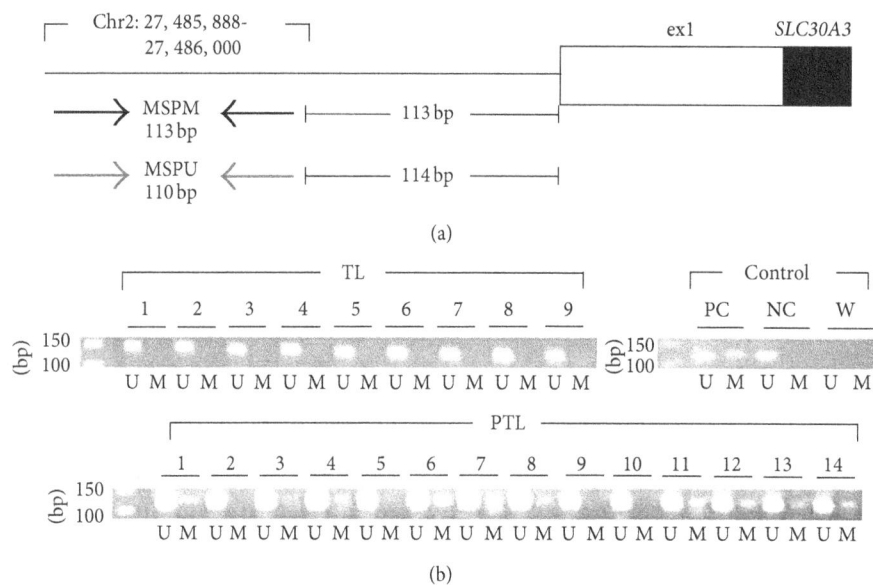

(a)

(b)

FIGURE 4: Methylation-specific PCR (MSP) analysis of *SLC30A3*. (a) Schematic representation of MSP primer binding sites. Black horizontal arrows: methylated-specific primer (MSPM) binding sites; gray horizontal arrows: unmethylated-specific primer (MSPU) binding sites. The expected PCR product sizes and positions of the primer annealing sites (chromosome and base count, NCBI Build GRCh37/hg19) are indicated. Solid box: coding region; open box: untranslated region. (b) Agarose gel electrophoresis of MSP products. M: product amplified with MSPM; U: product amplified with MSPU. PC: positive control, human methylated DNA standard; NC: negative control, human unmethylated DNA standard; W: water control (details can be found in Section 2). Note that the TL samples show only unmethylated PCR products, while many of the PTL samples show both methylated and unmethylated PCR products, indicative of partial methylation.

samples than those observed in the TNL pooled sample, suggesting the potential role of methylation in the regulation of labor, independent from GA. We identified a relatively small number of DMGs between the TNL and TL groups and the TL and PTL groups (65 genes each) at the $P < 0.001$ significance level. This observation may be attributed to the small sample size and the sample-to-sample variability

related to GA. Gene set enrichment analysis of those genes revealed significant overrepresentation of pathways that appear to be functionally relevant (Table 3). The enrichment of pathways related to ion transport, ion channel activity, and cytokine production among the DMGs between the TNL and TL groups reflects biochemical and molecular events associated with the onset of labor, which, along with hormonal factors, help to initiate parturition. These results are at least partially in line with previous gene expression profiling studies reporting labor-associated cytokine-related gene signatures in human amniotic [36] and chorioamniotic [37] membranes. The overrepresentation of heart- (development and contraction) related gene sets may be explained by the presence of myofibroblasts in the connective tissue of the amnion [38], which have contractile ability [39], and hence are involved in heart rhythm regulation [40] and, possibly, prevention of excessive distension of the amniotic membrane [38]. The DMGs between the TL and PTL groups were enriched in gene sets involved in cell adhesion, cell-cell and cell-extracellular matrix interactions, which have crucial roles in the modulation of cellular behavior and tissue maintenance and organization [41]. This observation confirms the importance of intact fetal membranes as a critical factor in the maintenance of pregnancy. Another overrepresented gene set was the negative regulation of transferase activity. Given the versatile roles of transferases, differential methylation of this group of genes (including *HEXIM1*, *SFN*, *CBLC*, and *DUSP2*) may influence a wide range of cellular processes in a way that interferes with timely onset of labor and parturition. Among these genes, *DUSP2* has previously been documented as being significantly upregulated following interleukin-1β (IL-1β) stimulation in myometrial cells [42], suggesting its potential role in the mediation of uterine contractions. It would be intriguing to examine how the activity of *DUSP2* in the amnion may contribute to the process of parturition.

Our study at the individual gene level using BS revealed three CpG sites (CpGs-934, -959, and -1084) in *OXTR* that exhibit significant DM among the three groups of amnion and decidua tissues, which are of fetal and maternal origin, respectively [4]. Subsequent gene expression analysis demonstrated no correlation between gene expression and methylation and therefore, the functional significance of the observed DM remains undetermined. Previous work showed that site-specific methylation can result in transcriptional alterations through its effects on the interaction of transcription factors (TFs) with its cognate DNA sequence [43]. Currently, there are no known TF binding sites around CpG-959, which was previously identified as significantly hypermethylated in peripheral blood mononuclear cells from autistic patients compared with those from control patients [19]. However, Gregory et al. [19] have indicated that CpG-934, whose differential methylation has also been associated with autism, falls within predicted binding domains for v-rel reticuloendotheliosis viral oncogene homolog (avian) (c-Rel), zinc fingers and homeoboxes 2 (ZHX2), and lectin, galactoside-binding, soluble, 4 (LGALS4). Using ENCODE ChIP-seq data available in the UCSC genome browser, we also found that CpG-1084 falls within putative binding sites for *SUZ12* and *Pol2* (see Section 2 for more details), which warrants future studies to dissect the impact of the methylation status at this specific dinucleotide on the interactions between these TFs and their binding sites.

Despite the lack of any significant difference in *UCN* methylation levels between the TL and PTL groups, our observation of a significant, more than 2-fold increase in *UCN* mRNA levels in the PTL amnion tissues compared with the term tissues suggests a potential role of this gene in the etiology of PTB, which encodes an endogenous ligand for corticotropin releasing hormone receptor (CRHR) that mediates the action of CRH, one of the major endocrine factors in parturition [44]. Given that there are several putative binding sites for TFs (such as C7EBP, GATA, and MyoD) [45] upstream of the region examined in this study, it would be intriguing to investigate whether the methylation status of CpG dinucleotides encompassing those sites correlates with the observed gene expression patterns. It would also be worthwhile to examine if mechanisms other than methylation underlie the transcriptional regulation of *UCN* in the amnion.

Our MSP analysis identified another gene (*SLC30A3*) that might play a role in pathogenic processes of PTB. This gene, also known as *ZNT3*, encodes a zinc transporter responsible for zinc efflux from the cytoplasm to extracellular spaces or intracellular organelles [46]. Given the differential expression of *SLC30A3* in relation to dietary zinc and/or glucose supply in mouse placenta [47] and beta cells [48], it is postulated that its dysregulated expression due to aberrant methylation in human amnion may influence nutritional homeostasis during pregnancy, ultimately, leading to PTB.

Our work was limited by the small sample size and the lack of control for gender-specific methylation differences [49, 50]. Another major limitation is that the PTL tissues were examined as a pooled sample, not individually. Previous studies have demonstrated that pooled DNA samples can be used to provide a reliable estimate of average group methylation when analyzed using high-throughput techniques such as MALDI-TOF mass spectrometry [51, 52]. Therefore, a DNA pooling approach using such systems could be employed in future studies for large-scale assessment of methylation variations in maternal and fetal tissues. Very recently, it has been shown that neonatal DNA exhibits a considerable degree of GA-associated variability in DNA methylation patterns [53]. Given this finding, a precisely stratified analysis based on GA may allow a more accurate characterization of DNA methylation profiles associated with term and preterm pregnancies.

5. Conclusion

This work provides preliminary evidence that DNA methylation changes may play at least a partial role in physiologic labor and the etiology of PTB, and suggests that DNA methylation profiles, together with other types of biological data, hold a promise for the identification of genes involved in normal parturition and preterm birth.

Conflict of Interests

The authors declare that there is no conflict of interests.

Acknowledgments

The authors thank all the families who participated in the study, and acknowledge the great efforts of Laura Knosp,

Susan Berends, Mercedes López, Mirta Gladys Leguiza-món, Silvina Argañaraz, Marta Padilla, and Azucena Singh who were critical in contacting and enrolling families and screening samples. They are grateful to the W.M. Keck Foundation Biotechnology Resource Laboratory and the Keck Biostatistics Resource at Yale University for help with conducting the genome-wide methylation experiment and analyzing the raw data. This research was supported by Grants from the National Institutes of Health (R01 HD052953 and HD57192) and the March of Dimes Foundation (FY2006-575 and FY2008-260).

References

[1] K. Benirschke, *Pathology of the Human Placenta*, Springer, New York, NY, USA, 2012.

[2] H. Niknejad, H. Peirovi, M. Jorjani, A. Ahmadiani, J. Ghanavi, and A. M. Seifalian, "Properties of the amniotic membrane for potential use in tissue engineering," *European Cells and Materials*, vol. 15, pp. 88–99, 2008.

[3] A. Toda, M. Okabe, T. Yoshida, and T. Nikaido, "The potential of amniotic membrane/amnion-derived cells for regeneration of various tissues," *Journal of Pharmacological Sciences*, vol. 105, no. 3, pp. 215–228, 2007.

[4] B. M. Carlson, *Human Embryology and Developmental Biology*, Mosby/Elsevier, Philadelphia, PA, USA, 2009.

[5] J. A. Keelan, T. Sato, and M. D. Mitchell, "Regulation of inter-leukin (IL)-6 and IL-8 production in an amnion-derived cell line by cytokines, growth factors, glucocorticoids, and phorbol esters," *American Journal of Reproductive Immunology*, vol. 38, no. 4, pp. 272–278, 1997.

[6] J. A. Keelan, T. Sato, and M. D. Mitchell, "Interleukin (IL)-6 and IL-8 production by human amnion: regulation by cytokines, growth factors, glucocorticoids, phorbol esters, and bacterial lipopolysaccharide," *Biology of Reproduction*, vol. 57, no. 6, pp. 1438–1444, 1997.

[7] W. E. Ackerman, T. L. S. Summerfield, D. D. Vandre, J. M. Robinson, and D. A. Kniss, "Nuclear factor-kappa B regu-lates inducible prostaglandin E synthase expression in human amnion mesenchymal cells," *Biology of Reproduction*, vol. 78, no. 1, pp. 68–76, 2008.

[8] P. Bernstein, N. Leyland, P. Gurland, and D. Gare, "Cervical ripening and labor induction with prostaglandin E2 gel: a placebo-controlled study," *American Journal of Obstetrics and Gynecology*, vol. 156, no. 2, pp. 336–340, 1987.

[9] W. F. Rayburn, "Prostaglandin E2 gel for cervical ripening and induction of labor: a critical analysis," *American Journal of Obstetrics and Gynecology*, vol. 160, no. 3, pp. 529–534, 1989.

[10] W. Gibb, "The role of prostaglandins in human parturition," *Annals of Medicine*, vol. 30, no. 3, pp. 235–241, 1998.

[11] G. J. Haluska, C. A. Kaler, M. J. Cook, and M. J. Novy, "Prostaglandin production during spontaneous labor and after treatment with RU486 in pregnant rhesus macaques," *Biology of Reproduction*, vol. 51, no. 4, pp. 760–765, 1994.

[12] J. C. Schellenberg and W. Kirkby, "Production of prostaglandin F(2α) and E2 in explants of intrauterine tissues of guinea pigs during late pregnancy and labor," *Prostaglandins*, vol. 54, no. 3, pp. 625–638, 1997.

[13] M. L. Oyen, S. E. Calvin, and D. V. Landers, "Premature rupture of the fetal membranes: is the amnion the major determinant?" *American Journal of Obstetrics and Gynecology*, vol. 195, no. 2, pp. 510–515, 2006.

[14] B. M. Mercer, "Preterm premature rupture of the membranes," *Obstetrics and Gynecology*, vol. 101, no. 1, pp. 178–193, 2003.

[15] H. Wang, M. Ogawa, J. R. Wood et al., "Genetic and epigenetic mechanisms combine to control MMP1 expression and its association with preterm premature rupture of membranes," *Human Molecular Genetics*, vol. 17, no. 8, pp. 1087–1096, 2008.

[16] T. Fujimoto, S. Parry, M. Urbanek et al., "A single nucleotide polymorphism in the matrix metalloproteinase-1 (MMP-1) promoter influences amnion cell MMP-1 expression and risk for preterm premature rupture of the fetal membranes," *The Journal of Biological Chemistry*, vol. 277, no. 8, pp. 6296–6302, 2002.

[17] G. Gimpl and F. Fahrenholz, "The oxytocin receptor system: structure, function, and regulation," *Physiological Reviews*, vol. 81, no. 2, pp. 629–683, 2001.

[18] V. Terzidou, A. M. Blanks, S. H. Kim, S. Thornton, and P. R. Bennett, "Labor and inflammation increase the expression of oxytocin receptor in human amnion," *Biology of Reproduction*, vol. 84, no. 3, pp. 546–552, 2011.

[19] S. G. Gregory, J. J. Connelly, A. J. Towers et al., "Genomic and epigenetic evidence for oxytocin receptor deficiency in autism," *BMC Medicine*, vol. 7, article 62, 2009.

[20] H. J. Larsson, W. W. Eaton, K. M. Madsen et al., "Risk factors for autism: perinatal factors, parental psychiatric history, and socioeconomic status," *American Journal of Epidemiology*, vol. 161, no. 10, pp. 916–925, 2005.

[21] C. Limperopoulos, H. Bassan, N. R. Sullivan et al., "Positive screening for autism in ex-preterm infants: prevalence and risk factors," *Pediatrics*, vol. 121, no. 4, pp. 758–765, 2008.

[22] M. Bibikova, J. Le, B. Barnes et al., "Genome-wide DNA methyl-ation profiling using Infinium assay," *Epigenomics*, vol. 1, no. 1, pp. 177–200, 2009.

[23] S. Chowdhury, S. W. Erickson, S. L. MacLeod et al., "Maternal genome-wide DNA methylation patterns and congenital heart defects," *PLoS ONE*, vol. 6, no. 1, Article ID e16506, 2011.

[24] E. Chudin, S. Kruglyak, S. C. Baker, S. Oeser, D. Barker, and T. K. McDaniel, "A model of technical variation of microarray signals," *Journal of Computational Biology*, vol. 13, no. 4, pp. 996–1003, 2006.

[25] O. Y. Naumova, M. Lee, R. Koposov, M. Szyf, M. Dozier, and E. L. Grigorenko, "Differential patterns of whole-genome DNA methylation in institutionalized children and children raised by their biological parents," *Development and Psychopathology*, vol. 24, no. 1, pp. 143–155, 2012.

[26] A. Subramanian, P. Tamayo, V. K. Mootha et al., "Gene set enrichment analysis: a knowledge-based approach for inter-preting genome-wide expression profiles," *Proceedings of the National Academy of Sciences of the United States of America*, vol. 102, no. 43, pp. 15545–15550, 2005.

[27] L. C. Li and R. Dahiya, "MethPrimer: designing primers for methylation PCRs," *Bioinformatics*, vol. 18, no. 11, pp. 1427–1431, 2002.

[28] C. Kusui, T. Kimura, K. Ogita et al., "DNA methylation of the human oxytocin receptor gene promoter regulates tissue-specific gene suppression," *Biochemical and Biophysical Research Communications*, vol. 289, no. 3, pp. 681–686, 2001.

[29] E. Birney, J. A. Stamatoyannopoulos, A. Dutta et al., "Identifi-cation and analysis of functional elements in 1% of the human genome by the ENCODE pilot project," *Nature*, vol. 447, no. 7146, pp. 799–816, 2007.

[30] K. R. Rosenbloom, T. R. Dreszer, M. Pheasant et al., "ENCODE whole-genome data in the UCSC genome browser," *Nucleic Acids Research*, vol. 38, no. 1, Article ID gkp961, pp. D620–D625, 2010.

[31] K. Ahn, J. W. Huh, S. J. Park et al., "Selection of internal reference genes for SYBR green qRT-PCR studies of rhesus monkey (*Macaca mulatta*) tissues," *BMC Molecular Biology*, vol. 9, article 78, 2008.

[32] K. J. Livak and T. D. Schmittgen, "Analysis of relative gene expression data using real-time quantitative PCR and the 2-ΔΔCT method," *Methods*, vol. 25, no. 4, pp. 402–408, 2001.

[33] S. Johnson, C. Hollis, P. Kochhar, E. Hennessy, D. Wolke, and N. Marlow, "Autism spectrum disorders in extremely preterm children," *Journal of Pediatrics*, vol. 156, no. 4, pp. 525.e2–531.e2, 2010.

[34] S. Johnson and N. Marlow, "Preterm birth and childhood psychiatric disorders," *Pediatric Research*, vol. 69, no. 5, pp. 11R–18R, 2011.

[35] Q. Gu, V. L. Clifton, J. Schwartz, G. Madsen, J. Y. Sha, and R. Smith, "Characterization of urocortin in human pregnancy," *Chinese Medical Journal*, vol. 114, no. 6, pp. 618–622, 2001.

[36] Y. M. Han, R. Romero, J. S. Kim et al., "Region-specific gene expression profiling: novel evidence for biological heterogeneity of the human amnion," *Biology of Reproduction*, vol. 79, no. 5, pp. 954–961, 2008.

[37] R. Haddad, G. Tromp, H. Kuivaniemi et al., "Human spontaneous labor without histologic chorioamnionitis is characterized by an acute inflammation gene expression signature," *American Journal of Obstetrics and Gynecology*, vol. 195, no. 2, pp. 394.e12–405.e12, 2006.

[38] T. Wang and J. Schneider, "Myofibroblasten im bindegewebe des menschlichen amnions," *Zeitschrift für Geburtshilfe und Perinatologie*, vol. 186, pp. 164–169, 1982.

[39] J. J. Tomasek, G. Gabbiani, B. Hinz, C. Chaponnier, and R. A. Brown, "Myofibroblasts and mechano: regulation of connective tissue remodelling," *Nature Reviews Molecular Cell Biology*, vol. 3, no. 5, pp. 349–363, 2002.

[40] S. Rohr, "Myofibroblasts in diseased hearts: new players in cardiac arrhythmias?" *Heart Rhythm*, vol. 6, no. 6, pp. 848–856, 2009.

[41] C. C. DuFort, M. J. Paszek, and V. M. Weaver, "Balancing forces: architectural control of mechanotransduction," *Nature Reviews Molecular Cell Biology*, vol. 12, no. 5, pp. 308–319, 2011.

[42] G. Chevillard, A. Derjuga, D. Devost, H. H. Zingg, and V. Blank, "Identification of interleukin-1β regulated genes in uterine smooth muscle cells," *Reproduction*, vol. 134, no. 6, pp. 811–822, 2007.

[43] A. S. Bélanger, J. Tojcic, M. Harvey, and C. Guillemette, "Regulation of UGT1A1 and HNF1 transcription factor gene expression by DNA methylation in colon cancer cells," *BMC Molecular Biology*, vol. 11, article 9, 2010.

[44] P. Florio, W. Vale, and F. Petraglia, "Urocortins in human reproduction," *Peptides*, vol. 25, no. 10, pp. 1751–1757, 2004.

[45] L. Zhao, C. J. Donaldson, G. W. Smith, and W. W. Vale, "The structures of the mouse and human urocortin genes (Ucn and UCN)," *Genomics*, vol. 50, no. 1, pp. 23–33, 1998.

[46] R. J. Cousins, J. P. Liuzzi, and L. A. Lichten, "Mammalian zinc transport, trafficking, and signals," *The Journal of Biological Chemistry*, vol. 281, no. 34, pp. 24085–24089, 2006.

[47] R. M. Helston, S. R. Phillips, J. A. McKay, K. A. Jackson, J. C. Mathers, and D. Ford, "Zinc transporters in the mouse placenta show a coordinated regulatory response to changes in dietary zinc intake," *Placenta*, vol. 28, no. 5-6, pp. 437–444, 2007.

[48] K. Smidt, N. Jessen, A. B. Petersen et al., "SLC30A3 responds to glucose- and zinc variations in β-cells and is critical for insulin production and in vivo glucose-metabolism during β-cell stress," *PLoS ONE*, vol. 4, no. 5, Article ID e5684, 2009.

[49] R. K. C. Yuen, M. S. Peñaherrera, P. von Dadelszen, D. E. McFadden, and W. P. Robinson, "DNA methylation profiling of human placentas reveals promoter hypomethylation of multiple genes in early-onset preeclampsia," *European Journal of Human Genetics*, vol. 18, no. 9, pp. 1006–1012, 2010.

[50] A. M. Cotton, L. Avila, M. S. Penaherrera, J. G. Affleck, W. P. Robinson, and C. J. Brown, "Inactive X chromosome-specific reduction in placental DNA methylation," *Human Molecular Genetics*, vol. 18, no. 19, pp. 3544–3552, 2009.

[51] S. J. Docherty, O. S. P. Davis, C. M. A. Haworth, R. Plomin, and J. Mill, "DNA methylation profiling using bisulfite-based epityping of pooled genomic DNA," *Methods*, vol. 52, no. 3, pp. 255–258, 2010.

[52] S. J. Docherty, O. S. Davis, C. M. Haworth, R. Plomin, and J. Mill, "Bisulfite-based epityping on pooled genomic DNA provides an accurate estimate of average group DNA methylation," *Epigenetics Chromatin*, vol. 2, article 3, 2009.

[53] J. W. Schroeder, K. N. Conneely, J. C. Cubells et al., "Neonatal DNA methylation patterns associate with gestational age," *Epigenetics*, vol. 6, no. 12, pp. 1498–1504, 2011.

6

Evaluation of Allele Frequency Estimation Using Pooled Sequencing Data Simulation

Yan Guo,[1] **David C. Samuels,**[2] **Jiang Li,**[1] **Travis Clark,**[3] **Chung-I Li,**[1] **and Yu Shyr**[1]

[1] *Vanderbilt Ingram Cancer Center, Center for Quantitative Sciences, Nashville, TN, USA*
[2] *Center for Human Genetics Research, Vanderbilt University Medical Center, Nashville, TN, USA*
[3] *VANTAGE, Vanderbilt University, Nashville, TN, USA*

Correspondence should be addressed to Yu Shyr; yu.shyr@vanderbilt.edu

Academic Editors: L. Han, X. Li, and Z. Su

Next-generation sequencing (NGS) technology has provided researchers with opportunities to study the genome in unprecedented detail. In particular, NGS is applied to disease association studies. Unlike genotyping chips, NGS is not limited to a fixed set of SNPs. Prices for NGS are now comparable to the SNP chip, although for large studies the cost can be substantial. Pooling techniques are often used to reduce the overall cost of large-scale studies. In this study, we designed a rigorous simulation model to test the practicability of estimating allele frequency from pooled sequencing data. We took crucial factors into consideration, including pool size, overall depth, average depth per sample, pooling variation, and sampling variation. We used real data to demonstrate and measure reference allele preference in DNAseq data and implemented this bias in our simulation model. We found that pooled sequencing data can introduce high levels of relative error rate (defined as error rate divided by targeted allele frequency) and that the error rate is more severe for low minor allele frequency SNPs than for high minor allele frequency SNPs. In order to overcome the error introduced by pooling, we recommend a large pool size and high average depth per sample.

1. Introduction

Over the last decade, large-scale genome-wide association studies (GWAS) based on genotyping arrays have helped researchers to identify hundreds of loci harboring common variants that are associated with complex traits. However, multiple disadvantages have limited genotyping arrays' ability for disease association detection. A major disadvantage of genotyping arrays is the limited power for detecting rare disease variance. Rare variants with minor allele frequency (MAF) less than 1% are not sufficiently captured by GWAS [1]. Such low MAF variants may have substantial effect sizes without showing Mendelian segregation. The lack of a functional link between the majority of the putative risk variants and the disease phenotypes is another major drawback for genotyping array-based GWAS [2]. The most popular genotyping chip, the Affymetrix 6.0 array, contains nearly 1 million SNPs, yet only one-third of these SNPs resides in the coding regions. Even though many GWAS-identified statistically significant SNPs lie in the intron or intergenic regions, [3–5] their biological function remains difficult to explain. Another limitation of genotyping arrays is that, because the SNPs are predetermined on the array, no finding of novel SNPs is possible.

Most of the above limitations can be overcome by using high throughput NGS technology [6]. NGS can target a specific region of interest, such as the exome or the mitochondria. Often, the functions of variants identified in coding regions of interest are much easier to explain than those of variants identified in the intron or intergenic regions. Also, by targeting the exome, we can effectively examine nearly 30 million base pairs in the coding region rather than just 0.3 million SNPs on the Affymetrix 6.0 array. Sequencing technology has been used to detect rare variants in many studies [7–10], with rare variants defined as 1%–5% frequency. Due to the large sample size needed to detect such low frequency variants, detection of rare variants less than 1% can still pose a significant challenge for NGS technology. One way to overcome this limitation is by doing a massive genotyping catalogue such as the 1000 Genomes Project [11]. Researchers are often too limited financially to conduct a genotyping study on such a large scale. DNA pooling is a strategy often used to reduce the financial burden in such cases.

The concept of pooling in genetic studies began in 1985 with the first genetic study to apply a pooling strategy [12]. Since then, pooling has been extensively applied in linkage studies in plants [13], allele frequency measurements of microsatellite markers and single nucleotide polymorphisms (SNPs) [10, 14–18], homozygosity mapping of recessive diseases in inbred populations [19–22], and mutation detection [23]. Even though pooling has also been used widely with NGS technology [24–26], the effectiveness of the pooling strategy has long been debated. On the one hand, several studies have claimed that data generated from pooling studies are accurate and reliable. For example, Huang et al. claimed that the minor allele odds ratio estimated from pooled DNA agreed fairly well with the minor allele odds ratio estimated from individual genotyping [27]. Docherty et al. demonstrated that pooling can be effectively applied to the genome-wide Affymetrix GeneChip Mapping 500 K Array [28]. Some studies have even found that pooling designs have an advantage in the detection of rare alleles and mutations, such as the study by Amos et al., which suggested that mutations in individuals could be more efficiently detected using pools [23]. On the other hand, several studies have argued that, when compared with individual sequencing, pooled sequencing can generate variant calls with high false-positive rates [29]. Other studies also found that the ability to accurately estimate the allele frequency from pooled sequencing is limited [30, 31].

Usually two different kinds of pooling paradigms are involved. The first is multiplexing (also known as barcoding). On an Illumina HiSeq 2000 sequencer, one lane can generate, on average, from 100 to 150 million reads per run. For exome sequencing, from 30 to 40 million reads per sample are needed to generate reliable coverage in the exome for variant detection. Thus, the common practice is to multiplex from 3 to 4 samples per lane to reduce cost. Using multiplexing with barcode technology, we are able to identify each read's origination. The disadvantage of multiplexing with barcoding is the extra cost of barcoding and labor. The cheaper alternative to pooling with multiplexing is pooling without multiplexing, which prevents us from identifying the origin of each read.

In this study, we focused on pooling without multiplexing. By using comprehensive and thorough simulations, we tried to determine the effectiveness of estimating allele frequency from pooled sequencing data. In our simulation model we considered important factors of pooled sequencing, including overall depth, the average depth per sample, pooling variation, sampling variation, and targeted minor allele frequency (MAF). Another important issue we addressed in our simulation is the reference allele preferential bias, which is a phenomenon during alignment when there is preference toward the reference allele. We used real data to show the effect of reference allele bias and adjusted our simulation model accordingly. We describe our simulation model in detail and present the results from the simulation.

2. Materials and Methods

We designed a thorough simulation model to closely reflect the real-world pooled sequencing situation. Our simulation model includes notations which we have defined as follows: let $\hat{F}$ be the allele frequency estimated from pooled sequencing data, and let F be the true allele frequency in the pool. Under the ideal assumption, all samples' contributions to the pool are equal. However, in practice, factors such as human error and library preparation variation can affect a sample's contribution to the pool. Very likely, each time a sample is added to the pool, an error is introduced. We let ε_i denote the error of sample i during the pooling process, and ε_i should follow a normal distribution $N(\mu, \sigma^2)$, where $\mu = 0$, and σ^2 denotes the variance of error in the pool. We assume that the amount of DNA added to the pool for each sample $c + \varepsilon_i$ follows a normal distribution $N(c, \sigma^2)$, where c is a constant and denotes the ideal constant contribution to the pool by each sample. The probability that a read is contributed by sample i can be represented as $p_i = (c + \varepsilon_i)/\sum_{i=1}^{N_s}(c + \varepsilon_i)$, where N_s denotes the number of samples in the pool. The contribution of each sample in the pool to a SNP can be modeled as a multinomial distribution $R_1, R_2, R_3, \ldots, R_{N_s} \sim Multinomial(D, p_1, p_2, p_3, \ldots, p_{N_s})$, where D equals the depth at this SNP and R_i represents the reads contributed by sample i for this SNP. The depth D follows a possion distribution $Poisson(\lambda_D)$, where λ_D equals the average depth for the exome regions. For sequencing data, the reads at heterozygous SNPs should have an allele balance of 50%, meaning 50% of the read should support the reference allele while the other 50% of the read should support the alternative allele. Thus the reads that support the alternative allele should follow a binomial distribution $Binomial(D, 0.5)$.

In our study we estimated the average depth for the exome regions as follows:

Average depth

$$= \frac{\text{Lane} \times \text{Reads per lane} \times \text{Capture efficiency}}{\text{number of exons}}. \tag{1}$$

In general the read output for 1 lane on an Illumina HiSeq 2000 sequencer is around 120 million reads. The most popular exome capture kits including Illumina TruSeq, Agilent SureSelect, and NimbleGen SeqCap EZ capture almost 100 percent of all known exons (about 30 million base pairs). Most capture kits claim that they have capture efficiency of at least 70 percent, but, in practice, it has been shown that the capture efficiency of all these capture kits are only around 50 percent [32], which implies that if a sample is sequenced for 120 million reads, only around 60 million reads will be aligned to exome regions. After filtering for mapping quality, the number of reads aligned to exome regions will be even smaller. However, to simplify, we ignored the reads that failed the mapping quality filter. There are about 180,000 exons [33]. Based on (1), for exome sequencing on 1 Illumina HiSeq lane, the average depth is expected to be roughly 400.

TABLE 1: Allele balance for 3 independent datasets.

Dataset	Sample	Min.	1st Qu.	Median	Mean	3rd Qu.	Max.	Mean 95% conf. lo.	Mean 95% conf. hi.
	1055QC0003	0.091	0.423	0.48	0.48	0.536	0.862	0.476	0.483
	1055QC0004	0.1	0.427	0.477	0.48	0.53	0.826	0.477	0.483
	1055QC0005	0.046	0.429	0.481	0.482	0.536	0.939	0.479	0.486
	1055QC0006	0.1	0.418	0.478	0.481	0.542	0.909	0.477	0.485
	1055QC0007	0.156	0.417	0.476	0.475	0.536	0.879	0.472	0.479
	1055QC0008	0.148	0.421	0.481	0.482	0.542	0.905	0.479	0.486
	1055QC0009	0.148	0.422	0.478	0.48	0.536	0.963	0.476	0.483
	1055QC0011	0.1	0.421	0.481	0.48	0.538	0.952	0.477	0.484
	1055QC0012	0.095	0.429	0.478	0.48	0.531	1	0.477	0.483
	1055QC0013	0.165	0.424	0.482	0.482	0.541	0.9	0.479	0.486
SureSelect	1055QC0014	0.103	0.429	0.481	0.483	0.538	0.818	0.48	0.487
	1055QC0016	0.13	0.425	0.48	0.482	0.54	0.909	0.478	0.485
	1055QC0017	0.136	0.422	0.481	0.48	0.536	0.9	0.477	0.483
	1055QC0018	0.182	0.424	0.48	0.48	0.537	0.987	0.477	0.483
	1055QC0020	0.2	0.432	0.483	0.485	0.536	0.815	0.482	0.488
	1055QC0021	0.12	0.429	0.481	0.484	0.538	1	0.48	0.487
	1055QC0022	0.091	0.424	0.478	0.479	0.533	0.905	0.476	0.482
	1055QC0024	0.077	0.422	0.478	0.478	0.535	0.857	0.474	0.481
	1055QC0025	0.13	0.429	0.481	0.484	0.54	0.897	0.481	0.488
	1055QC0026	0.13	0.42	0.478	0.479	0.539	0.793	0.476	0.482
	1055QC0028	0.039	0.419	0.477	0.476	0.531	0.938	0.472	0.479
	10009	0.044	0.447	0.5	0.499	0.55	1	0.496	0.501
	10244	0.091	0.444	0.5	0.497	0.55	0.909	0.495	0.499
TruSeq	10290	0.065	0.444	0.5	0.497	0.55	0.917	0.495	0.499
	20007	0.077	0.447	0.5	0.498	0.55	0.923	0.496	0.5
	20017	0.044	0.447	0.5	0.498	0.55	0.921	0.496	0.5
	20301	0.077	0.449	0.5	0.499	0.55	0.967	0.497	0.501
	ERR004043	0.04	0.376	0.44	0.447	0.511	0.986	0.44	0.453
	ERR004047	0.125	0.391	0.447	0.451	0.503	1	0.446	0.457
Array based	SRR013908	0.081	0.37	0.475	0.481	0.584	0.977	0.472	0.489
	SRR013909	0.071	0.372	0.476	0.484	0.591	0.95	0.476	0.492
	SRR015428	0.093	0.389	0.488	0.49	0.586	0.909	0.483	0.498
	SRR015429	0.1	0.426	0.496	0.497	0.564	0.913	0.491	0.503
All	Mean	0.103	0.421	0.482	0.483	0.543	0.919	0.479	0.487

To measure the accuracy of the allele frequency $\widehat{F}$ estimated from pooled sequencing data, we computed the relative root mean square error (RMSE) as follows:

$$\frac{\sqrt[2]{\sum_1^n \left(\widehat{F}_i - F\right)^2 / n}}{F}, \quad (2)$$

where n is the number of simulations we performed to estimate the target allele frequency F. In our simulations, we set $n = 10,000$. Unlike the traditional RMSE, we divided it by the target allele frequency F to make the result relative to the allele frequency we were simulating, so we could compare RMSE for allele frequencies as small as 0.5% and as large as 50%.

Reference allele preferential bias is a phenomenon during alignment when there is preference toward the reference allele. Degner et al. described such bias in RNA-seq data [34]. To examine whether this bias also exists in DNAseq data, we measured allele balance (defined as reads that support the alternative allele divided by total reads) of three independent DNA sequencing datasets. The three datasets were sequenced at different facilities (Broad Institute, HudsonAlpha, Illumina), at different time points, and using different capture methods (Agilent SureSelect, Illumina TruSeq, and Array Based Capture). The theoretical allele balance for heterozygous SNPs should be around 50%. In real data, we observed that the mean allele balance for all heterozygous SNPs for all samples is 0.483 (range: 0.447–0.499) (Table 1). Thus, we modified our previously defined

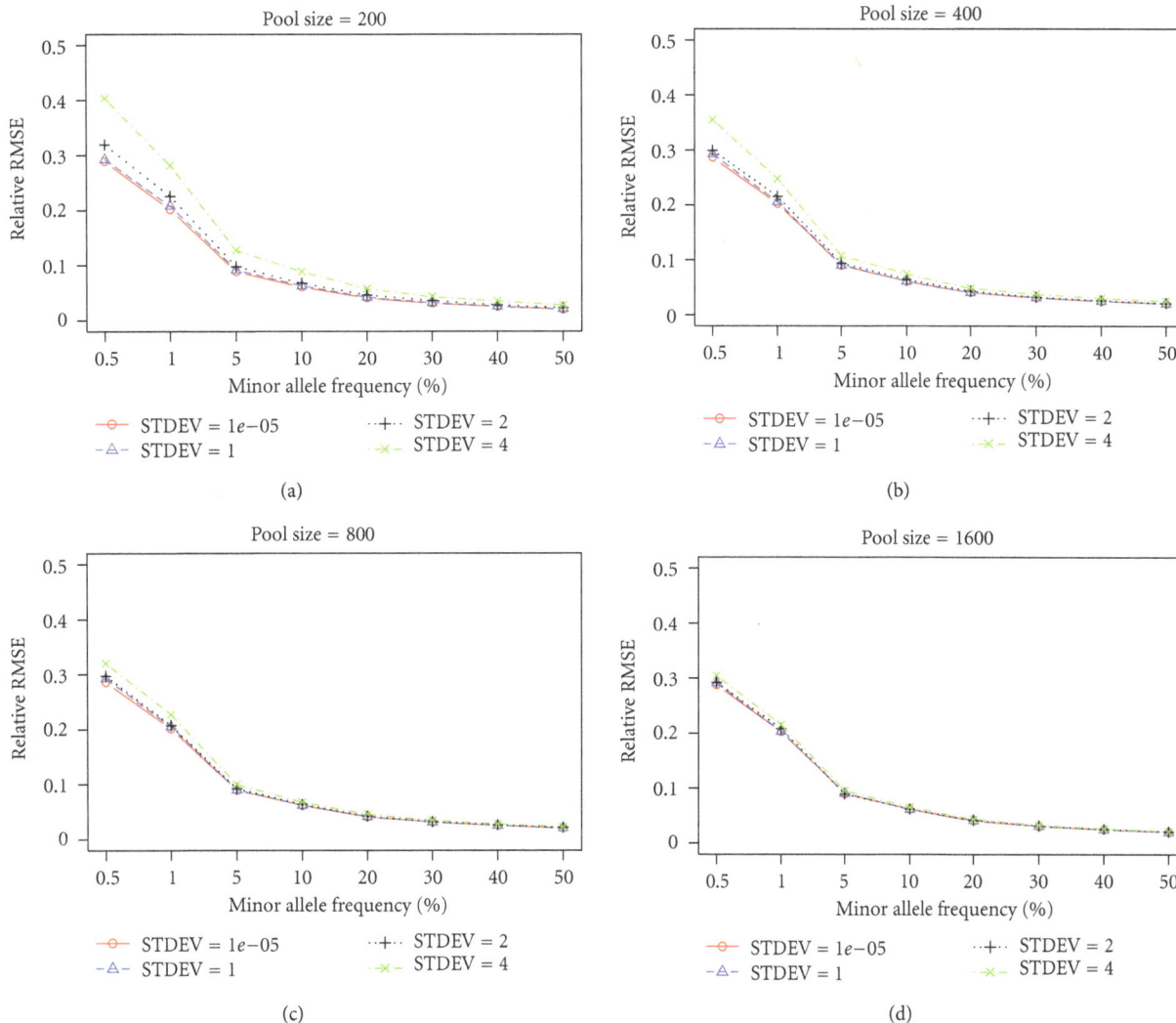

FIGURE 1: Relative RMSE for different pool sizes and MAFs under different standard deviations.

read distribution at heterozygous site $Binomial(D, 0.5)$ to $Binomial(D, P)$ where P follows a normal distribution $N(\mu_P, \sigma_p^2)$, where μ_P and σ_p^2 are estimated by the empirical mean allele balance we observed in real data.

Three simulations were conducted to evaluate the accuracy of allele frequency estimation from pooled sequencing data. The detailed descriptions of the three simulations are as follows.

Simulation 1. The goal of Simulation 1 was to study the relationship between different levels of ε and relative RMSE under different pool sizes ($N_s = 200, 400, 800,$ and 1600) and different MAF (MAF = 0.5%, 1%, %5, 10%, 20%, 30%, 40%, and 50%). Each sample's DNA contribution $c + \varepsilon_i$ to the pool follows a normal distribution $N(c, \sigma^2)$. For simulation purpose, we set an arbitrary value $c = 10$ units; the actual value of c does not affect the outcome of the simulations, because the simulation merely scales around it. To best

represent the scenario in practice, we used several different standard deviations values for the distribution of sample contribution to the pool. For the ideal situation, we set σ^2 to a very small number (10^{-5}); then, we increase σ^2 to 1, 2, and 4 (10%, 20%, and 40% of c) to see the effect of larger error variance on the accuracy of allele frequency estimation using pooled sequencing data. Each allele frequency was simulated 10,000 times.

Simulation 2. The goal of Simulation 2 was to study the relationship between depth and relative RMSE. The average depth of exome coverage can be estimated using the number of lanes. Instead of looking directly at average depth in the exome regions, we looked at average depth per sample $\lambda_{ps} = \lambda_D / N_s$ (i.e., average depth divided by pool size). If the average depth of exome regions for the pool of 200 people is 600x, then the depth per sample is 3x. In this simulation, we used $\lambda_{ps} = 0.5, 1, 2, 4, 6, 8, 10, 12, 14, 16, 18,$ and 20, pool size

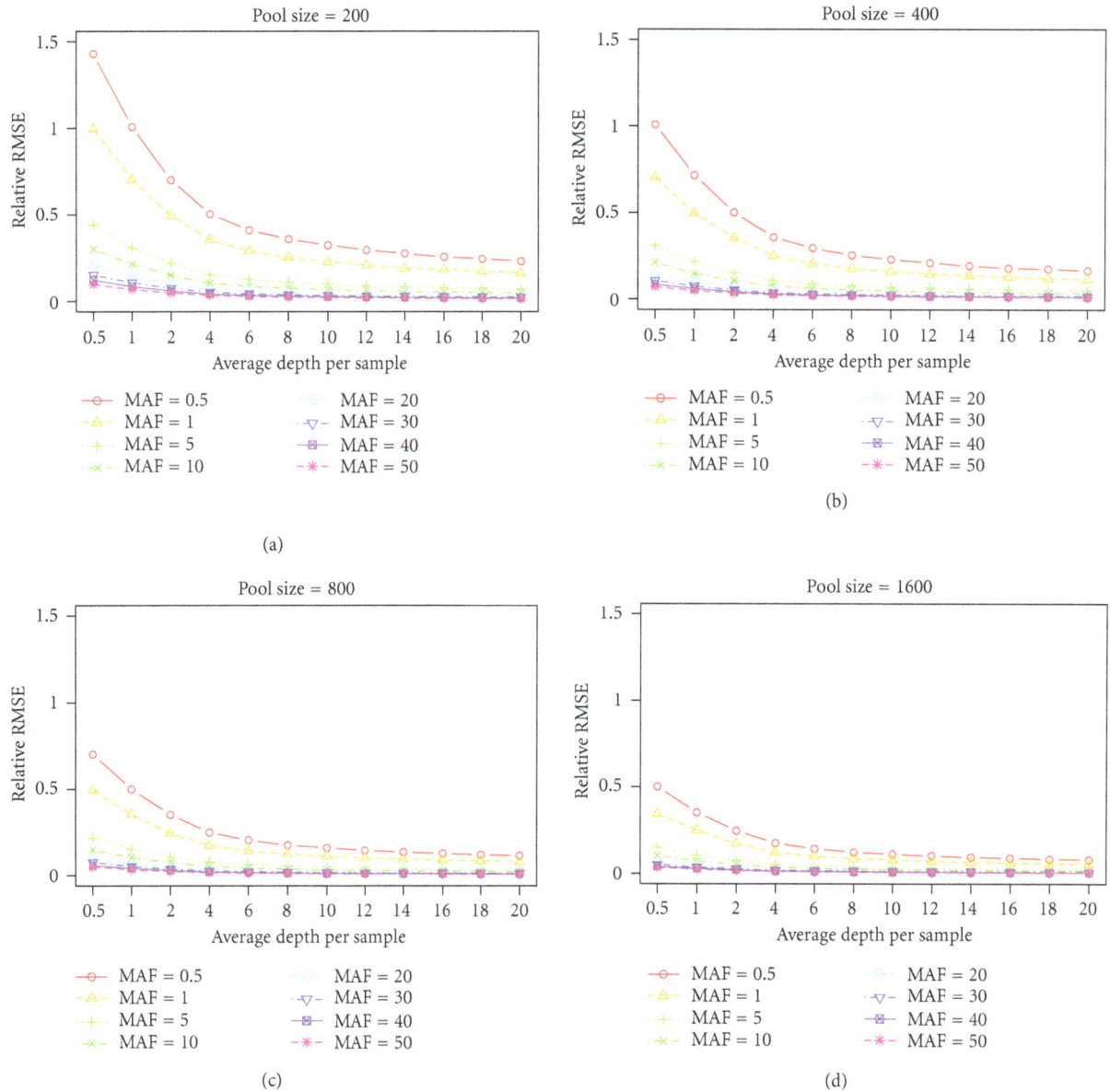

FIGURE 2: Relative RMSE for different pool sizes and MAFs under different average per sample depths.

N_s = 200, 400, 800, and 1600, and MAF = 0.5%, 1%, %5, 10%, 20%, 30%, 40%, and 50%. Each allele frequency was simulated 10,000 times.

Simulation 3. The goal of Simulation 3 was to determine the overall performance of a pooled exome sequencing study. In practice, we cannot measure a SNP 10,000 times and then compute the average allele frequency as we did in Simulations 1 and 2. We are limited with one measurement only at a given SNP. It is important that we look at the overall performance too rather than just at a single SNP. Based on the released data of the 1000 Genomes Project, we built an empirical MAF distribution. This distribution should represent an overall picture of MAF distribution in the population. A typical exome study will yield 10,000–100,000 SNPs after filtering,

with the number of SNPs heavily dependent on the number of samples sequenced in the study. Following the empirical distribution of the MAF, we randomly drew 10,000 SNPs from this distribution to simulate an exome sequencing dataset and computed an overall error rate. The error rate is defined as $|\widehat{F} - F|/((\widehat{F} + F)/2)$. We further repeated this simulation 1000 times and computed the median error rates.

3. Results

Simulation 1. We assume that each sample's DNA contribution to the pool follows a normal distribution $N(c, \sigma^2)$. In an ideal situation, σ^2 is small, and if we fix overall depth, the pool size does not make a significant difference for the RMSE. For example, in an ideal situation, for MAF = 0.5, the relative

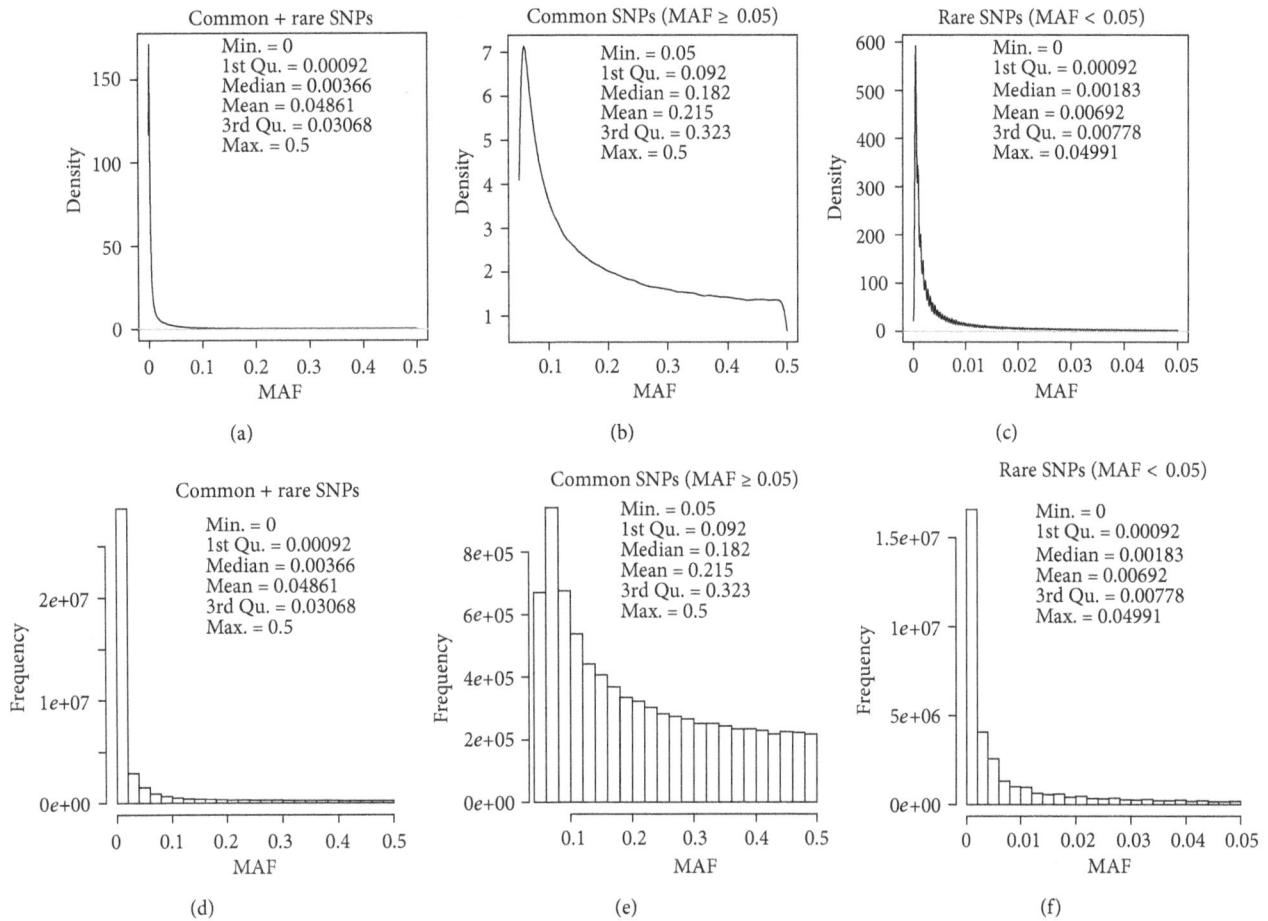

FIGURE 3: 1000 Genome MAF distributions.

TABLE 2: Statistics for doing 10,000 simulations at different MAFs.

MAF	Min.	1st Qu.	Median	Mean	3rd Qu.	Max.	Var.	Relative RMSE
0.5	0.0000	0.0036	0.0049	0.0050	0.0064	0.0162	0.0000	0.5037
1	0.0000	0.0075	0.0098	0.0100	0.0124	0.0264	0.0000	0.3552
5	0.0256	0.0448	0.0500	0.0500	0.0551	0.0795	0.0001	0.1540
10	0.0615	0.0928	0.0999	0.1000	0.1071	0.1401	0.0001	0.1070
20	0.1444	0.1904	0.2000	0.2001	0.2098	0.2558	0.0002	0.0716
30	0.2449	0.2889	0.2997	0.2998	0.3106	0.3619	0.0003	0.0537
40	0.3397	0.3879	0.3998	0.4000	0.4118	0.4707	0.0003	0.0442
50	0.4348	0.4877	0.5000	0.4998	0.5116	0.5675	0.0003	0.0359

TABLE 3: Pooled and individual sequencing pricing.

Sequencing per pool	200	400	600	800	1000
2 lanes	$3,650	$4,050	$4,450	$4,850	$5,250
4 lanes	$6,650	$7,050	$7,450	$7,850	$8,250
6 lanes	$9,650	$10,050	$10,450	$10,850	$11,250
8 lanes	$12,650	$13,050	$13,450	$13,850	$14,250
10 lanes	$15,650	$16,050	$16,450	$16,850	$17,250
12 lanes	$18,650	$19,050	$19,450	$19,850	$20,250
16 lanes	$24,650	$25,050	$25,450	$25,850	$26,250
Individual prep.	$125,000	$250,000	$375,000	$500,000	$625,000

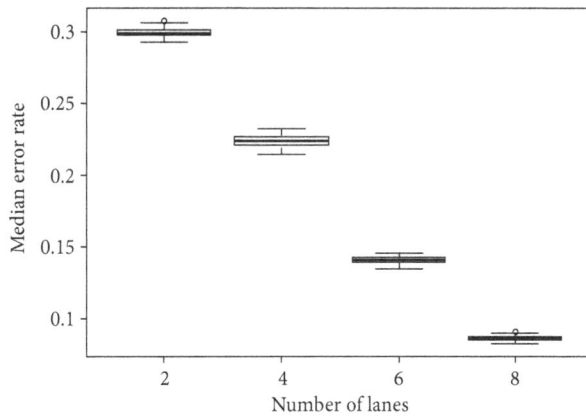

FIGURE 4: Median error rates for simulating 1000 exome sequences using different numbers of lanes. Simulation on 2 lanes shows nearly 30% error, and only around 5% error rate is observed for 16 lanes simulation.

RMSE for pool size N_s = 200, 400, 800, and 1600 equals 0.023, 0.024, 0.024, and 0.020, respectively. However, if we increase σ^2, pools with greater size tend to have lower RMSEs. For example, when σ^2 is increased to 4, for MAF = 50%, the relative RMSEs for pool size N_s = 200, 400, 800, and 1600 equals 0.028, 0.0250, 0.023, and 0.022, respectively. Increasing σ^2 clearly also increased relative RMSE for all MAFs and for all pool sizes. For example, for pool size N_s = 200, MAF = 50%, σ^2 = 0.00001, 1, 2, and 4, the relative RMSE are 0.020, 0.021, 0.023, and 0.028, respectively. Also lower MAF tended to have high relative RMSE than high MAF. For example, in an ideal situation, for pool size N_s = 200, the relative RMSEs for MAF = 0.5%, 1%, 5%, 10%, 20%, 30%, 40%, and 50% are equal to 0.289, 0.202, 0.088, 0.061, 0.041, 0.031, 0.030, and 0.020, respectively. The results of Simulation 1 can be viewed in Figure 1.

Simulation 2. In this simulation, the goal was to examine the relationship between average depth per sample λ_{ps} and pool size N_s. We found that, with the same average depth per sample λ_{ps}, higher pool sizes will generate lower relative RMSEs. Also, as the MAF increases, the relative RMSE decreases. For example, for MAF = 50% and average depth per sample λ_{ps} = 1, the relative RMSEs for N_s = 200, 400, 800, and 1600 are 0.071, 0.050, 0.036, and 0.025, respectively. If we can infinitely increase pool size or average depth per sample while fixing the other, the RMSE will reach zero. The result of Simulation 2 can be viewed in Figure 2.

In our study, we performed simulations at each MAF 10,000 times. However, in practice, we do not have the resources to measure a SNP 10,000 times and then take the average. In real exome sequencing, each SNP is only measured one time. Table 2 shows the quantile information for simulating MAF = 0.5%, 1%, 5%, 10%, 20%, 30%, 40%, and 50% 10,000 times. The mean and median of the estimated MAF are very close to the targeted MAF value. When MAF increases, the variance also increases. If we account for

relative RMSE, the simulations still produced more accurate results for larger MAFs.

Simulation 3. In this simulation, we simulated the scenario of pooled exome sequencing. Using data from the 1000 Genomes Project as prior information that contains genotyping data from 1092 individuals, we built an empirical distribution of MAF (Figure 3). Based on this empirical distribution, we simulated the pooled exome sequencing with pool size N_s = 1092 1000 times and computed the median error rate for each simulation (Figure 4). The results clearly indicate that higher depth is required to produce an acceptable error rate (>5%). For standard exome sequencing, pooled DNA from 1000 subjects will require roughly 16 Illumina HiSeq lanes to produce results with an acceptable error rate.

Financial Implication. The ultimate goal of pooling is to ease the financial burden on large association studies. Based on the most up-to-date pricing information on NGS, we compared the total cost of conducting association studies using pooling at different pool sizes with individual sequencing using Illumina HighSeq 2000 sequencer, which contains 2 flow cells, and each flow cell contains 16 lanes. Table 3 shows the price difference between pooling and individual sequencing. The savings using pooling is more substantial when pool sample size is large. When using all 16 lanes, the savings for 200 samples is roughly 500% over individual sequencing and, for 1000 samples, a 2300% saving.

4. Discussion

Our simulation showed that there are several important factors to consider when designing a pooling study. Those factors include sample size, targeted MAF, and, most importantly, the depth. The sample size directly affects the ability to detect rare SNPs. Larger pool size will increase the accuracy of MAF estimation with the same per sample depth but will not have much effect with the same overall depth. Similarly, with the same pool size, increasing depth will decrease relative RMSE. Our simulation also showed that pooled sequencing is not ideal for estimating the MAF of rare SNPs. The relative RMSE is much higher for SNPs with MAF < 1% compared to SNPs with MAF > 5% (Figure 1).

Sequencing pooled DNA will ease financial burdens and make large association possible. At the same time, however, pooling introduces additional errors. A majority of the errors are caused by the unequal representation of each sample's DNA in the pool. This unequal representation could be due to human or machine error, which we have considered in our simulation. There are other factors which can also cause the unequal representation, such as a sample's DNA quality and variation introduced in the PCR/amplification stage. Unfortunately, we can only minimize such errors and variation using more sophisticated lab techniques. Even if every sample is equally represented in the pool, the sequenced data still do not truly reflect the equality due to sampling variance. Based on our simulation results, when designing a pooling study, we recommend the following: larger pool size is better, and higher depth is better. More elaborately, it is better to keep balance between pool size and depth. We recommend keeping the average depth per sample at 10 minimum if rare SNPs are not of interest; otherwise, average depth per sample at 20 minimum is highly recommended.

Acknowledgments

The authors would like to thank Peggy Schuyler and Margot Bjoring for their editorial support.

References

[1] M. I. McCarthy and J. N. Hirschhorn, "Genome-wide association studies: potential next steps on a genetic journey," *Human Molecular Genetics*, vol. 17, no. 2, pp. R156–R165, 2008.

[2] J. McClellan and M. C. King, "Genetic heterogeneity in human disease," *Cell*, vol. 141, no. 2, pp. 210–217, 2010.

[3] D. B. Hancock, I. Romieu, M. Shi et al., "Genome-wide association study implicates chromosome 9q21.31 as a susceptibility locus for asthma in Mexican children," *PLoS Genetics*, vol. 5, no. 8, Article ID e1000623, 2009.

[4] F. A. Wright, L. J. Strug, V. K. Doshi et al., "Genome-wide association and linkage identify modifier loci of lung disease severity in cystic fibrosis at 11p13 and 20q13.2," *Nature Genetics*, vol. 43, no. 6, pp. 539–546, 2011.

[5] E. Einarsdottir, M. R. Bevova, A. Zhernakova et al., "Multiple independent variants in 6q21-22 associated with susceptibility to celiac disease in the Dutch, Finnish and Hungarian populations," *European Journal of Human Genetics*, vol. 19, no. 6, pp. 682–686, 2011.

[6] T. A. Manolio, F. S. Collins, N. J. Cox et al., "Finding the missing heritability of complex diseases," *Nature*, vol. 461, no. 7265, pp. 747–753, 2009.

[7] W. Ji, J. N. Foo, B. J. O'Roak et al., "Rare independent mutations in renal salt handling genes contribute to blood pressure variation," *Nature Genetics*, vol. 40, no. 5, pp. 592–599, 2008.

[8] S. Nejentsev, N. Walker, D. Riches, M. Egholm, and J. A. Todd, "Rare variants of IFIH1, a gene implicated in antiviral responses, protect against type 1 diabetes," *Science*, vol. 324, no. 5925, pp. 387–389, 2009.

[9] S. Romeo, W. Yin, J. Kozlitina et al., "Rare loss-of-function mutations in ANGPTL family members contribute to plasma triglyceride levels in humans," *The Journal of Clinical Investigation*, vol. 119, no. 1, pp. 70–79, 2009.

[10] M. A. Rivas, M. Beaudoin, A. Gardet et al., "Deep resequencing of GWAS loci identifies independent rare variants associated with inflammatory bowel disease," *Nature Genetics*, vol. 43, pp. 1066–1073, 2011.

[11] R. M. Durbin, D. L. Altshuler, G. R. Abecasis et al., "A map of human genome variation from population-scale sequencing," *Nature*, vol. 467, pp. 1061–1073, 2010.

[12] N. Arnheim, C. Strange, and H. Erlich, "Use of pooled DNA samples to detect linkage disequilibrium of polymorphic restriction fragments and human disease: studies of the HLA class II loci," *Proceedings of the National Academy of Sciences of the United States of America*, vol. 82, no. 20, pp. 6970–6974, 1985.

[13] R. W. Michelmore, I. Paran, and R. V. Kesseli, "Identification of markers linked to disease-resistance genes by bulked segregant analysis: a rapid method to detect markers in specific genomic regions by using segregating populations," *Proceedings of the National Academy of Sciences of the United States of America*, vol. 88, no. 21, pp. 9828–9832, 1991.

[14] P. Pacek, A. Sajantila, and A. C. Syvanen, "Determination of allele frequencies at loci with length polymorphism by quantitative analysis of DNA amplified from pooled samples," *PCR Methods and Applications*, vol. 2, no. 4, pp. 313–317, 1993.

[15] L. F. Barcellos, W. Klitz, L. L. Field et al., "Association mapping of disease loci, by use of a pooled DNA genomic screen," *American Journal of Human Genetics*, vol. 61, no. 3, pp. 734–747, 1997.

[16] J. Daniels, P. Holmans, N. Williams et al., "A simple method for analyzing microsatellite allele image patterns generated from DNA pools and its application to allelic association studies," *American Journal of Human Genetics*, vol. 62, no. 5, pp. 1189–1197, 1998.

[17] S. H. Shaw, M. M. Carrasquillo, C. Kashuk, E. G. Puffenberger, and A. Chakravarti, "Allele frequency distributions in pooled DNA samples: applications to mapping complex disease genes," *Genome Research*, vol. 8, no. 2, pp. 111–123, 1998.

[18] M. Krumbiegel, F. Pasutto, U. Schlötzer-Schrehardt et al., "Genome-wide association study with DNA pooling identifies variants at CNTNAP2 associated with pseudoexfoliation syndrome," *European Journal of Human Genetics*, vol. 19, no. 2, pp. 186–193, 2011.

[19] V. C. Sheffield, R. Carmi, A. Kwitek-Black et al., "Identification of a Bardet—Biedl syndrome locus on chromosome 3 and evaluation of an efficient approach to homozygosity mapping," *Human Molecular Genetics*, vol. 3, no. 8, pp. 1331–1335, 1994.

[20] R. Carmi, T. Rokhlina, A. E. Kwitek-Black et al., "Use of a DNA pooling strategy to identify a human obesity syndrome locus on chromosome 15," *Human Molecular Genetics*, vol. 4, no. 1, pp. 9–13, 1995.

[21] A. Nystuen, P. J. Benke, J. Merren, E. M. Stone, and V. C. Sheffield, "A cerebellar ataxia locus identified by DNA pooling to search for linkage disequilibrium in an isolated population from the Cayman Islands," *Human Molecular Genetics*, vol. 5, no. 4, pp. 525–531, 1996.

[22] D. A. Scott, R. Carmi, K. Elbedour, S. Yosefsberg, E. M. Stone, and V. C. Sheffield, "An autosomal recessive nonsyndromic-hearing-loss locus identified by DNA pooling using two inbred bedouin kindreds," *American Journal of Human Genetics*, vol. 59, no. 2, pp. 385–391, 1996.

[23] C. I. Amos, M. L. Frazier, and W. Wang, "DNA pooling in mutation detection with reference to sequence analysis," *American Journal of Human Genetics*, vol. 66, no. 5, pp. 1689–1692, 2000.

[24] P. Benaglio, T. L. Mcgee, L. P. Capelli, S. Harper, E. L. Berson, and C. Rivolta, "Next generation sequencing of pooled samples reveals new SNRNP200 mutations associated with retinitis pigmentosa," *Human Mutation*, vol. 32, no. 6, pp. E2246–E2258, 2011.

[25] A. A. Out, I. J. H. M. van Minderhout, J. J. Goeman et al., "Deep sequencing to reveal new variants in pooled DNA samples," *Human Mutation*, vol. 30, no. 12, pp. 1703–1712, 2009.

[26] M. A. Rivas, M. Beaudoin, A. Gardet et al., "Deep resequencing of GWAS loci identifies independent rare variants associated with inflammatory bowel disease," *Nature Genetics*, vol. 43, pp. 1066–1073, 2011.

[27] Y. Huang, D. A. Hinds, L. Qi, and R. L. Prentice, "Pooled versus individual genotyping in a breast cancer genome-wide association study," *Genetic Epidemiology*, vol. 34, no. 6, pp. 603–612, 2010.

[28] S. J. Docherty, L. M. Butcher, L. C. Schalkwyk, and R. Plomin, "Applicability of DNA pools on 500 K SNP microarrays for cost-effective initial screens in genomewide association studies," *BMC Genomics*, vol. 8, article 214, 2007.

[29] M. Harakalova, I. J. Nijman, J. Medic et al., "Genomic DNA pooling strategy for next-generation sequencing-based rare variant discovery in abdominal aortic aneurysm regions of interest—challenges and limitations," *Journal of Cardiovascular Translational Research*, vol. 4, no. 3, pp. 271–280, 2011.

[30] A. G. Day-Williams, K. McLay, E. Drury et al., "An evaluation of different target enrichment methods in pooled sequencing designs for complex disease association studies," *PLoS One*, vol. 6, Article ID e26279, 2011.

[31] X. Chen, J. B. Listman, F. J. Slack, J. Gelernter, and H. Zhao, "Biases and errors on Allele frequency estimation and disease association tests of next-generation sequencing of Pooled samples," *Genetic Epidemiology*, vol. 36, no. 6, pp. 549–560, 2012.

[32] Y. Guo, J. Long, J. He et al., "Exome sequencing generates high quality data in non-target regions," *BMC Genomics*, vol. 13, article 194, 2012.

[33] S. B. Ng, E. H. Turner, P. D. Robertson et al., "Targeted capture and massively parallel sequencing of 12 human exomes," *Nature*, vol. 461, no. 7261, pp. 272–276, 2009.

[34] J. F. Degner, J. C. Marioni, A. A. Pai et al., "Effect of read-mapping biases on detecting allele-specific expression from RNA-sequencing data," *Bioinformatics*, vol. 25, no. 24, pp. 3207–3212, 2009.

7

Comparison of Nasal Epithelial Smoking-Induced Gene Expression on Affymetrix Exon 1.0 and Gene 1.0 ST Arrays

Xiaoling Zhang,[1,2] **Marc E. Lenburg,**[1] **and Avrum Spira**[1,3]

[1] *Division of Computational Biomedicine, Boston University School of Medicine, 72 East Concord Street, E631, Boston, MA 02118, USA*
[2] *Division of Intramural Research, National Heart, Lung and Blood Institute, The NHLBI's Framingham Heart Study,*
  73 Mt. Wayte Avenue Suite 2, Framingham, MA 01702, USA
[3] *Pulmonary Center, Boston University Medical Center, 715 Albany Street, Boston, MA 02118, USA*

Correspondence should be addressed to Xiaoling Zhang; shirley0818@gmail.com

Academic Editors: X. Li, Z. Su, and X. Xu

We have previously defined the impact of tobacco smoking on nasal epithelium gene expression using Affymetrix Exon 1.0 ST arrays. In this paper, we compared the performance of the Affymetrix GeneChip Human Gene 1.0 ST array with the Human Exon 1.0 ST array for detecting nasal smoking-related gene expression changes. RNA collected from the nasal epithelium of five current smokers and five never smokers was hybridized to both arrays. While the intersample correlation within each array platform was relatively higher in the Gene array than that in the Exon array, the majority of the genes most changed by smoking were tightly correlated between platforms. Although neither array dataset was powered to detect differentially expressed genes (DEGs) at a false discovery rate (FDR) < 0.05, we identified more DEGs than expected by chance using the Gene ST array. These findings suggest that while both platforms show a high degree of correlation for detecting smoking-induced differential gene expression changes, the Gene ST array may be a more cost-effective platform in a clinical setting for gene-level genomewide expression profiling and an effective tool for exploring the host response to cigarette smoking and other inhaled toxins.

1. Introduction

Cigarette smoking is well recognized as the major cause of lung cancer and chronic obstructive pulmonary disease (COPD) [1]; however, only 10%–20% of smokers actually develop these diseases [2]. Further, it is unclear why some smokers remain healthy while others are still at high risk decades even after they have quit [3]. Unfortunately, there are currently no effective tools for identifying current and former smokers at highest risk for developing tobacco-related lung diseases.

Based on the concept that cigarette smoking creates a "field of injury" in epithelial cells throughout the respiratory tract, we have previously measured smoking-induced gene expression changes in bronchial airway epithelial cells obtained via bronchoscopy among healthy never, former, and current smokers using Affymetrix HG-U133A Array [4, 5]. Further, we developed a profile of bronchial airway gene expression that can distinguish smokers with and without

lung cancer and could serve as an early diagnostic biomarker for disease [6]. However, the invasiveness of bronchoscopy prevents it from being used as a screening tool for assessing smoking-induced lung cancer risk in large population studies. Most recently, utilizing Affymetrix Human Exon 1.0 ST (sense target) Array, we have demonstrated that smoking induces mostly similar gene-expression changes in both nasal and bronchial epithelium [7], suggesting that nasal epithelium could be a relatively noninvasive surrogate to measure physiological responses to cigarette smoking in several scenarios, for example, to estimate lung cancer risk of smokers in large-scale epidemiological studies, to detect second-hand smoking effects among children and adults, and to examine short- and long-term smoking damage in smoking cessation studies which require repeatedly collecting multiple samples from the same individuals.

In the previously mentioned comparison study, the Human Exon 1.0 ST array, the first in the whole-transcripts- (WTs-) based array family, was applied. This array contains

>5 million 25 mer probes, interrogating 1 million known and predicted exons [8], resulting in a comprehensive gene-level analysis, alternative splicing analysis, and novel exon and transcript detection. However, this platform is cost prohibitive, and the majority of expression biomarker applications only focus on known and manually curated genes. The Human Gene 1.0 ST array (Affymetrix, Santa Clara, CA) uses a subset of the same probes on the Human Exon 1.0 ST array to interrogate the more focused, well-annotated content at the gene level. Probes are also designed across the entire length of the genomic locus to provide a robust and accurate representation of total transcription activity for genes from RefSeq, Ensembl, and putative complete CDS GenBank transcripts. Predicted and discovery-oriented content from the Human Exon 1.0 ST array has been dropped; together this permits the use of a smaller, more affordable chip format for the Human Gene 1.0 ST array. The end result is a single focused gene-level expression array interrogating 28,869 well-annotated full-length genes with 764,885 distinct probes (on average, 28 probes per gene). Although the Gene ST array uses sparser probe coverage than the Exon array, it provides the same advantages of the Exon array for gene-level analysis and for partially degraded mRNA since both are WT-based arrays (*probes are distributed along the whole length of the gene*). Furthermore, the Gene ST array offers several advantages over the Exon array: (1) lower cost, easier to analyze due to having more than 6.5 times less, and more focused content (764,885 probes compared to 5 million on Exon arrays); (2) requiring 10 times less starting RNA materials (100 ng of total RNAs compared to 1 μg for Exon arrays), which will be tremendously beneficial for clinical studies where limited amounts of RNAs are available.

There are very limited data comparing the robustness and reproducibility of the Gene ST and Exon arrays, particularly in the setting of clinical samples. Previously, we have shown that there was a high correspondence of smoking-induced gene expression changes between the Human Exon and U133 arrays by hybridizing the same bronchial epithelial samples to both platforms [8]. Other studies have demonstrated a reasonable correlation in signals for genes that were differentially expressed between tissue types (heart versus brain) by assaying those tissue samples in parallel on both Exon and Gene ST arrays [9]. In this study, in order to estimate the performance of these two platforms on detecting differentially expressed genes in our nasal epithelial cells, we first performed a systematic comparison of the gene signal estimations from the Human Exon and Gene ST arrays by hybridizing the same nasal epithelial RNA samples obtained from smokers and nonsmokers to both arrays. Then, utilizing different chip description files (CDFs) for the preprocessing, we evaluated the impact of preprocessing and probe selection on the performance of these two array systems. Our data suggest a high degree of correlation between both platforms for detecting smoking-induced differential gene expression changes, with the Gene ST array being a more cost-effective and flexible platform in a clinical setting for the genomewide study of gene-level expression profiling.

2. Material and Methods

2.1. Study Population. We recruited 5 healthy never smokers and 5 current smokers for the study at Boston Medical Center. Nonsmokers with a history of significant second-hand environmental cigarette exposure, individuals with respiratory symptoms, or regular use of inhaled medications was excluded. For each participant, a detailed smoking history was obtained including, for smokers, cumulative tobacco exposure (measured in pack-years), age when they began smoking, and for all participants the extent of second-hand tobacco exposure. All individuals were screened with routine chest X-ray and spirometry and were excluded if there was evidence of pulmonary pathology. The study was approved by the Institutional Review Board of Boston Medical Center, and all participants provided written informed consent. Of note, there were no significant differences ($P > 0.05$) in age, race, and gender between 5 nonsmokers and 5 smokers in this study.

2.2. Sample Collection. Nasal epithelial cells were collected by brushing the inferior turbinate of the nose as previously described [7]. Briefly, the right nare was lavaged with 1 cc of 1% lidocaine. A nasal speculum (Bionix, Toledo, OH) then spread the nare while a standard cytology brush was inserted underneath the inferior nasal turbinate. The brush was rotated in place for 3 seconds, removed, and immediately placed in 1 mL RNAlater (Qiagen, Valencia, CA). After storage at 4°C, RNA was isolated via Qiagen RNeasy Mini Kits as per the manufacturer protocol. Integrity of the RNA samples was assessed by Agilent BioAnalyzer, and purity of the RNA was confirmed using a NanoDrop spectrophotometer.

4-5 mL of blood was obtained from the study participants for determination of plasma cotinine. Samples were centrifuged, and 2.2 mL of plasma was stored at −80°C, and then shipped on dry ice to the San Francisco Division of Clinical Pharmacology and Experimental Therapeutics, University of California. Gas chromatography (quantitation limit = 10 ng/mL) or liquid chromatography-tandem mass spectrometry (quantitation limit = .02 ng/mL) was used to analyze samples for the presence of this nicotine metabolite from self-reported current and never smokers, respectively.

2.3. Microarray Data Acquisition and Preprocessing

2.3.1. Affymetrix Human Exon 1.0 ST Array. 1 μg of total RNA from the nasal epithelium samples was used as the starting material. Ribosomal RNA was first removed using the RiboMinus Human/Mouse Transcriptome Isolation Kit (Invitrogen, Carlsbad, CA). This treated RNA was then converted to cDNA and subsequently processed, labeled, and hybridized onto the Exon arrays as previously described [7]. Following hybridization, each array was washed and stained according to the standard Affymetrix protocol. The stained array was scanned using an Affymetrix GeneChip Scanner 3000, resulting in a raw data CEL file for each array. The approximately 17,800 empirically supported transcripts (RefSeq and full-length GenBank mRNAs) were used for gene-level analysis. About 230,000 "core" exon-level probesets on

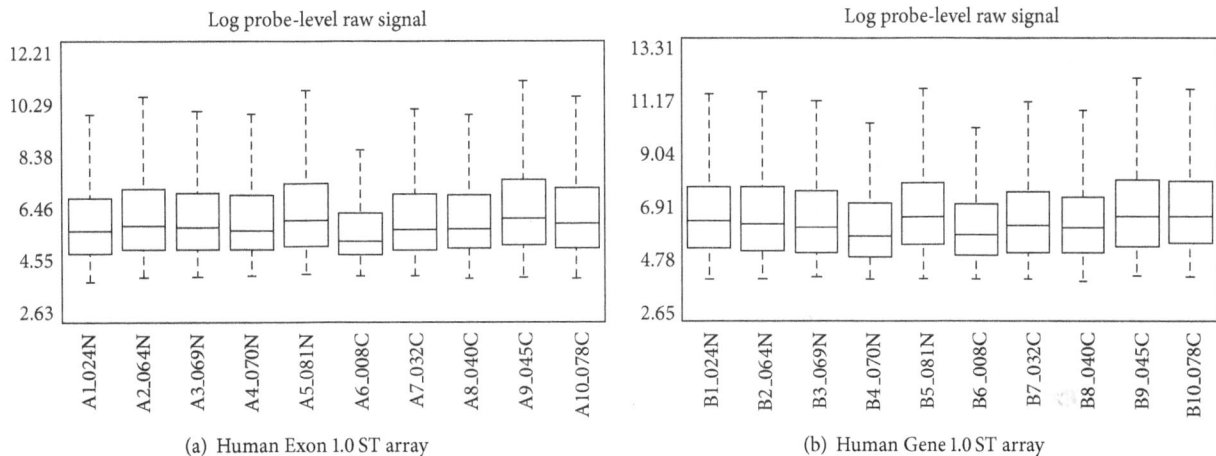

FIGURE 1: Distribution (boxplot) of the probe-level raw signal derived from (a) the Human Exon 1.0 ST array and (b) the Human Gene 1.0 ST. The raw signal distribution of each sample is similar across two platforms; for example, sample no. 6 has the lowest signal intensity distribution.

the Exon array were mapped to these core transcripts with a high degree of confidence. Gene-level expression values were derived from CEL files by quantile sketch normalization using the model-based Robust Multichip Average (RMA) method [10] as implemented in the Exon Array Computational Tool (ExACT) software package (Affymetrix, Santa Clara, CA). The gene annotations used for each probe set were from the annotation file obtained from Affymetrix (http://www.affymetrix.com/).

2.3.2. Affymetrix Human Gene 1.0 ST Array. Between 100 and 300 ng of total RNA was processed, labeled, and hybridized to Gene 1.0 ST arrays. The protocol for the Gene ST array was essentially the same as that for the Exon array, so they were combined in one protocol. The RMA method in the ExACT package was used for background adjustment, normalization, and probe-level summarization of the microarray samples. Although the Gene ST array contains 28,869 genes with support from RefSeq, Ensembl, and putative complete CDS GenBank, to be consistent with the previous analysis for the Exon array, we used the CDF provided by Affymetrix (the default parameter in ExACT) to preprocess the raw data. This led to 19,734 gene-level probe sets with putative full-length transcript support in the GenBank and RefSeq databases (http://www.affymetrix.com/estore/browse/products.jsp;jsessionid=7F717F80BE253ABBF6155B16AC95C6F9?navMode=34000&productId=131453&navAction=jump&aId=productsNav#1_1).

2.4. Microarray Data Analysis. All of these 10 samples were processed in one batch for each array type. After preprocessing, we applied the principal components analysis (PCA) and the relative log expression (RLE) signal to check the quality of the arrays. Not surprisingly, all of the samples passed the quality metrics. For comparing the fold change of smoking-induced gene expression changes on both arrays, we considered the RefSeq mappings as a nonredundant and relatively complete database of transcripts. In total, 17,881 core

transcripts with annotation on the Exon array were mapped to 19,734 transcripts with annotation on the Gene ST array. Only those probe sets for genes in common across both array types were used in the analysis. This produced a set of 16,482 transcripts, and these probe sets in general exhibited a higher signal than the other probe sets on each of the arrays.

After obtaining transcripts in common across the Exon and Gene ST arrays, we first compared smoking-induced gene expression changes in nasal epithelial cells between these two array systems. In order to identify genes that were differentially expressed between the 5 smokers and 5 nonsmokers, first, a simple Student's *t*-test was applied for data from each array platform, respectively. Then, fold changes between the log2 mean values for the smoker and nonsmoker replicates were calculated independently for each array platform. Using the matched 16,482 transcripts, we then characterized the sample correlation between these two platforms for each matched sample.

Finally, in order to estimate the impact of different chip description files (CDFs) on gene expression measurements in our nasal clinical samples, we modified the Exon array CDF by removing probes not found in Gene ST arrays. As a result, 19,802 transcripts were left for comparison of smoking-induced gene expression changes by examining the fold changes between smokers and nonsmokers in nasal epithelium.

2.5. Additional Information. All statistical analyses described previously were performed with R 2.11.0 (available at http://r-project.org/) and Bioconductor [11]. All microarray data from this study has been deposited in Gene Expression Omnibus (GEO).

3. Results and Discussion

3.1. Distribution of the Probe-Level Raw Signal on Exon and Gene ST Arrays. Each RNA sample collected from nasal epithelium was hybridized to both the Exon and Gene ST

Pearson's correlation (RMA gene signal)

(a) Human Exon 1.0 ST array

Pearson's correlation (RMA gene signal)

(b) Human Gene 1.0 ST array

(c) A cross-platform comparison of samples. For each paired samples (from the Exon array and the Gene ST array, resp.), the sample correlation was calculated using Pearson's Correlation for 16,482 genes found on two platforms. The number in each box represents the correlation coefficient. The Exon array data is on x-axis, and Gene ST array data is on y-axis

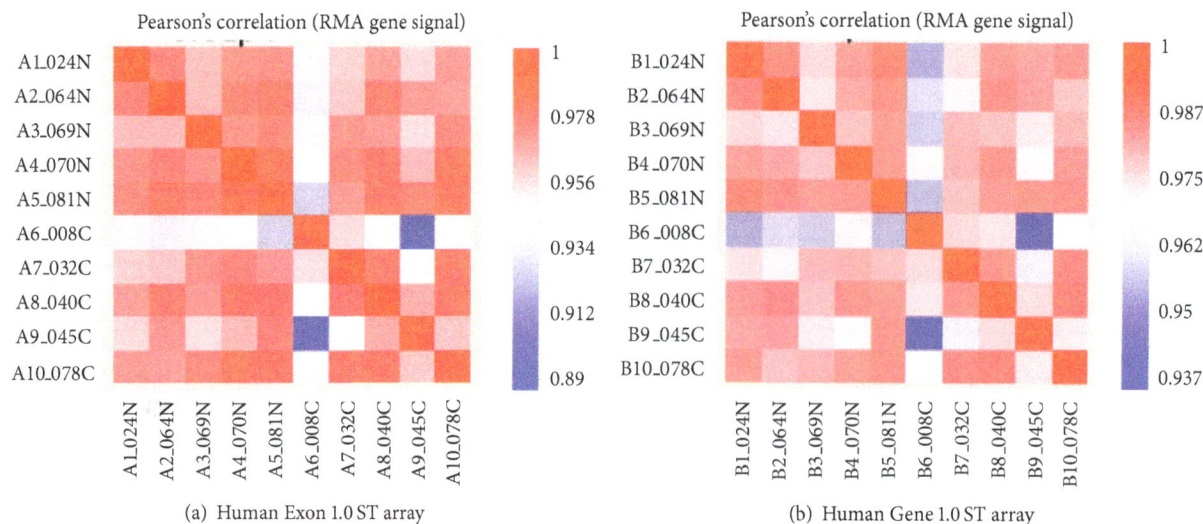

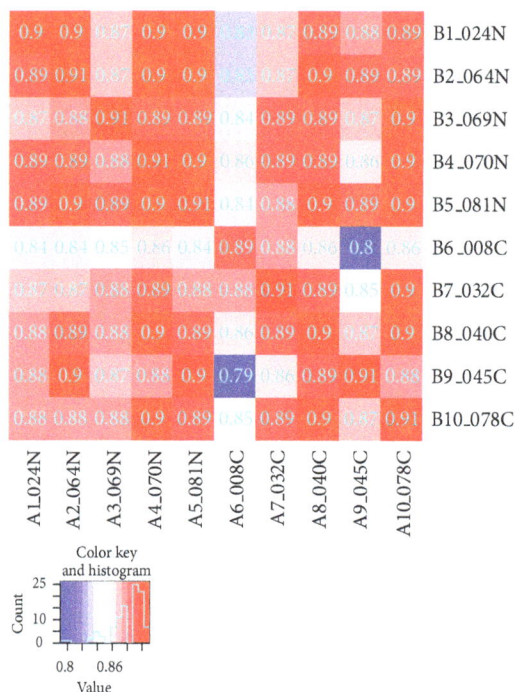

FIGURE 2: Sample correlation using RMA-normalized gene-level signals. (a) Within the Human Exon 1.0 ST array; (b) within the Human Gene 1.0 ST;(c) cross-platform sample correlation of the Exon and Gene ST arrays for 16,482 matched genes.

arrays. The distribution of the raw signal for each sample is shown in Figure 1(a) (the Exon array) and Figure 1(b) (the Gene array), respectively. Although there is a time difference for the processing and labeling between the two array systems, overall, the raw intensity for these 10 samples (5 current and 5 never smokers) is similar across the two platforms; for example, sample no. 6 has the lowest signal

intensity measured by both arrays compared to the other 9 samples.

3.2. Sample Correlation within and between Exon Arrays and Gene ST Arrays. We first calculated the sample correlation within each platform using normalized gene level signal for

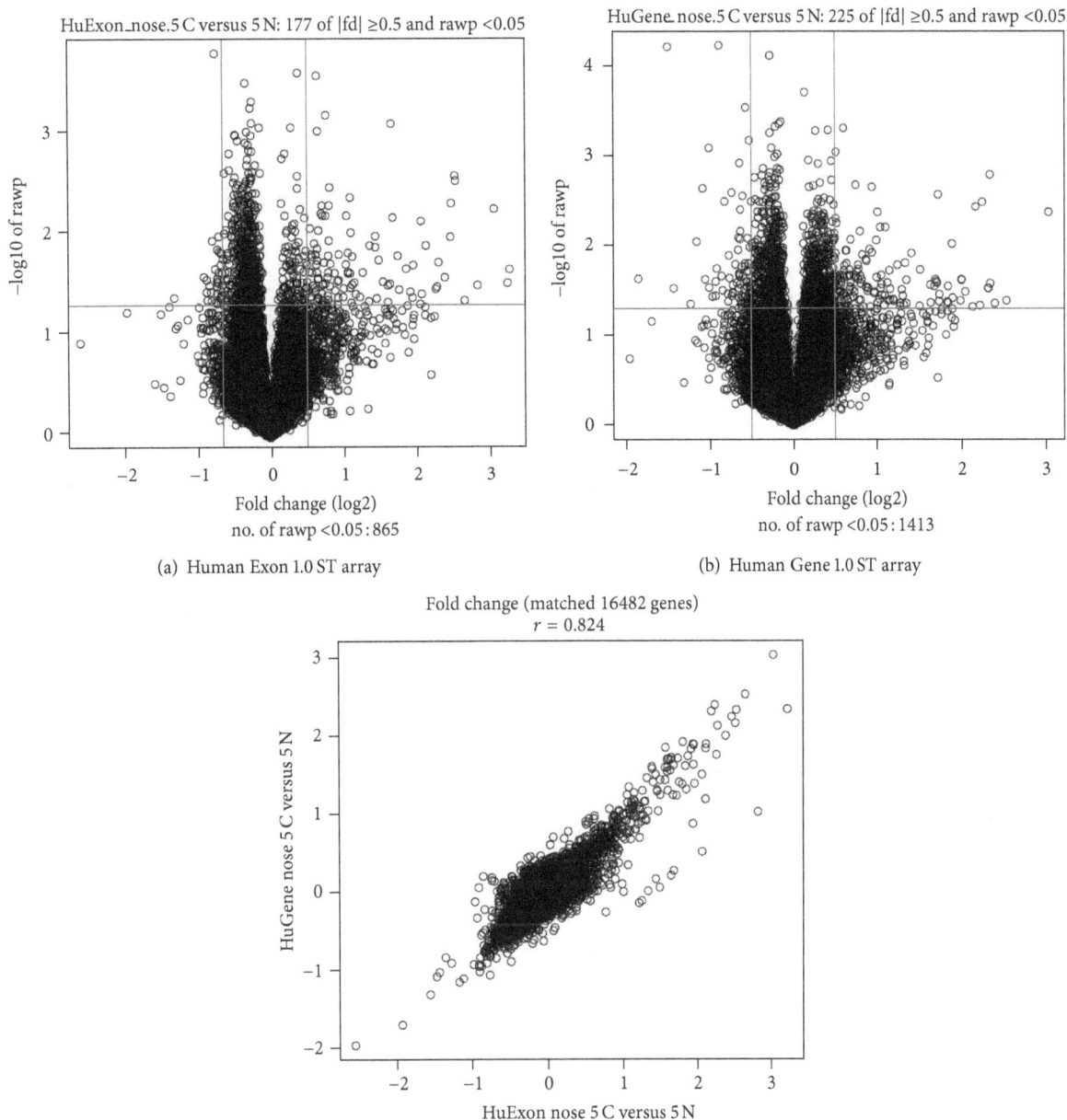

(a) Human Exon 1.0 ST array

(b) Human Gene 1.0 ST array

(c) Fold changes between the log2 mean values for 5 current smokers (C) and 5 never smokers (N) were calculated independently for the Exon and Gene ST arrays for 16,482 best matched genes. This fold change correlation ($r = 0.82$) is higher than the correlation between the Exon and U133A arrays ($r = 0.62$) [8]

FIGURE 3: Smoking-induced differentially expressed genes using a Student's t-test (a) within the Human Exon 1.0 ST array and (b) within the Human Gene 1.0 ST. (c) Correlation between smoking-induced gene-expression differences (fold change) of the Exon and Gene ST arrays for 16,482 best matched gene-level probe sets.

all core transcripts on the Exon arrays and all transcripts on the Gene arrays. As can be seen in Figure 2, the correlation is relatively high ($r > 0.95$) for all samples with the exception of sample no.6 which is slightly less correlated with the other samples ($r = 0.9$). This could be due to its relatively low normalized signal compared to the other samples. Furthermore, the correlation is relatively higher in the Gene arrays than in the Exon arrays, perhaps due to more noisy probes on the Exon arrays (Figures 2(a) and 2(b)).

Then, in order to directly compare the correlation between the same samples hybridized on both arrays, we first identified 16,482 genes found on both platforms. The matched sample correlation between the Exon arrays and Gene ST arrays is high (≥ 0.9 except for sample no. 6) (Figure 2(c)).

These results suggest that, compared to the Exon array, the Gene ST array is a comparable platform that is a much more cost-effective choice for well-annotated gene-level analysis.

3.3. Gene-Level Differential Expression Changes on Exon and Gene ST Arrays. For both Exon and Gene ST arrays, "gene-level" analysis of multiple probes on different exons is summarized into an expression value representing all transcripts from the same gene. This approach allowed us to compare genes differentially expressed between the same 5 current and 5 never smokers assayed on both arrays using a simple Student's t-test. Very few differentially expressed genes (DEGs) pass multiple testing (FDR < 0.05), which might be due to the small sample size in this study. However, at $P < 0.05$, many more genes are discovered from the Gene arrays (1,413 genes) than from the Exon arrays (865 genes), as shown in Figure 3(a) (the Exon array) and Figure 3(b) (the Gene array). At a significance level of 0.05, 894 (17,881*0.05) genes are expected to be detected on the Exon array and 987 genes (19,734*0.05) on the Gene ST array. We can see that the number of DEGs is more than expected by chance on the Gene ST array, while the number of DEGs is less than expected by chance on the Exon array using the same samples as the Gene ST array. This suggests that the Gene ST array has higher signal-to-noise ratio for gene-level analysis compared to the Exon array. This might be due to the design of the Gene ST array, on which 90% of the gene-level probe sets contain only probes that match uniquely to the genome [12], indicating less noisy probe design/contents in the Gene ST array.

After comparing statistical results of differentially expressed genes on the Exon and Gene ST arrays, we then estimated the similarities and differences in the magnitude of the gene-level fold changes between smoking status across both array platforms (Figure 3(c)). In this scatter plot, each point corresponds to a pair of transcripts for which a successful cross-chip mapping could be found ($n = 16,482$). The majority of detected gene-expression differences correlate tightly between the Exon and Gene ST arrays ($r = 0.82; P < 4 \times 10^{-18}$), which is significantly higher than the fold change correlation observed between the Exon array and the U133A array in bronchial smoking gene expression data ($r = 0.62$) [8]. This is not surprising since U133A array is one of the traditional 3' arrays, which may lack probe sets to measure the expression of some particular transcripts. However, the Gene ST arrays, whenever possible, use a subset of the same probes on the Exon array and are similar in other ways, like having probes targeting the whole transcripts (WTs), compatible with WT Sense Target Labeling and Control Reagent Kits for maximum coverage of the entire gene.

Finally, we applied a modified CDF to preprocess the data including background correction, normalization, and summarization, to see whether there is a difference of the impact on smoking-induced gene expression changes between utilizing different CDFs for preprocessing. As shown in Figure 4, the gene-level fold change correlation between Exon arrays and Gene ST arrays is higher with a modified

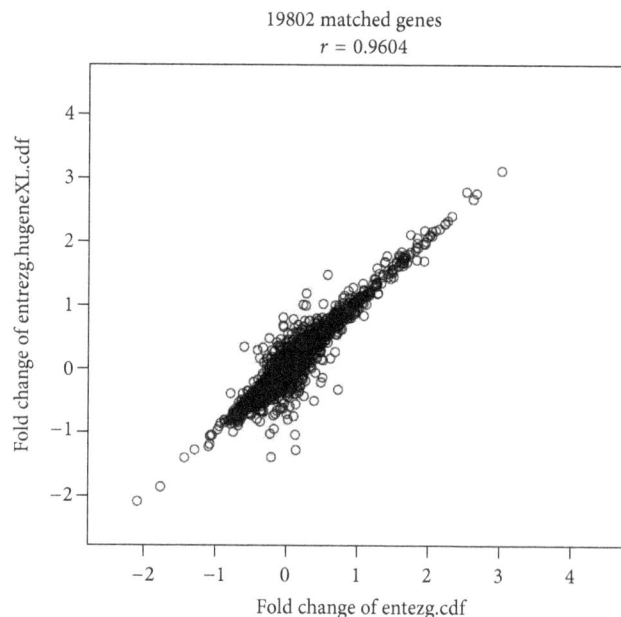

FIGURE 4: The fold change comparison of smoking-induced gene expression changes for a total of 19,802 matched genes between two different CDFs.

CDF using a total of 19,802 matched genes ($r = 0.96$). Furthermore, 68% of DEGs defined at $P < 0.05$ overlapped, which is much higher than expected by chance.

4. Conclusion

In this study, we compared the performance of Exon and Gene ST arrays in our nasal epithelial samples at gene-level since the Gene ST array is designed as a focused gene level expression microarray and covers only well-characterized full-length genes by using sparser probe coverage than the Exon array. Both are WT-based arrays with a comprehensive coverage of the entire gene locus, so the Gene ST array therefore provides similar power to the Exon array for gene-level analysis in a more affordable format. But without the discovery content and full exon-level coverage, the Gene array is not designed for the high-resolution study of known and predicted alternative splicing. However, there is at least 1 probe per known exon on Gene ST arrays, so it could also be used to detect alternative splicing (AS) events. In a recent study [9], Ha et al. have shown a comparable performance of the Gene ST arrays for detecting transcript isoforms expressed differently between brain and other tissues, when compared to Exon arrays even though there are 4 probes targeting each exon on the Exon arrays.

On the other hand, the Exon array includes substantially more discovery contents, such as predicted exons and transcripts. RNA-Seq is an alternate approach for this type of discovery, but RNA-Seq has its own limitations which prevent it from having as wide an application as expression arrays, for example, much higher cost and computational and storage challenges. As one possible way to address this need, Affymetrix recently released the Whole Transcriptome Arrays, with probes targeting exon-exon junctions, noncoding RNAs, and so forth. Additional studies are needed to compare Whole Transcriptome Arrays and Gene ST arrays for well-annotated gene-level applications.

In summary, we compared the performance of the Human Gene 1.0 ST array with the Exon 1.0 ST array for detecting smoking-related gene expression changes in nasal epithelium. The Gene ST array appears to be a reproducible platform capable of working with smaller amounts (100–300 ng) of RNAs. In a gene-level fold change comparison, we found a strong correlation between the two platforms for smoking-related gene expression changes even though the Gene ST array contains much fewer probes. These findings suggest that the Gene ST array can serve as a clinically relevant and more flexible tool (in terms of cost and input RNAs) for exploring host response to tobacco smoking in large-scale population-based studies.

Disclosure

None of the authors has a financial relationship with a commercial entity that has an interest in the subject of this paper.

Authors' Contribution

M. E. Lenburg and A. Spira contributed equally as cosenior authors.

Acknowledgment

This work was supported by NIH/NIEHS UO1ES016035 (Genes, Environment and Health Initiative).

References

[1] R. T. Greenlee, M. B. Hill-Harmon, T. Murray, and M. Thun, "Cancer statistics, 2001," *CA—A Cancer Journal for Clinicians*, vol. 51, no. 1, pp. 15–36, 2001.

[2] P. G. Shields, "Molecular epidemiology of lung cancer," *Annals of Oncology*, vol. 10, supplement 5, pp. S7–S11, 1999.

[3] J. O. Ebbert, P. Yang, C. M. Vachon et al., "Lung cancer risk reduction after smoking cessation: observations from a prospective cohort of women," *Journal of Clinical Oncology*, vol. 21, no. 5, pp. 921–926, 2003.

[4] A. Spira, J. Beane, V. Shah et al., "Effects of cigarette smoke on the human airway epithelial cell transcriptome," *Proceedings of the National Academy of Sciences of the United States of America*, vol. 101, no. 27, pp. 10143–10148, 2004.

[5] J. Beane, P. Sebastiani, G. Liu, J. S. Brody, M. E. Lenburg, and A. Spira, "Reversible and permanent effects of tobacco smoke exposure on airway epithelial gene expression," *Genome Biology*, vol. 8, no. 9, article R201, 2007.

[6] A. Spira, J. E. Beane, V. Shah et al., "Airway epithelial gene expression in the diagnostic evaluation of smokers with suspect lung cancer," *Nature Medicine*, vol. 13, no. 3, pp. 361–366, 2007.

[7] X. Zhang, P. Sebastiani, G. Liu et al., "Similarities and differences between smoking-related gene expression in nasal and bronchial epithelium," *Physiological Genomics*, vol. 41, no. 1, pp. 1–8, 2010.

[8] X. Zhang, G. Liu, M. E. Lenburg, and A. Spira, "Comparison of smoking-induced gene expression on Affymetrix Exon and 3'-based expression arrays," *Genome Informatics*, vol. 18, pp. 247–257, 2007.

[9] K. C. H. Ha, J. Coulombe-Huntington, and J. Majewski, "Comparison of Affymetrix Gene Array with the Exon Array shows potential application for detection of transcript isoform variation," *BMC Genomics*, vol. 10, article 519, 2009.

[10] R. A. Irizarry, B. Hobbs, F. Collin et al., "Exploration, normalization, and summaries of high density oligonucleotide array probe level data," *Biostatistics*, vol. 4, no. 2, pp. 249–264, 2003.

[11] R. C. Gentleman, V. J. Carey, D. M. Bates et al., "Bioconductor: open software development for computational biology and bioinformatics," *Genome Biology*, vol. 5, no. 10, article R80, 2004.

[12] Human Gene 1.0 ST Array Performance, White Paper, http://www.affymetrix.com/support/technical/whitepapers/hugene_perf_whitepaper.pdf.

8

Evidence for Directed Evolution of Larger Size Motif in *Arabidopsis thaliana* Genome

Rajesh Mehrotra,[1,2] **Amit Yadav,**[1] **Purva Bhalothia,**[1] **Ratna Karan,**[2] **and Sandhya Mehrotra**[1]

[1]*Department of Biological Sciences, Birla Institute of Technology and Science, Pilani, Rajasthan 333031, India*
[2]*Department of Plant Environmentand Soil Sciences, Louisiana State University, Baton Rouge, LA 70894, USA*

Correspondence should be addressed to Rajesh Mehrotra, rajmeh25@hotmail.com

Academic Editors: Z.-G. Han, Y. Y. Shugart, and W. A. Thompson

Transcription control of gene expression depends on a variety of interactions mediated by the core promoter region, sequence specific DNA-binding proteins, and their cognate promoter elements. The prominent group of *cis* acting elements in plants contains an ACGT core. The *cis* element with this core has been shown to be involved in abscisic acid, salicylic acid, and light response. In this study, genome-wide comparison of the frequency of occurrence of two ACGT elements without any spacers as well as those separated by spacers of different length was carried out. In the first step, the frequency of occurrence of the *cis* element sequences across the whole genome was determined by using BLAST tool. In another approach the spacer sequence was randomized before making the query. As expected, the sequence ACGTACGT had maximum occurrence in *Arabidopsis thaliana* genome. As we increased the spacer length, one nucleotide at a time, the probability of its occurrence in genome decreased. This trend continued until an unexpectedly sharp rise in frequency of (ACGT)N25(ACGT). The observation of higher probability of bigger size motif suggests its directed evolution in *Arabidopsis thaliana* genome.

1. Introduction

Gene expression in eukaryotic organisms has been a topic of great interest. Careful regulation and recruitment of transcription factors (TFs) to *cis* regulatory elements in promoter regions lead to generation of specificity and diversity [1] in genetic regulation. Promoters are arrays of *cis* regulatory elements present upstream of a gene arranged with other specific *cis* elements. At present 469 *cis* elements have been reported in the plant *cis* regulatory element (PLACE) database. The prominent group of *cis* acting elements in plants contains an ACGT core. Several *cis* elements with this core have been shown to be responding to abscisic acid [2–4], salicylic acid [5], and light signals [6]. It has been reported by Foster et al. [7] that bZIP class of transcription factors binds to this core motif. In an elegant study Krawczyk et al. [8] showed deletion of two base pairs between activator sequence-1 (as1) palindromes does not affect binding of activator sequence binding factor (ASF-1) and TGA factors (which binds to TGACG sequence), whereas insertion decreases factor binding *in vitro*. In their

study the distance between palindromic centers was 12 base pairs. Mehrotra et al. [9, 10] have shown that this motif functions even when they are placed out of the native context. R. Mehrotra and S. Mehrotra [11] have shown that promoter activation by ACGT in response to salicylic and abscisic acids is differentially regulated by the spacing between these motifs. It contributes synergistically to gene expression by stabilising the transcription complex formed on minimal promoter [10]. The present study is an extension of aforementioned work. In this study, genome-wide comparison of the frequency of occurrence of two ACGT elements without any spacers and also separated by spacers of different lengths was done. Based on the data obtained we report that there is a directed evolution of bigger size of motif in the *Arabidopsis thaliana* genome.

2. Materials and Methods

The objective was to find out the frequency of the recurring sequences and then use these recurring sequences with

TABLE 1: Frequency of occurrence of the various promoter sequences in which spacer sequence length between two ACGT palindromes is gradually increased from 5 to 25 nucleotides.

	Cis element	Chromosome 1	Chromosome 2	Chromosome 3	Chromosome 4	Chromosome 5	Total
(ACGT)$_2$	ACGTACGT	469	312	367	327	410	1885
(ACGT)$_8$	ACGTACGTACGTACGTAC-GTACGTACGTACGT	70	31	12	28	59	200
(ACGT)$_{N5}$(ACGT)	ACGTGGCTAACGT	16	11	13	13	19	72
(ACGT)$_{N10}$(ACGT)	ACGTGGCTATGGCGACGT	8	5	10	4	12	39
(ACGT)$_{N25}$(ACGT)	ACGTGGCTATGGCGGAGC-AAGATTCACTCACGT	15	12	13	9	13	62
(ACGT)$_{RN5}$(ACGT)	ACGT--GCTAG--ACGT	7	5	5	2	4	23
(ACGT)$_{RN10}$(ACGT)	ACGT--TGGGGCCGAT--ACGT	2	2	4	3	3	14
(ACGT)$_{RN25}$(ACGT)	ACGTAGACACGTTGGGGG-AACTTACTGCCACGT	3	1	7	5	5	21
(ACGT)$_{RN25}$(ACGT)	ACGT-ATATGAGATCGGCGCT-TCACGGAGC-ACGT	4	14	6	4	4	32
(ACGT)$_{N5}$(ACGT) randomized	GGAATCCTTGGCA	41	24	30	19	23	137
(ACGT)$_{N10}$(ACGT) randomized	GCGGGCTATCGGTAGCAT	2	5	2	0	1	10
(ACGT)$_{N25}$(ACGT) randomized	TAAGGCTTAGCCACGCTT-AGGGTGTGAGCACAC	6	6	3	0	3	18
(TGCA)$_{N25}$(TGCA)	TGCAGGCTATGGCGGAGC-AAGATTCACTCTGCA	13	12	9	12	9	55

N5, N10, N25 denote sequence length between two ACGT palindromes. RN5, RN10, RN25—signify only spacer sequence being randomized. (ACGT)$_N$_(ACGT) randomized—signify complete sequence being randomized.

a random minimal promoter to predict transcription factors likely to interact with them.

The genomic sequence database of *Arabidopsis thaliana* at http://www.arabidopsis.org/ (The *Arabidopsis* Information Resource, TAIR) was analyzed using software BLASTn (available at NCBI website). All sequences were run in BLASTn against whole *Arabidopsis thaliana* genome to find their frequency of occurrence. Accession numbers of *Arabidopsis thaliana* chromosomes are as follows: chromosome 1: NC_003070.9, chromosome 2: NC_003071.7, chromosome 3: NC_003074.8, chromosome 4: NC_003075.7, and chromosome 5: NC_003076.8.

Randomization of the sequence was carried out using SHUFFLE program [12]. Different sequences obtained are listed in Table 1. In the next step we found the transcription factors binding to these *cis* elements separated by different length of nucleotides. A 139 bp long minimal promoter *Pmec* [13] was used in this study. The minimal promoter sequence as shown below was suffixed to the sequences shown in Table 1;

TCACTATATATAGGAAGTTCATTTCATTTGGAA-TGGACACGTGTTGTCATTTCTCAACAATTACCAACA-ACAACAAACAACAAACAACATTATACAATTACTATT-TACAATTACATCTAGATAAACAATGGCTTCCTCC.

These extended sequences were used in JASPAR core database [14] to scan for transcription factors and then these TFs were crosschecked with results obtained from CONSITE [15].

3. Results and Discussion

3.1. Promoters with Greater Length between ACGT Motifs Are More Frequent. It has been reported that ACGT *cis* elements function even when they are placed out of native sequence context [9, 10]. When the distance of separation between two ACGT elements are 5 base pairs, and 10 base pairs, they are induced in response to salicylic acid (SA) and abscisic acid (ABA), respectively. Interestingly, SA mimics biotic stress response and ABA mimics abiotic stress response in plants and thus is of great interest to plant biologists. Paixão and Azevedo [16] showed that multiplicity of *cis* element evolved through transitional forms showing redundant *cis* regulation. In this study, when the frequency of occurrence of two ACGT elements without any spacers and also separated by the spacer of different lengths was observed, we found that the total frequency of occurrence of two ACGT element in tandem is 1885 (Table 1), while the e value was same for all alignments obtained on a particular chromosome. When two ACGT elements were separated by spacer of 5, 10, and 25 nucleotides their frequency of occurrence was 72, 39, and 62, respectively. An unexpectedly high frequency of occurrence was observed when two ACGT elements were separated by 25 base pairs. According to the rule of probability the frequency of two ACGT elements separated by 25 base pairs should be less than when they are separated by 10 base pairs or lesser. Hobo et al. [17] have earlier reported that in ABA responsive promoters the distance between ACGT elements

TABLE 2: Frequency of occurrence of nitrogenous bases when spacer sequence length between two ACGT palindromes is gradually increased from 5 to 25 nucleotides.

		A	C	G	T	Seq. used	Gap	Count
$(ACGT)_{N5}(ACGT)$	ACGTGGCT_ACGT	72	42	33	34	72	5	690
$(ACGT)_{N6}(ACGT)$	ACGTGGCTA_ACGT	98	65	45	44	44	6	611
$(ACGT)_{N7}(ACGT)$	ACGTGGCTAT_ACGT	92	91	77	80	77	7	824
$(ACGT)_{N8}(ACGT)$	ACGTGGCTATG_ACGT	97	30	64	55	64	8	852
$(ACGT)_{N9}(ACGT)$	ACGTGGCTATGG_ACGT	39	32	22	32	32	9	602
$(ACGT)_{N10}(ACGT)$	ACGTGGCTATGGC_ACGT	34	36	39	66	39	10	600
$(ACGT)_{N11}(ACGT)$	ACGTGGCTATGGCG_ACGT	36	23	38	29	38	11	681
$(ACGT)_{N12}(ACGT)$	ACGTGGCTATGGCGG_ACGT	56	54	65	45	56	12	638
$(ACGT)_{N13}(ACGT)$	ACGTGGCTATGGCGGA_ACGT	78	50	77	59	77	13	652
$(ACGT)_{N14}(ACGT)$	ACGTGGCTATGGCGGAG_ACGT	86	53	96	52	53	14	841
$(ACGT)_{N15}(ACGT)$	ACGTGGCTATGGCGGAGC_ACGT	56	67	44	66	56	15	709
$(ACGT)_{N16}(ACGT)$	ACGTGGCTATGGCGGAGCA_ACGT	60	34	52	34	60	16	843
$(ACGT)_{N17}(ACGT)$	ACGTGGCTATGGCGGAGCAA_ACGT	39	41	42	39	42	17	830
$(ACGT)_{N18}(ACGT)$	ACGTGGCTATGGCGGAGCAAG_ACGT	49	47	58	48	49	18	719
$(ACGT)_{N19}(ACGT)$	ACGTGGCTATGGCGGAGCAAGA_ACGT	50	38	49	44	44	19	695
$(ACGT)_{N20}(ACGT)$	ACGTGGCTATGGCGGAGCAAGAT_ACGT	34	30	44	37	37	20	821
$(ACGT)_{N21}(ACGT)$	ACGTGGCTATGGCGGAGCAAGATT_ACGT	36	40	42	43	40	21	717
$(ACGT)_{N22}(ACGT)$	ACGTGGCTATGGCGGAGCAAGATTC_ACGT	53	42	42	46	53	22	726
$(ACGT)_{N23}(ACGT)$	ACGTGGCTATGGCGGAGCAAGATTCA_ACGT	91	55	60	61	55	23	771
$(ACGT)_{N24}(ACGT)$	ACGTGGCTATGGCGGAGCAAGATTCAC_ACGT	77	64	57	53	53	24	1171
$(ACGT)_{N25}(ACGT)$	ACGTGGCTATGGCGGAGCAAGATTCACT_ACGT	76	62	58	69	62	25	708

TABLE 3: Alterations in transcription factor binding sites when spacer sequence length between two ACGT palindromes is gradually increased from 5 to 25 nucleotides.

	Minimal promoter sequence (MPS)	(ACGT)	(ACGT) (MPS)	$(ACGT)_2$ (MPS)	$(ACGT)_{N5}$ (ACGT)(MPS)	$(ACGT)_{N10}$ (ACGT)(MPS)	$(ACGT)_{N25}$ (ACGT)(MPS)
Model name	Frequency						
ARR10	0	0	0	0	0	0	1
AGL3	2	0	2	2	2	2	2
ATHB-5	1	0	1	2	1	1	1
bZIP910	0	0	0	0	1	1	1
Dof3	1	0	1	1	1	1	2
EmBP-1	2	0	2	1	2	2	2
Gamyb	5	0	5	5	5	5	5
HAT5	2	0	2	2	2	2	2
HMG-1	6	0	6	6	6	6	6
HMG-I/Y	6	0	6	6	6	6	6
id1	5	0	5	5	5	5	5
myb.Ph3	1	0	1	1	2	1	1
PEND	1	0	1	1	1	1	1
squamosa	2	0	3	3	3	3	3
TGA1A	1	0	1	1	2	2	2
	35	0	36	36	39	38	40

is 30 base pairs. To address this discrepancy in the data obtained, we randomized the spacer sequence keeping the ACGT motif unchanged. The logic of this randomization was to identify how important is the distance between the binding sites for transcription factors. After randomization of the spacer there was a drop in the frequency of occurrence to 23, 14, and 21 from 72, 39, and 62 for $(ACGT)_{N5}(ACGT)$, $(ACGT)_{N10}(ACGT)$, and $(ACGT)_{N25}(ACGT)$, respectively. This means that along with the distance between binding motifs there has been a positive selection for the sequence of the spacer in transcriptional regulation. In the next step we completely randomized the sequence and we observed that there is a drop in frequency of occurrence of two ACGT elements when separated by 10 and 25 base pairs while there was an unexpected increase in the frequency when ACGT elements were separated by five base pairs. This happened because randomization generated a motif that has been positively selected in evolution.

3.2. A and G Are the Preferred Bases. We increased the spacer length one residue at a time and looked for the frequency of each resultant sequence in the database. As shown in Table 2, there has been preference for A and G in the spacer region between two ACGT sequences.

3.3. Increasing Spacing between Motifs Increases Transcription Factor Binding Sites. Potential transcription factor binding sites for all experimental sequences when predicted using JASPAR CORE software and subsequently crosschecked with CONSITE revealed the minimal promoter sequence to be possessing 35 potential TF binding sites (Table 3, MPS). Interestingly the sequence ACGT as such has no site for binding of transcription factors but when minimal promoter is suffixed to it, an extra site for squamosa is generated and the total transcription factor binding site increases from 35 to 36 in minimal promoter alone (Table 3, (ACGT)(MPS)). When two ACGT elements in tandem are placed over minimal promoter sequence no extra site for binding of transcription factor is generated (Table 3, $(ACGT)_2(MPS)$). However, when ACGT elements are separated by five base pairs (Table 3, $(ACGT)_{N5}(ACGT)(MPS)$), four additional transcriptional binding sites are generated while ATHB-5 binding site which existed in the earlier cases is lost. The new sites generated are for transcription factors bzip9-10, EmBP-1, myb.Ph3, and TGA1a. Placement of two ACGT elements separated by 10 base pairs, however, resulted in loss of one myb.Ph3 site and the total transcriptional binding site decreased to 38 (Table 3, $(ACGT)_{N10}(ACGT)(MPS)$). In case when ACGT elements are separated by 25 base pairs followed by minimal promoter an additional site for ARR10 and dof3 was generated (Table 3, $(ACGT)_{N25}(ACGT)(MPS)$).

Based on the data obtained in this study, we report here that there has been directed evolution of bigger size of the motif in the *Arabidopsis thaliana* genome.

4. Conclusions

The central question in promoter evolution is to know how does *cis* regulatory element multiplicity evolved. The promoter regions of many genes contains multiple binding sites for the same transcription factor. Multiplicity may have evolved through transitional forms showing redundant *cis* regulation. In this paper, we focused on multiplicity of ACGT *cis* element and the distances between them which occurs in natural promoters. We found that ACGT element separated by 25 base pairs is more frequent than those by 10 base pairs which is against the law of probability. It signifies that under some evolutionary forces this interval was favoured since this distance may cause changes in the level of gene expression or in its robustness against variation in transcription factor concentration. Selection for different levels of expression of certain genes in certain environment could, over time, generates a positive association between *cis* element multiplicity and expression level.

Acknowledgments

The authors are grateful to the Department of Science and Technology, New Delhi, India for Grant-in-Aid and financial support to carry out this work bearing the file no. SR/FT/LS-126/2008. The authors are grateful to the administration of Birla Institute of Technology and Sciences, Pilani, Rajasthan for providing logistic support. They are thankful to Professor C. Gatz for critically reading the paper.

References

[1] G. A. Wray, M. W. Hahn, E. Abouheif et al., "The evolution of transcriptional regulation in eukaryotes," *Molecular Biology and Evolution*, vol. 20, no. 9, pp. 1377–1419, 2003.

[2] M. J. Guiltinan, W. R. Marcotte, and R. S. Quatrano, "A plant leucine zipper protein that recognizes an abscisic acid response element," *Science*, vol. 250, no. 4978, pp. 267–271, 1990.

[3] Q. Shen and T. H. Ho, "Functional dissection of an abscisic acid (ABA)-inducible gene reveals two independent ABA-responsive complexes each containing a G-box and a novel cis-acting element," *Plant Cell*, vol. 7, no. 3, pp. 295–307, 1995.

[4] P. K. Busk and M. Pagès, "Regulation of abscisic acid-induced transcription," *Plant Molecular Biology*, vol. 37, no. 3, pp. 425–435, 1998.

[5] I. Jupin and N. H. Chua, "Activation of the CaMV as-1 cis-element by salicylic acid: differential DNA-binding of a factor related to TGA1a," *EMBO Journal*, vol. 15, no. 20, pp. 5679–5689, 1996.

[6] R. G. K. Donald and A. R. Cashmore, "Mutation of either G box or I box sequences profoundly affects expression from the *Arabidopsis* rbcS-1A promoter," *EMBO Journal*, vol. 9, no. 6, pp. 1717–1726, 1990.

[7] R. Foster, T. Izawa, and N. H. Chua, "Plant basic leucine zipper proteins gather at ACGT elements," *FASEB Journal*, vol. 8, pp. 192–200, 1994.

[8] S. Krawczyk, C. Thurow, R. Niggeweg, and C. Gatz, "Analysis of the spacing between the two palindromes of activation sequence-1 with respect to binding to different TGA factors and transcriptional activation potential," *Nucleic Acids Research*, vol. 30, no. 3, pp. 775–781, 2002.

[9] R. Mehrotra, K. Kiran, C. P. Chaturvedi et al., "Effect of copy number and spacing of the ACGT and GT cis elements on transient expression of minimal promoter in plants," *Journal of Genetics*, vol. 84, no. 2, pp. 183–187, 2005.

[10] S. V. Sawant, K. Kiran, R. Mehrotra et al., "A variety of synergistic and antagonistic interactions mediated by cis-acting DNA motifs regulate gene expression in plant cells and modulate stability of the transcription complex formed on a basal promoter," *Journal of Experimental Botany*, vol. 56, no. 419, pp. 2345–2353, 2005.

[11] R. Mehrotra and S. Mehrotra, "Promoter activation by ACGT in response to salicylic and abscisic acids is differentially regulated by the spacing between two copies of the motif," *Journal of Plant Physiology*, vol. 167, no. 14, pp. 1214–1218, 2010.

[12] L. Doelz, *BioCompanion*, Biocomputing Essentials Series, Dr. Ing. U. Doelz, Basel, Switzerland, 1990.

[13] S. Sawant, P. K. Singh, R. Madanala, and R. Tuli, "Designing of an artificial expression cassette for the high-level expression of transgenes in plants," *Theoretical and Applied Genetics*, vol. 102, no. 4, pp. 635–644, 2001.

[14] J. C. Bryne, E. Valen, M. H. E. Tang et al., "JASPAR, the open access database of transcription factor-binding profiles: new content and tools in the 2008 update," *Nucleic Acids Research*, vol. 36, no. 1, pp. D102–D106, 2008.

[15] A. Sandelin, W. W. Wasserman, and B. Lenhard, "ConSite: web-based prediction of regulatory elements using cross-species comparison," *Nucleic Acids Research*, vol. 32, pp. W249–W252, 2004.

[16] T. Paixão and R. B. R. Azevedo, "Redundancy and the evolution of cis-regulatory element multiplicity," *PLoS Computational Biology*, vol. 6, no. 7, Article ID e1000848, 2010.

[17] T. Hobo, M. Asada, Y. Kowyama, and T. Hattori, "ACGT-containing abscisic acid response element (ABRE) and coupling element 3 (CE3) are functionally equivalent," *Plant Journal*, vol. 19, no. 6, pp. 679–689, 1999.

9

Comparison of Next-Generation Sequencing Systems

Lin Liu, Yinhu Li, Siliang Li, Ni Hu, Yimin He, Ray Pong, Danni Lin, Lihua Lu, and Maggie Law

NGS Sequencing Department, Beijing Genomics Institute (BGI), 4th Floor, Building 11, Beishan Industrial Zone, Yantian District, Guangdong, Shenzhen 518083, China

Correspondence should be addressed to Lin Liu, linda.liu79@gmail.com

Academic Editor: P. J. Oefner

With fast development and wide applications of next-generation sequencing (NGS) technologies, genomic sequence information is within reach to aid the achievement of goals to decode life mysteries, make better crops, detect pathogens, and improve life qualities. NGS systems are typically represented by SOLiD/Ion Torrent PGM from Life Sciences, Genome Analyzer/HiSeq 2000/MiSeq from Illumina, and GS FLX Titanium/GS Junior from Roche. Beijing Genomics Institute (BGI), which possesses the world's biggest sequencing capacity, has multiple NGS systems including 137 HiSeq 2000, 27 SOLiD, one Ion Torrent PGM, one MiSeq, and one 454 sequencer. We have accumulated extensive experience in sample handling, sequencing, and bioinformatics analysis. In this paper, technologies of these systems are reviewed, and first-hand data from extensive experience is summarized and analyzed to discuss the advantages and specifics associated with each sequencing system. At last, applications of NGS are summarized.

1. Introduction

(Deoxyribonucleic acid) DNA was demonstrated as the genetic material by Oswald Theodore Avery in 1944. Its double helical strand structure composed of four bases was determined by James D. Watson and Francis Crick in 1953, leading to the central dogma of molecular biology. In most cases, genomic DNA defined the species and individuals, which makes the DNA sequence fundamental to the research on the structures and functions of cells and the decoding of life mysteries [1]. DNA sequencing technologies could help biologists and health care providers in a broad range of applications such as molecular cloning, breeding, finding pathogenic genes, and comparative and evolution studies. DNA sequencing technologies ideally should be fast, accurate, easy-to-operate, and cheap. In the past thirty years, DNA sequencing technologies and applications have undergone tremendous development and act as the engine of the genome era which is characterized by vast amount of genome data and subsequently broad range of research areas and multiple applications. It is necessary to look back on the history of sequencing technology development to review the NGS systems (454, GA/HiSeq, and SOLiD), to compare their advantages and disadvantages, to discuss the various

applications, and to evaluate the recently introduced PGM (personal genome machines) and third-generation sequencing technologies and applications. All of these aspects will be described in this paper. Most data and conclusions are from independent users who have extensive first-hand experience in these typical NGS systems in BGI (Beijing Genomics Institute).

Before talking about the NGS systems, we would like to review the history of DNA sequencing briefly. In 1977, Frederick Sanger developed DNA sequencing technology which was based on chain-termination method (also known as Sanger sequencing), and Walter Gilbert developed another sequencing technology based on chemical modification of DNA and subsequent cleavage at specific bases. Because of its high efficiency and low radioactivity, Sanger sequencing was adopted as the primary technology in the "first generation" of laboratory and commercial sequencing applications [2]. At that time, DNA sequencing was laborious and radioactive materials were required. After years of improvement, Applied Biosystems introduced the first automatic sequencing machine (namely AB370) in 1987, adopting capillary electrophoresis which made the sequencing faster and more accurate. AB370 could detect 96 bases one time, 500 K bases a day, and the read length could reach 600 bases.

The current model AB3730xl can output 2.88 M bases per day and read length could reach 900 bases since 1995. Emerged in 1998, the automatic sequencing instruments and associated software using the capillary sequencing machines and Sanger sequencing technology became the main tools for the completion of human genome project in 2001 [3]. This project greatly stimulated the development of powerful novel sequencing instrument to increase speed and accuracy, while simultaneously reducing cost and manpower. Not only this, X-prize also accelerated the development of next-generation sequencing (NGS) [4]. The NGS technologies are different from the Sanger method in aspects of massively parallel analysis, high throughput, and reduced cost. Although NGS makes genome sequences handy, the followed data analysis and biological explanations are still the bottle-neck in understanding genomes.

Following the human genome project, 454 was launched by 454 in 2005, and Solexa released Genome Analyzer the next year, followed by (Sequencing by Oligo Ligation Detection) SOLiD provided from Agencourt, which are three most typical massively parallel sequencing systems in the next-generation sequencing (NGS) that shared good performance on throughput, accuracy, and cost compared with Sanger sequencing (shown in Table 1(a)). These founder companies were then purchased by other companies: in 2006 Agencourt was purchased by Applied Biosystems, and in 2007, 454 was purchased by Roche, while Solexa was purchased by Illumina. After years of evolution, these three systems exhibit better performance and their own advantages in terms of read length, accuracy, applications, consumables, man power requirement and informatics infrastructure, and so forth. The comparison of these three systems will be focused and discussed in the later part of this paper (also see Tables 1(a), 1(b), and 1(c)).

2. Roche 454 System

Roche 454 was the first commercially successful next generation system. This sequencer uses pyrosequencing technology [5]. Instead of using dideoxynucleotides to terminate the chain amplification, pyrosequencing technology relies on the detection of pyrophosphate released during nucleotide incorporation. The library DNAs with 454-specific adaptors are denatured into single strand and captured by amplification beads followed by emulsion PCR [6]. Then on a picotiter plate, one of dNTP (dATP, dGTP, dCTP, dTTP) will complement to the bases of the template strand with the help of ATP *sulfurylase*, *luciferase*, luciferin, DNA *polymerase*, and adenosine 5′ phosphosulfate (APS) and release pyrophosphate (PPi) which equals the amount of incorporated nucleotide. The ATP transformed from PPi drives the luciferin into oxyluciferin and generates visible light [7]. At the same time, the unmatched bases are degraded by *apyrase* [8]. Then another dNTP is added into the reaction system and the pyrosequencing reaction is repeated.

The read length of Roche 454 was initially 100–150 bp in 2005, 200000+ reads, and could output 20 Mb per run

[9, 10]. In 2008 454 GS FLX Titanium system was launched; through upgrading, its read length could reach 700 bp with accuracy 99.9% after filter and output 0.7 G data per run within 24 hours. In late 2009 Roche combined the GS Junior a bench top system into the 454 sequencing system which simplified the library preparation and data processing, and output was also upgraded to 14 G per run [11, 12]. The most outstanding advantage of Roche is its speed: it takes only 10 hours from sequencing start till completion. The read length is also a distinguished character compared with other NGS systems (described in the later part of this paper). But the high cost of reagents remains a challenge for Roche 454. It is about 12.56×10^{-6} per base (counting reagent use only). One of the shortcomings is that it has relatively high error rate in terms of poly-bases longer than 6 bp. But its library construction can be automated, and the emulsion PCR can be semiautomated which could reduce the manpower in a great extent. Other informatics infrastructure and sequencing advantages are listed and compared with HiSeq 2000 and SOLiD systems in Tables 1(a), 1(b), and 1(c).

2.1. 454 GS FLX Titanium Software. GS RunProcessor is the main part of the GS FLX Titanium system. The software is in charge of picture background normalization, signal location correction, cross-talk correction, signals conversion, and sequencing data generation. GS RunProcessor would produce a series of files including SFF (standard flowgram format) files each time after run. SFF files contain the basecalled sequences and corresponding quality scores for all individual, high-quality reads (filtered reads). And it could be viewed directly from the screen of GS FLX Titanium system. Using GS De Novo Assembler, GS Reference Mapper and GS Amplicon Variant Analyzer provided by GS FLX Titanium system, SFF files can be applied in multiaspects and converted into fastq format for further data analyzing.

3. AB SOLiD System

(Sequencing by Oligo Ligation Detection) SOLiD was purchased by Applied Biosystems in 2006. The sequencer adopts the technology of two-base sequencing based on ligation sequencing. On a SOLiD flowcell, the libraries can be sequenced by 8 base-probe ligation which contains ligation site (the first base), cleavage site (the fifth base), and 4 different fluorescent dyes (linked to the last base) [10]. The fluorescent signal will be recorded during the probes complementary to the template strand and vanished by the cleavage of probes' last 3 bases. And the sequence of the fragment can be deduced after 5 round of sequencing using ladder primer sets.

The read length of SOLiD was initially 35 bp reads and the output was 3 G data per run. Owing to two-base sequencing method, SOLiD could reach a high accuracy of 99.85% after filtering. At the end of 2007, ABI released the first SOLiD system. In late 2010, the SOLiD 5500xl sequencing system was released. From SOLiD to SOLiD 5500xl, five upgrades were released by ABI in just three years. The SOLiD 5500xl realized improved read length, accuracy,

TABLE 1: (a) Advantage and mechanism of sequencers. (b) Components and cost of sequencers. (c) Application of sequencers.

(a)

Sequencer	454 GS FLX	HiSeq 2000	SOLiDv4	Sanger 3730xl
Sequencing mechanism	Pyrosequencing	Sequencing by synthesis	Ligation and two-base coding	Dideoxy chain termination
Read length	700 bp	50SE, 50PE, 101PE	50 + 35 bp or 50 + 50 bp	400~900 bp
Accuracy	99.9%*	98%, (100PE)	99.94% *raw data	99.999%
Reads	1 M	3 G	1200~1400 M	—
Output data/run	0.7 Gb	600 Gb	120 Gb	1.9~84 Kb
Time/run	24 Hours	3~10 Days	7 Days for SE 14 Days for PE	20 Mins~3 Hours
Advantage	Read length, fast	High throughput	Accuracy	High quality, long read length
Disadvantage	Error rate with polybase more than 6, high cost, low throughput	Short read assembly	Short read assembly	High cost low throughput

(b)

Sequencers	454 GS FLX	HiSeq 2000	SOLiDv4	3730xl
Instrument price	Instrument $500,000, $7000 per run	Instrument $690,000, $6000/(30x) human genome	Instrument $495,000, $15,000/100 Gb	Instrument $95,000, about $4 per 800 bp reaction
CPU	2* Intel Xeon X5675	2* Intel Xeon X5560	8* processor 2.0 GHz	Pentium IV 3.0 GHz
Memory	48 GB	48 GB	16 GB	1 GB
Hard disk	1.1 TB	3 TB	10 TB	280 GB
Automation in library preparation	Yes	Yes	Yes	No
Other required device	REM e system	cBot system	EZ beads system	No
Cost/million bases	$10	$0.07	$0.13	$2400

(c)

Sequencers	454 GS FLX	HiSeq 2000	SOLiDv4	3730xl
Resequencing		Yes	Yes	
De novo	Yes	Yes		Yes
Cancer	Yes	Yes	Yes	
Array	Yes	Yes	Yes	Yes
High GC sample	Yes	Yes	Yes	
Bacterial	Yes	Yes	Yes	
Large genome	Yes	Yes		
Mutation detection	Yes	Yes	Yes	Yes

(1) All the data is taken from daily average performance runs in BGI. The average daily sequence data output is about 8 Tb in BGI when about 80% sequencers (mainly HiSeq 2000) are running.

(2) The reagent cost of 454 GS FLX Titanium is calculated based on the sequencing of 400 bp; the reagent cost of HiSeq 2000 is calculated based on the sequencing of 200 bp; the reagent cost of SOLiDv4 is calculated based on the sequencing of 85 bp.

(3) HiSeq 2000 is more flexible in sequencing types like 50SE, 50PE, or 101PE.

(4) SOLiD has high accuracy especially when coverage is more than 30x, so it is widely used in detecting variations in resequencing, targeted resequencing, and transcriptome sequencing. Lanes can be independently run to reduce cost.

and data output of 85 bp, 99.99%, and 30 G per run, respectively. A complete run could be finished within 7 days. The sequencing cost is about 40×10^{-9} per base estimated from reagent use only by BGI users. But the short read length and resequencing only in applications is still its major shortcoming [13]. Application of SOLiD includes whole genome resequencing, targeted resequencing, transcriptome research (including gene expression profiling, small RNA analysis, and whole transcriptome analysis), and epigenome (like ChIP-Seq and methylation). Like other NGS systems, SOLiD's computational infrastructure is expensive and not trivial to use; it requires an air-conditioned data center, computing cluster, skilled personnel in computing, distributed memory cluster, fast networks, and batch queue system. Operating system used by most researchers is GNU/LINUX. Each solid sequencer run takes 7 days and generates around 4 TB of raw data. More data will be generated after bioinformatics analysis. This information is listed and compared with other NGS systems in Tables 1(a), 1(b), and 1(c). Automation can be used in library preparations, for example, Tecan system which integrated a Covaris A and Roche 454 REM e system [14].

3.1. SOLiD Software. After the sequencing with SOLiD, the original sequence of color coding will be accumulated. According to double-base coding matrix, the original color sequence can be decoded to get the base sequence if we knew the base types for one of any position in the sequence. Because of a kind of color corresponding four base pair, the color coding of the base will directly influence the decoding of its following base. It said that a wrong color coding will cause a chain decoding mistakes. BioScope is SOLiD data analysis package which provides a validated, single framework for resequencing, ChIP-Seq, and whole transcriptome analysis. It depends on reference for the follow-up data analysis. First, the software converts the base sequences of references into color coding sequence. Second, the color-coding sequence of references is compared with the original sequence of color-coding to get the information of mapping with newly developed mapping algorithm MaxMapper.

4. Illumina GA/HiSeq System

In 2006, Solexa released the Genome Analyzer (GA), and in 2007 the company was purchased by Illumina. The sequencer adopts the technology of sequencing by synthesis (SBS). The library with fixed adaptors is denatured to single strands and grafted to the flowcell, followed by bridge amplification to form clusters which contains clonal DNA fragments. Before sequencing, the library splices into single strands with the help of linearization enzyme [10], and then four kinds of nucleotides (ddATP, ddGTP, ddCTP, ddTTP) which contain different cleavable fluorescent dye and a removable blocking group would complement the template one base at a time, and the signal could be captured by a (charge-coupled device) CCD.

At first, solexa GA output was 1 G/run. Through improvements in polymerase, buffer, flowcell, and software, in 2009 the output of GA increased to 20 G/run in August (75PE), 30 G/run in October (100PE), and 50 G/run in December (Truseq V3, 150PE), and the latest GAIIx series can attain 85 G/run. In early 2010, Illumina launched HiSeq 2000, which adopts the same sequencing strategy with GA, and BGI was among the first globally to adopt the HiSeq system. Its output was 200 G per run initially, improved to 600 G per run currently which could be finished in 8 days. In the foreseeable future, it could reach 1 T/run when a personal genome cost could drop below $1 K. The error rate of 100PE could be below 2% in average after filtering (BGI's data). Compared with 454 and SOLiD, HiSeq 2000 is the cheapest

in sequencing with $0.02/million bases (reagent counted only by BGI). With multiplexing incorporated in P5/P7 primers and adapters, it could handle thousands of samples simultaneously. HiSeq 2000 needs (HiSeq control software) HCS for program control, (real-time analyzer software) RTA to do on-instrument base-calling, and CASAVA for secondary analysis. There is a 3 TB hard disk in HiSeq 2000. With the aid of Truseq v3 reagents and associated softwares, HiSeq 2000 has improved much on high GC sequencing. MiSeq, a bench top sequencer launched in 2011 which shared most technologies with HiSeq, is especially convenient for amplicon and bacterial sample sequencing. It could sequence 150PE and generate 1.5 G/run in about 10 hrs including sample and library preparation time. Library preparation and their concentration measurement can both be automated with compatible systems like Agilent Bravo, Hamilton Banadu, Tecan, and Apricot Designs.

4.1. HiSeq Software. HiSeq control system (HCS) and real-time analyzer (RTA) are adopted by HiSeq 2000. These two softwares could calculate the number and position of clusters based on their first 20 bases, so the first 20 bases of each sequencing would decide each sequencing's output and quality. HiSeq 2000 uses two lasers and four filters to detect four types of nucleotide (A, T, G, and C). The emission spectra of these four kinds of nucleotides have cross-talk, so the images of four nucleotides are not independent and the distribution of bases would affect the quality of sequencing. The standard sequencing output files of the HiSeq 2000 consist of *bcl files, which contain the base calls and quality scores in each cycle. And then it is converted into *_qseq.txt files by BCL Converter. The ELAND program of CASAVA (offline software provided by Illumina) is used to match a large number of reads against a genome.

In conclusion, of the three NGS systems described before, the Illumina HiSeq 2000 features the biggest output and lowest reagent cost, the SOLiD system has the highest accuracy [11], and the Roche 454 system has the longest read length. Details of three sequencing system are list in Tables 1(a), 1(b), and 1(c).

5. Compact PGM Sequencers

Ion Personal Genome Machine (PGM) and MiSeq were launched by Ion Torrent and Illumina. They are both small in size and feature fast turnover rates but limited data throughput. They are targeted to clinical applications and small labs.

5.1. Ion PGM from Ion Torrent. Ion PGM was released by Ion Torrent at the end of 2010. PGM uses semiconductor sequencing technology. When a nucleotide is incorporated into the DNA molecules by the polymerase, a proton is released. By detecting the change in pH, PGM recognized whether the nucleotide is added or not. Each time the chip was flooded with one nucleotide after another, if it is not the correct nucleotide, no voltage will be found; if there is 2 nucleotides added, there is double voltage detected [15].

PGM is the first commercial sequencing machine that does not require fluorescence and camera scanning, resulting in higher speed, lower cost, and smaller instrument size. Currently, it enables 200 bp reads in 2 hours and the sample preparation time is less than 6 hours for 8 samples in parallel.

An exemplary application of the Ion Torrent PGM sequencer is the identification of microbial pathogens. In May and June of 2011, an ongoing outbreak of exceptionally virulent Shiga-toxin- (Stx) producing *Escherichia coli* O104:H4 centered in Germany [16, 17], there were more than 3000 people infected. The whole genome sequencing on Ion Torrent PGM sequencer and HiSeq 2000 helped the scientists to identify the type of *E. coli* which would directly apply the clue to find the antibiotic resistance. The strain appeared to be a hybrid of two *E. coli* strains—entero aggregative *E. coli* and entero hemorrhagic *E. coli*—which may help explain why it has been particularly pathogenic. From the sequencing result of *E. coli* TY2482 [18], PGM shows the potential of having a fast, but limited throughput sequencer when there is an outbreak of new disease.

In order to study the sequencing quality, mapping rate, and GC depth distribution of Ion Torrent and compare with HiSeq 2000, a high GC *Rhodobacter* sample with high GC content (66%) and 4.2 Mb genome was sequenced in these two different sequencers (Table 2). In another experiment, *E. coli* K12 DH10B (NC_010473.1) with GC 50.78% was sequenced by Ion Torrent for analysis of quality value, read length, position accuracies, and GC distribution (Figure 1).

5.1.1. Sequencing Quality.
The quality of Ion Torrent is more stable, while the quality of HiSeq 2000 decreases noticeably after 50 cycles, which may be caused by the decay of fluorescent signal with increasing the read length (shown in Figure 1).

5.1.2. Mapping.
The insert size of library of Rhodobacter was 350 bp, and 0.5 Gb data was obtained from HiSeq. The sequencing depth was over 100x, and the contig and scaffold N50 were 39530 bp and 194344 bp, respectively. Based on the assembly result, we used 33 Mb which is obtained from ion torrent with 314 chip to analyze the map rate. The alignment comparison is Table 2.

The map rate of Ion Torrent is higher than HiSeq 2000, but it is incomparable because of the different alignment methods used in different sequencers. Besides the significant difference on data including mismatch rate, insertion rate, and deletion rate, HiSeq 2000 and Ion Torrent were still incomparable because of the different sequencing principles. For example, the polynucleotide site could not be indentified easily in Ion Torrent. But it is shown that Ion Torrent has a stable quality along sequencing reads and a good performance on mismatch accuracies, but rather a bias in detection of indels. Different types of accuracy are analyzed and shown in Figure 1.

5.1.3. GC Depth Distribution.
The GC depth distribution is better in Ion Torrent from Figure 1. In Ion Torrent, the sequencing depth is similar while the GC content is from

TABLE 2: Comparison in alignment between Ion Torrent and HiSeq 2000.

	Ion Torrent[a]	HiSeq 2000[b]
Total reads num	165518	205683
Total bases num	18574086	18511470
Max read length	201	90
Min read length	15	90
Map reads num	157258	157511
Map rate	95%	76.57%
Covered rate	96.50%	93.11%
Total map length	15800258	14176420
Total mismatch base	53475	142425
Total insertion base	109550	1397
Total insertion num	95740	1332
Total deletion base	152495	431
Total deletion num	139264	238
Ave mismatch rate	0.338%	1.004%
Ave insertion rate	0.693%	0.009%
Ave deletion rate	0.965%	0.003%

[a]: use TMAP to align; [b]: use SOAP2 to align.

63% to 73%. However in HiSeq 2000, the average sequencing depth is 4x when the GC content is 60%, while it is 3x with 70% GC content.

Ion Torrent has already released Ion 314 and 316 and planned to launch Ion 318 chips in late 2011. The chips are different in the number of wells resulting in higher production within the same sequencing time. The Ion 318 chip enables the production of >1 Gb data in 2 hours. Read length is expected to increase to >400 bp in 2012.

5.2. MiSeq from Illumina.
MiSeq which still uses SBS technology was launched by Illumina. It integrates the functions of cluster generation, SBS, and data analysis in a single instrument and can go from sample to answer (analyzed data) within a single day (as few as 8 hours). The Nextera, TruSeq, and Illumina's reversible terminator-based sequencing by synthesis chemistry was used in this innovative engineering. The highest integrity data and broader range of application, including amplicon sequencing, clone checking, ChIP-Seq, and small genome sequencing, are the outstanding parts of MiSeq. It is also flexible to perform single 36 bp reads (120 MB output) up to 2 × 150 paired-end reads (1–1.5 GB output) in MiSeq. Due to its significant improvement in read length, the resulting data performs better in contig assembly compared with HiSeq (data not shown). The related sequencing result of MiSeq is shown in Table 3. We also compared PGM with MiSeq in Table 4.

5.3. Complete Genomics.
Complete genomics has its own sequencer based on Polonator G.007, which is ligation-based sequencer. The owner of Polonator G.007, Dover, collaborated with the Church Laboratory of Harvard Medical School, which is the same team as SOLiD system, and introduced this cheap open system. The Polonator could

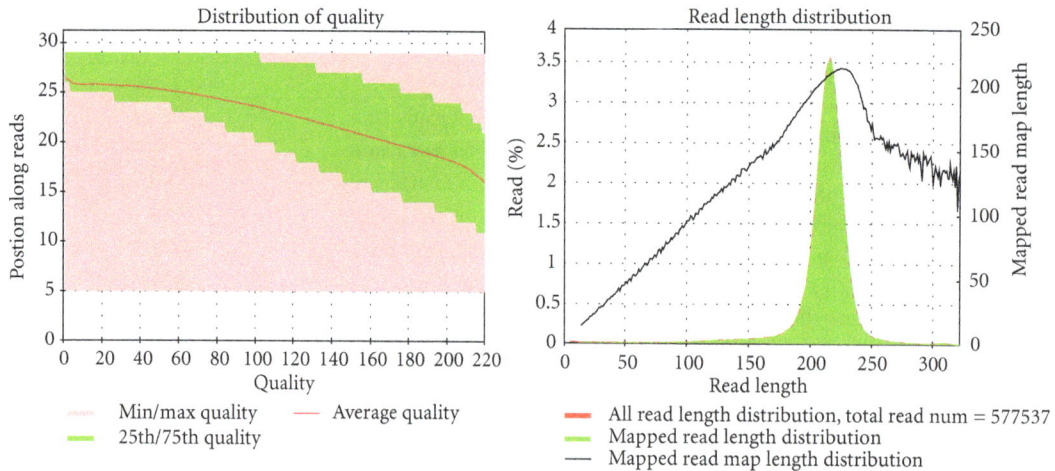

(a)

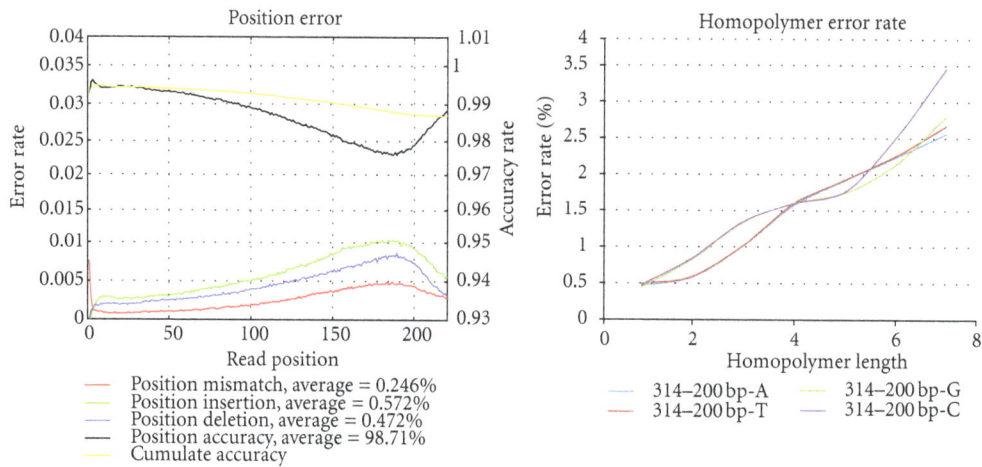

(b)

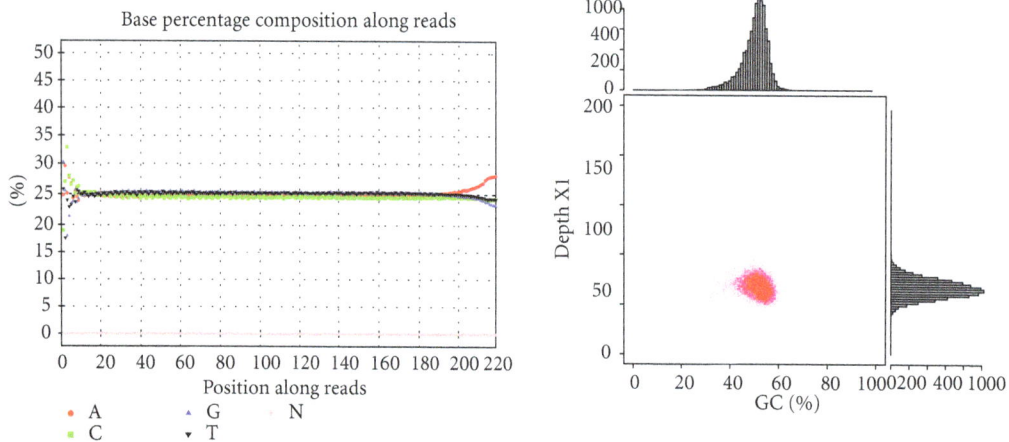

(c)

FIGURE 1: Ion Torrent sequencing quality. *E. coli* K12 DH10B (NC_010473.1) with GC 50.78% was used for this experiment. (a) is 314–200 bp from Ion Torrent. The left figure is quality value: pink range represents quality minimum and maximum values each position has. Green area represents the top and bottom quarter (1/4) reads of quality. Red line represents the average quality value in the position. The right figure is read length analysis: colored histogram represents the real read length. The black line represents the mapped length, and because it allows 3′ soft clipping, the length is different from the real read length. (b) is accuracy analysis. In each position, accuracy type including mismatch, insertion, and deletion is shown on the left y-axis. The average accuracy is shown the right y-axis. Accuracy of 200 bp sequencing could reach 99%. (c) is base composition along reads (left) and GC distribution analysis (right). The left figure is base composition in each position of reads. Base line splits after about 95 cycles indicating an inaccurate sequencing. The right one uses 500 bp window and the GC distribution is quite even. The data using high GC samples also indicates a good performance in Ion Torrent (data not shown).

TABLE 3: MiSeq 150PE data.

Sample	GC	Q20	Q30
Human HPV	33.57; 33.62	98.26; 95.52	93.64; 88.52
Bacteria	61.33; 61.43	90.84; 83.86	78.46; 69.04

(1) The data in the table includes both read 1 and read 2 from paired-end sequencing.
(2) GC represents the GC content of libraries.
(3) Q20 value is the average Q20 of all bases in a read, which represents the ratio of bases with probability of containing no more than one error in 100 bases. Q30 value is the average Q30 of all bases in a read, which represents the ratio of bases with probability of containing no more than one error in 1,000 bases.

combine a high-performance instrument at very low price and the freely downloadable, open-source software and protocols in this sequencing system. The Polonator G.007 is ligation detection sequencing, which decodes the base by the single-base probe in nonanucleotides (nonamers), not by dual-base coding [19]. The fluorophore-tagged nonamers will be degenerated by selectively ligate onto a series of anchor primers, whose four components are labeled with one of four fluorophores with the help of T4 DNA ligase, which correspond to the base type at the query position. In the ligation progress, T4 DNA ligase is particularly sensitive to mismatches on 3′-side of the gap which is benefit to improve the accuracy of sequencing. After imaging, the Polonator chemically strips the array of annealed primer-fluorescent probe complex; the anchor primer is replaced and the new mixture are fluorescently tagged nonamers is introduced to sequence the adjacent base [20]. There are two updates compared with Polonator G.007, DNA nanoball (DNB) arrays, and combinatorial probe-anchor ligation (cPAL). Compared with DNA cluster or microsphere, DNA nanoball arrays obtain higher density of DNA cluster on the surface of a silicon chip. As the seven 5-base segments are discontinuous, so the system of hybridization-ligation-detection cycle has higher fault-tolerant ability compared with SOLiD. Complete genomics claim to have 99.999% accuracy with 40x depth and could analyze SNP, indel, and CNV with price 5500$–9500$. But Illumina reported a better performance of HiSeq 2000 use only 30x data (Illumina Genome Network). Recently some researchers compared CG's human genome sequencing data with Illumina system [21], and there are notable differences in detecting SNVs, indels, and system-specific detections in variants.

5.4. The Third Generation Sequencer. While the increasing usage and new modification in next generation sequencing, the third generation sequencing is coming out with new insight in the sequencing. Third-generation sequencing has two main characteristics. First, PCR is not needed before sequencing, which shortens DNA preparation time for sequencing. Second, the signal is captured in real time, which means that the signal, no matter whether it is fluorescent (Pacbio) or electric current (Nanopore), is monitored during the enzymatic reaction of adding nucleotide in the complementary strand.

Single-molecule real-time (SMRT) is the third-generation sequencing method developed by Pacific Bioscience (Menlo Park, CA, USA), which made use of modified enzyme and direct observation of the enzymatic reaction in real time. SMRT cell consists of millions of zero-mode waveguides (ZMWs), embedded with only one set of enzymes and DNA template that can be detected during the whole process. During the reaction, the enzyme will incorporate the nucleotide into the complementary strand and cleave off the fluorescent dye previously linked with the nucleotide. Then the camera inside the machine will capture signal in a movie format in real-time observation [19]. This will give out not only the fluorescent signal but also the signal difference along time, which may be useful for the prediction of structural variance in the sequence, especially useful in epigenetic studies such as DNA methlyation [22].

Comparing to second generation, PacBio RS (the first sequencer launched by PacBio) has several advantages. First the sample preparation is very fast; it takes 4 to 6 hours instead of days. Also it does not need PCR step in the preparation step, which reduces bias and error caused by PCR. Second, the turnover rate is quite fast; runs are finished within a day. Third, the average read length is 1300 bp, which is longer than that of any second-generation sequencing technology. Although the throughput of the PacBioRS is lower than second-generation sequencer, this technology is quite useful for clinical laboratories, especially for microbiology research. A paper has been published using PacBio RS on the Haitian cholera outbreak [19].

We have run a *de novo* assembly of DNA fosmid sample from Oyster with PacBio RS in standard sequencing mode (using LPR chemistry and SMRTcells instead of the new version FCR chemistry and SMRTcells). An SMRT belt template with mean insert size of 7500 kb is made and run in one SMRT cell and a 120-minute movie is taken. After Post-QC filter, 22,373,400 bp reads in 6754 reads (average 2,566 bp) were sequenced with the average Read Score of 0.819. The Coverage is 324x with mean read score of 0.861 and high accuracy (~99.95). The result is exhibited in Figure 2.

Nanopore sequencing is another method of the third generation sequencing. Nanopore is a tiny biopore with diameter in nanoscale [23], which can be found in protein channel embedded on lipid bilayer which facilitates ion exchange. Because of the biological role of nanopore, any particle movement can disrupt the voltage across the channel. The core concept of nanopore sequencing involves putting a thread of single-stranded DNA across α-haemolysin (αHL) pore. αHL, a 33 kD protein isolated from *Staphylococcus aureus* [20], undergoes self-assembly to form a heptameric transmembrane channel [23]. It can tolerate extraordinary voltage up to 100 mV with current 100 pA [20]. This unique property supports its role as building block of nanopore. In nanopore sequencing, an ionic flow is applied continuously. Current disruption is simply detected by standard electrophysiological technique. Readout is relied on the size difference between all deoxyribonucleoside monophosphate (dNMP). Thus, for given dNMP, characteristic current modulation is shown for discrimination. Ionic current is resumed after trapped nucleotide entirely squeezing out.

TABLE 4: The comparison between PGM and MiSeq.

	PGM	MiSeq
Output	10 MB–100 MB	120 MB–1.5 GB
Read length	~200 bp	Up to 2 × 150 bp
Sequencing time	2 hours for 1 × 200 bp	3 hours for 1 × 36 single read 27 hours for 2 × 150 bp pair end read
Sample preparation time	8 samples in parallel, less than 6 hrs	As fast as 2 hrs, with 15 minutes hand on time
Sequencing method	semiconductor technology with a simple sequencing chemistry	Sequencing by synthesis (SBS)
Potential for development	Various parameters (read length, cycle time, accuracy, etc.)	Limited factors, major concentrate in flowcell surface size, insert sizes, and how to pack cluster in tighter
Input amount	μg	Ng (Nextera)
Data analysis	Off instrument	On instrument

Nanopore sequencing possesses a number of fruitful advantages over existing commercialized next-generation sequencing technologies. Firstly, it potentially reaches long read length >5 kbp with speed 1 bp/ns [19]. Moreover, detection of bases is fluorescent tag-free. Thirdly, except the use of exonuclease for holding up ssDNA and nucleotide cleavage [24], involvement of enzyme is remarkably obviated in nanopore sequencing [22]. This implies that nanopore sequencing is less sensitive to temperature throughout the sequencing reaction and reliable outcome can be maintained. Fourthly, instead of sequencing DNA during polymerization, single DNA strands are sequenced through nanopore by means of DNA strand depolymerization. Hence, hand-on time for sample preparation such as cloning and amplification steps can be shortened significantly.

6. Discussion of NGS Applications

Fast progress in DNA sequencing technology has made for a substantial reduction in costs and a substantial increase in throughput and accuracy. With more and more organisms being sequenced, a flood of genetic data is inundating the world every day. Progress in genomics has been moving steadily forward due to a revolution in sequencing technology. Additionally, other of types-large scale studies in exomics, metagenomics, epigenomics, and transcriptomics all become reality. Not only do these studies provide the knowledge for basic research, but also they afford immediate application benefits. Scientists across many fields are utilizing these data for the development of better-thriving crops and crop yields and livestock and improved diagnostics, prognostics, and therapies for cancer and other complex diseases.

BGI is on the cutting edge of translating genomics research into molecular breeding and disease association studies with belief that agriculture, medicine, drug development, and clinical treatment will eventually enter a new stage for more detailed understanding of the genetic components of all the organisms. BGI is primarily focused on three projects. (1) The Million Species/Varieties Genomes Project, aims to sequence a million economically and scientifically important plants, animals, and model organisms, including different breeds, varieties, and strains. This project is best represented by our sequencing of the genomes of the Giant panda, potato, macaca, and others, along with multiple resequencing projects. (2) The Million Human Genomes Project focuses on large-scale population and association studies that use whole-genome or whole-exome sequencing strategies. (3) The Million Eco-System Genomes Project has the objective of sequencing the metagenome and cultured microbiome of several different environments, including microenvironments within the human body [25]. Together they are called 3 M project.

In the following part, each of the following aspects of applications including *de novo* sequencing, mate-pair, whole genome or target-region resequencing, small RNA, transcriptome, RNA seq, epigenomics, and metagenomics, is briefly summarized.

In DNA *de novo* sequencing, the library with insert size below 800 bp is defined as DNA short fragment library, and it is usually applied in *de novo* and resequencing research. Skovgaard et al. [26] have applied a combination method of WGS (whole-genome sequencing) and genome copy number analysis to identify the mutations which could suppress the growth deficiency imposed by excessive initiations from the *E. coli* origin of replication, *oriC*.

Mate-pair library sequencing is significant beneficial for *de novo* sequencing, because the method could decrease gap region and extend scaffold length. Reinhardt et al. [27] developed a novel method for *de novo* genome assembly by analyzing sequencing data from high-throughput short read sequencing technology. They assembled genomes into large scaffolds at a fraction of the traditional cost and without using reference sequence. The assembly of one sample yielded an N50 scaffold size of 531,821 bp with >75% of the predicted genome covered by scaffolds over 100,000 bp.

Whole genome resequencing sequenced the complete DNA sequence of an organism's genome including the whole chromosomal DNA at a single time and alignment with the reference sequence. Mills et al. [28] constructed a map of unbalanced SVs (genomic structural variants) based on whole genome DNA sequencing data from 185 human genomes with SOLiD platform; the map encompassed 22,025

	Prefilter	Post-QC filter*
Number of bases	84, 110, 272 bp	22, 373, 400 bp
Number of reads	46, 861	6, 754
Mean read length	513 bp	2, 566 bp
Mean read score	0.144	0.819

* MinRL = 50, MinRS = 0.75

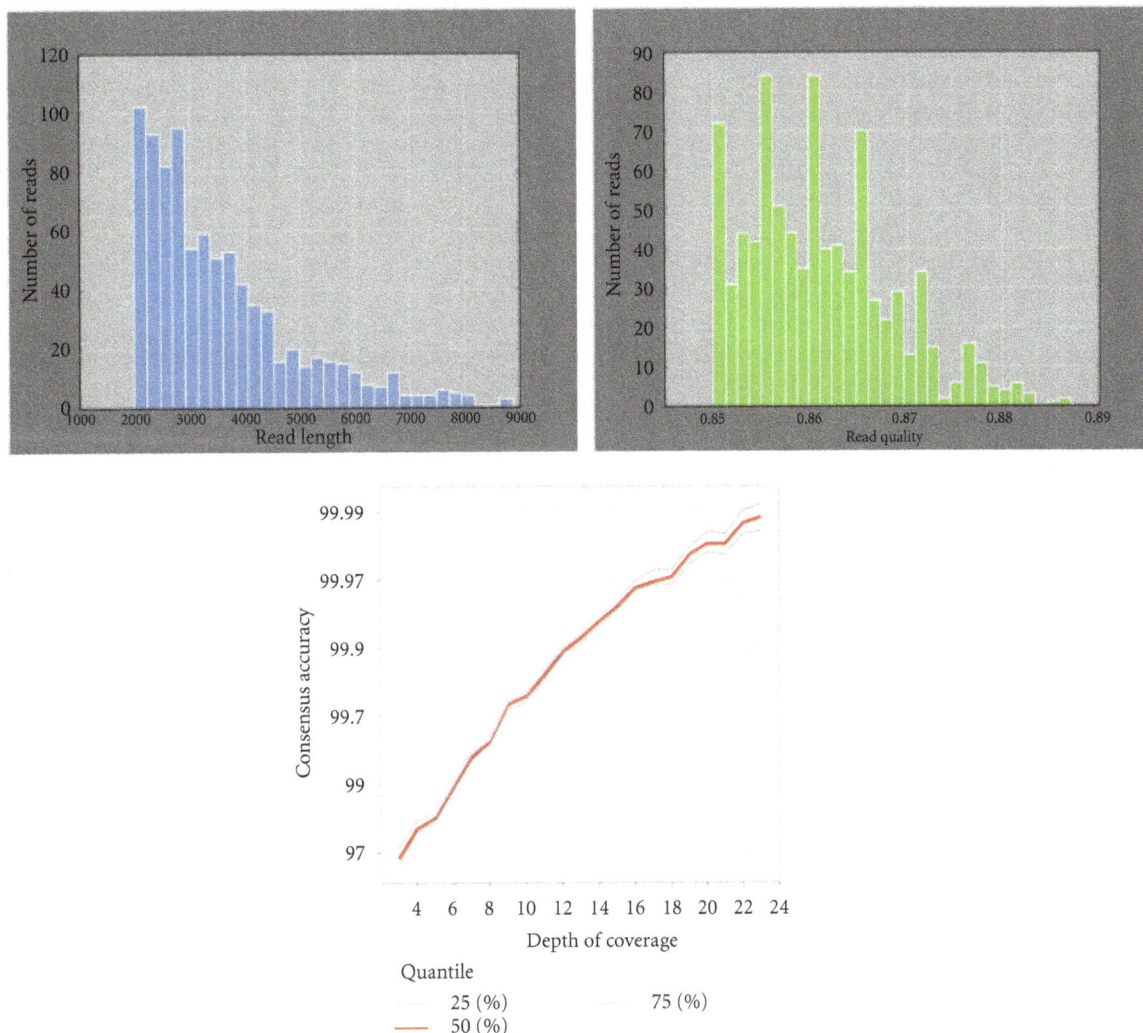

FIGURE 2: Sequencing of a fosmid DNA using Pacific Biosciences sequencer. With coverage, the accuracy could be above 97%. The figure was constructed by BGI's own data.

deletions and 6,000 additional SVs, including insertions and tandem duplications [28]. Most SVs (53%) were mapped to nucleotide resolution, which facilitated analyzing their origin and functional impact [28].

The whole genome resequencing is an effective way to study the functional gene, but the high cost and massive data are the main problem for most researchers. Target region sequencing is a solution to solve it. Microarray capture is a popular way of target region sequencing, which

uses hybridization to arrays containing synthetic oligonucleotides matching the target DNA sequencing. Gnirke et al. [29] developed a captured method that uses an RNA "baits" to capture target DNA fragments from the "pond" and then uses the Illumina platform to read out the sequence. About 90% of uniquely aligning bases fell on or near bait sequence; up to 50% lay on exons proper [29].

Fehniger et al. used two platforms, Illumina GA and ABI SOLiD, to define the miRNA transcriptomes of resting and cytokine-activated primary murine NK (natural killer) cells [30]. The identified 302 known and 21 novel mature miRNAs were analyzed by unique bioinformatics pipeline from small RNA libraries of NK cell. These miRNAs are overexpressed in broad range and exhibit isomiR complexity, and a subset is differentially expressed following cytokine activation, which were the clue to identify the identification of miRNAs by the Illumina GA and SOLiD instruments [30].

The transcriptome is the set of all RNA molecules, including mRNA, rRNA, tRNA, and other noncoding RNA produced in one or a population of cells. In these years, next-generation sequencing technology is used to study the transcriptome compares with DNA microarray technology in the past. The *S. mediterranea* transcriptome could be sequenced by an efficient sequencing strategy which designed by Adamidi et al. [31]. The catalog of assembled transcripts and the identified peptides in this study dramatically expand and refine planarian gene annotation, which is demonstrated by validation of several previously unknown transcripts with stem cell-dependent expression patterns.

RNA-seq is a new method in RNA sequencing to study mRNA expression. It is similar to transcriptome sequencing in sample preparation, except the enzyme. In order to estimate the technical variance, Marioni et al. [32] analyzed a kidney RNA samples on both Illumina platform and Affymetrix arrays. The additional analyses such as low-expressed genes, alternative splice variants, and novel transcripts were found on Illumina platform. Bradford et al. [33] compared the data of RNA-seq library on the SOLiD platform and Affymetrix Exon 1.0ST arrays and found a high degree of correspondence between the two platforms in terms of exon-level fold changes and detection. And the greatest detection correspondence was seen when the background error rate is extremely low in RNA-seq. The difference between RNA-seq and transcriptome on SOLiD is not so obvious as Illumina.

There are two kinds of application of epigenetic, Chromatin immunoprecipitation and methylation analysis. Chromatin immunoprecipitation (ChIP) is an immunoprecipitation technique which is used to study the interaction between protein and DNA in a cell, and the histone modifies would be found by the specific location in genome. Based on next-generation sequencing technology, Johnson et al. [34] developed a large-scale chromatin immunoprecipitation assay to identify motif, especially noncanonical NRSF-binding motif. The data displays sharp resolution of binding position (± 50 bp), which is important to infer new candidate interaction for the high sensitivity and specificity (ROC (receiver operator characteristic) area ≥ 0.96) and statistical confidence ($P < 10-4$). Another important application in epigenetic is DNA methylation analysis. DNA methylation exists typically in vertebrates at CpG sites; the methylation caused the conversion of the cytosine to 5-methylcytosine. Chung presented a whole methylome sequencing to study the difference between two kinds of bisulfite conversion methods (in solution versus in gel) by SOLiD platform [35].

The world class genome projects include the 1000 genome project, and the human ENCODE project, the human Microbiome (HMP) project, to name a few. BGI takes an active role in these and many more ongoing projects like 1000 Animal and Plant Genome project, the MetaHIT project, Yanhuang project, LUCAMP (Diabetes-associated Genes and Variations Study), ICGC (international cancer genome project), Ancient human genome, 1000 Mendelian Disorders Project, Genome 10 K Project, and so forth [25]. These internationally collaborated genome projects greatly enhanced genomics study and applications in healthcare and other fields.

To manage multiple projects including large and complex ones with up to tens of thousands of samples, a superior and sophisticated project management system is required handling information processing from the very beginning of sample labeling and storage to library construction, multiplexing, sequencing, and informatics analysis. Research-oriented bioinformatics analysis and followup experiment processed are not included. Although automation techniques' adoption has greatly simplified bioexperiment human interferences, all other procedures carried out by human power have to be managed. BGI has developed BMS system and Cloud service for efficient information exchange and project management. The behavior management mainly follows Japan 5S onsite model. Additionally, BGI has passed ISO9001 and CSPro (authorized by Illumina) QC system and is currently taking (Clinical Laboratory Improvement Amendments) CLIA and (American Society for Histocompatibility and Immunogenetics) AShI tests. Quick, standard, and open reflection system guarantees an efficient troubleshooting pathway and high performance, for example, instrument design failure of Truseq v3 flowcell resulting in bubble appearance (which is defined as "bottom-middle-swatch" phenomenon by Illumina) and random N in reads. This potentially hazards sequencing quality, GC composition as well as throughput. It not only effects a small area where the bubble locates resulting in reading N but also effects the focus of the place nearby, including the whole swatch, and the adjacent swatch. Filtering parameters have to be determined to ensure quality raw data for bioinformatics processing. Lead by the NGS tech group, joint meetings were called for analyzing and troubleshooting this problem, to discuss strategies to best minimize effect in terms of cost and project time, to construct communication channel, to statistically summarize compensation, in order to provide best project management strategies in this time. Some reagent QC examples are summaried in Liu et al. [36].

BGI is establishing their cloud services. Combined with advanced NGS technologies with multiple choices, a plug-and-run informatics service is handy and affordable. A series of softwares are available including BLAST, SOAP, and SOAP SNP for sequence alignment and pipelines for RNAseq data. Also SNP calling programs such as Hecate and Gaea are about to be released. Big-data studies from the whole spectrum of life and biomedical sciences now can be shared and published on a new journal GigaSicence cofounded by BGI and Biomed Central. It has a novel publication format: each piece of data links to a standard manuscript publication with an extensive database which hosts all associated data, data analysis tools, and cloud-computing resources. The scope covers not just omic type data and the fields of high-throughput biology currently serviced by large public repositories but also the growing range of more difficult-to-access data, such as imaging, neuroscience, ecology, cohort data, systems biology, and other new types of large-scale sharable data.

References

[1] G. M. Church and W. Gilbert, "Genomic sequencing," *Proceedings of the National Academy of Sciences of the United States of America*, vol. 81, no. 7, pp. 1991–1995, 1984.

[2] http://en.wikipedia.org/wiki/DNA_sequencing/.

[3] F. S. Collins, M. Morgan, and A. Patrinos, "The Human Genome Project: lessons from large-scale biology," *Science*, vol. 300, no. 5617, pp. 286–290, 2003.

[4] http://genomics.xprize.org/.

[5] http://my454.com/products/technology.asp.

[6] J. Berka, Y. J. Chen, J. H. Leamon et al., "Bead emulsion nucleic acid amplification," U.S. Patent Application, 2005.

[7] T. Foehlich et al., "High-throughput nucleic acid analysis," U.S. Patent, 2010.

[8] http://www.pyrosequencing.com/DynPage.aspx.

[9] http://www.roche-applied-science.com/.

[10] E. R. Mardis, "The impact of next-generation sequencing technology on genetics," *Trends in Genetics*, vol. 24, no. 3, pp. 133–141, 2008.

[11] S. M. Huse, J. A. Huber, H. G. Morrison, M. L. Sogin, and D. M. Welch, "Accuracy and quality of massively parallel DNA pyrosequencing," *Genome Biology*, vol. 8, no. 7, article R143, 2007.

[12] "The new GS junior sequencer," http://www.gsjunior.com/instrument-workflow.php.

[13] "SOLiD system accuray," http://www.appliedbiosystems.com/absite/us/en/home/applications-technologies/solid-next-generation-sequencing.html.

[14] http://www.tecan.com/platform/apps/product/index.asp?MenuID=3465&ID=7191&Menu=1&Item=33.52.2.

[15] B. A. Flusberg, D. R. Webster, J. H. Lee et al., "Direct detection of DNA methylation during single-molecule, real-time sequencing," *Nature Methods*, vol. 7, no. 6, pp. 461–465, 2010.

[16] A. Mellmann, D. Harmsen, C. A. Cummings et al., "Prospective genomic characterization of the german enterohemorrhagic Escherichia coli O104:H4 outbreak by rapid next generation sequencing technology," *PLoS ONE*, vol. 6, no. 7, Article ID e22751, 2011.

[17] H. Rohde, J. Qin, Y. Cui et al., "Open-source genomic analysis of Shiga-toxin-producing E. coli O104:H4," *New England Journal of Medicine*, vol. 365, no. 8, pp. 718–724, 2011.

[18] C. S. Chin, J. Sorenson, J. B. Harris et al., "The origin of the Haitian cholera outbreak strain," *New England Journal of Medicine*, vol. 364, no. 1, pp. 33–42, 2011.

[19] W. Timp, U. M. Mirsaidov, D. Wang, J. Comer, A. Aksimentiev, and G. Timp, "Nanopore sequencing: electrical measurements of the code of life," *IEEE Transactions on Nanotechnology*, vol. 9, no. 3, pp. 281–294, 2010.

[20] D. W. Deamer and M. Akeson, "Nanopores and nucleic acids: prospects for ultrarapid sequencing," *Trends in Biotechnology*, vol. 18, no. 4, pp. 147–151, 2000.

[21] "Performance comparison of whole-genome sequencing systems," *Nature Biotechnology*, vol. 30, pp. 78–82, 2012.

[22] D. Branton, D. W. Deamer, A. Marziali et al., "The potential and challenges of nanopore sequencing," *Nature Biotechnology*, vol. 26, no. 10, pp. 1146–1153, 2008.

[23] L. Song, M. R. Hobaugh, C. Shustak, S. Cheley, H. Bayley, and J. E. Gouaux, "Structure of staphylococcal α-hemolysin, a heptameric transmembrane pore," *Science*, vol. 274, no. 5294, pp. 1859–1866, 1996.

[24] J. Clarke, H. C. Wu, L. Jayasinghe, A. Patel, S. Reid, and H. Bayley, "Continuous base identification for single-molecule nanopore DNA sequencing," *Nature Nanotechnology*, vol. 4, no. 4, pp. 265–270, 2009.

[25] Website of BGI, http://www.genomics.org.cn.

[26] O. Skovgaard, M. Bak, A. Løbner-Olesen et al., "Genome-wide detection of chromosomal rearrangements, indels, and mutations in circular chromosomes by short read sequencing," *Genome Research*, vol. 21, no. 8, pp. 1388–1393, 2011.

[27] J. A. Reinhardt, D. A. Baltrus, M. T. Nishimura, W. R. Jeck, C. D. Jones, and J. L. Dangl, "De novo assembly using low-coverage short read sequence data from the rice pathogen Pseudomonas syringae pv. oryzae," *Genome Research*, vol. 19, no. 2, pp. 294–305, 2009.

[28] R. E. Mills, K. Walter, C. Stewart et al., "Mapping copy number variation by population-scale genome sequencing," *Nature*, vol. 470, no. 7332, pp. 59–65, 2011.

[29] A. Gnirke, A. Melnikov, J. Maguire et al., "Solution hybrid selection with ultra-long oligonucleotides for massively parallel targeted sequencing," *Nature Biotechnology*, vol. 27, no. 2, pp. 182–189, 2009.

[30] T. A. Fehniger, T. Wylie, E. Germino et al., "Next-generation sequencing identifies the natural killer cell microRNA transcriptome," *Genome Research*, vol. 20, no. 11, pp. 1590–1604, 2010.

[31] C. Adamidi, Y. Wang, D. Gruen et al., "De novo assembly and validation of planaria transcriptome by massive parallel sequencing and shotgun proteomics," *Genome Research*, vol. 21, no. 7, pp. 1193–1200, 2011.

[32] J. C. Marioni, C. E. Mason, S. M. Mane, M. Stephens, and Y. Gilad, "RNA-seq: an assessment of technical reproducibility and comparison with gene expression arrays," *Genome Research*, vol. 18, no. 9, pp. 1509–1517, 2008.

[33] J. R. Bradford, Y. Hey, T. Yates, Y. Li, S. D. Pepper, and C. J. Miller, "A comparison of massively parallel nucleotide sequencing with oligonucleotide microarrays for global transcription profiling," *BMC Genomics*, vol. 11, no. 1, article 282, 2010.

[34] D. S. Johnson, A. Mortazavi, R. M. Myers, and B. Wold, "Genome-wide mapping of in vivo protein-DNA interactions," *Science*, vol. 316, no. 5830, pp. 1497–1502, 2007.

[35] H. Gu, Z. D. Smith, C. Bock, P. Boyle, A. Gnirke, and A. Meissner, "Preparation of reduced representation bisulfite sequencing libraries for genome-scale DNA methylation profiling," *Nature Protocols*, vol. 6, no. 4, pp. 468–481, 2011.

[36] L. Liu, N. Hu, B. Wang et al., "A brief utilization report on the Illumina HiSeq 2000 sequencer," *Mycology*, vol. 2, no. 3, pp. 169–191, 2011.

10

Is BAC Transgenesis Obsolete? State of the Art in the Era of Designer Nucleases

J. Beil,[1] **L. Fairbairn,**[1] **P. Pelczar,**[2] **and T. Buch**[1]

[1] *Institute for Medical Microbiology, Immunology and Hygiene, Technische Universität München, Trogerstraße 30, 81679 Munich, Germany*
[2] *Institute of Animal Laboratory Sciences, VetSuisse Faculty, University of Zurich, Winterthurer Straße 190, 8057 Zurich, Switzerland*

Correspondence should be addressed to T. Buch, thorsten.buch@tum.de

Academic Editor: Masamitsu Yamaguchi

DNA constructs based on bacterial artificial chromosomes (BACs) are frequently used to generate transgenic animals as they reduce the influence of position effects and allow predictable expression patterns for genes whose regulatory sequences are not fully identified. Despite these advantages BAC transgenics suffer from drawbacks such as complicated vector construction, low efficiency of transgenesis, and some remaining expression variegation. The recent development of transcription activator-like effector nucleases (TALENs) and zinc finger nucleases (ZFNs) has resulted in new transgenic techniques which do not have the drawbacks associated with BAC transgenesis. Initial reports indicate that such designer nucleases (DNs) allow the targeted insertion of transgenes into endogenous loci by direct injection of the targeting vector and mRNA/DNA encoding the predesigned nucleases into oocytes. This results in the transgene being inserted at a specific locus in the mouse genome, thus circumventing the drawbacks associated with BAC transgenesis.

Sophisticated transgenic mouse models frequently require that the transgene expression is restricted to a particular tissue or cell type as defined by a specific promoter. BACs have been the most commonly used method for generation of transgenic animals when the transcriptional control elements of the gene of interest have not been clearly identified. Thus, in many applications BACs replaced "conventional" transgenesis using short constructs with well-defined promoter regions. BACs are based on the single-copy functional fertility plasmid (F factor) of *Escherichia coli*. In contrast to high copy number plasmids, replication of the F factor is tightly controlled and the plasmid exists as only one or two copies per cell. Unidirectional replication is determined by the regulatory genes oriS and repE of the F plasmid while the copy number is controlled by parA and parB genes. These regulatory elements, together with a resistance marker, constitute the BAC vector backbone that facilitates cloning and stable propagation of DNA fragments up to 300 kilobase pairs [1]. Originally developed for construction of genomic libraries used in the early genome sequencing projects, BAC constructs and BAC libraries have also proven to be very useful genetic tools in other aspects of basic research.

As mentioned before, the use of BACs enabled scientists to generate transgenes even when the promoter region of a gene is unknown. Furthermore, BAC-based transgenes more frequently yield the expected expression pattern among founder lines compared to conventional transgenes. This is most likely the result of reduced influence of position effects due to either the sheer size of the BAC transgenes which insulate the transgene cassette from the influence of the chromosomal environment or through the inclusion of elements such as enhancers, silencers, locus control regions, and matrix attachment regions. BAC transgenes can be rapidly constructed through recombination in *E. coli* and subsequently used to generate transgenic animals using standard techniques such as pronuclear injection in case of mice or rats (Figure 1(a)). This rapid and relatively straight

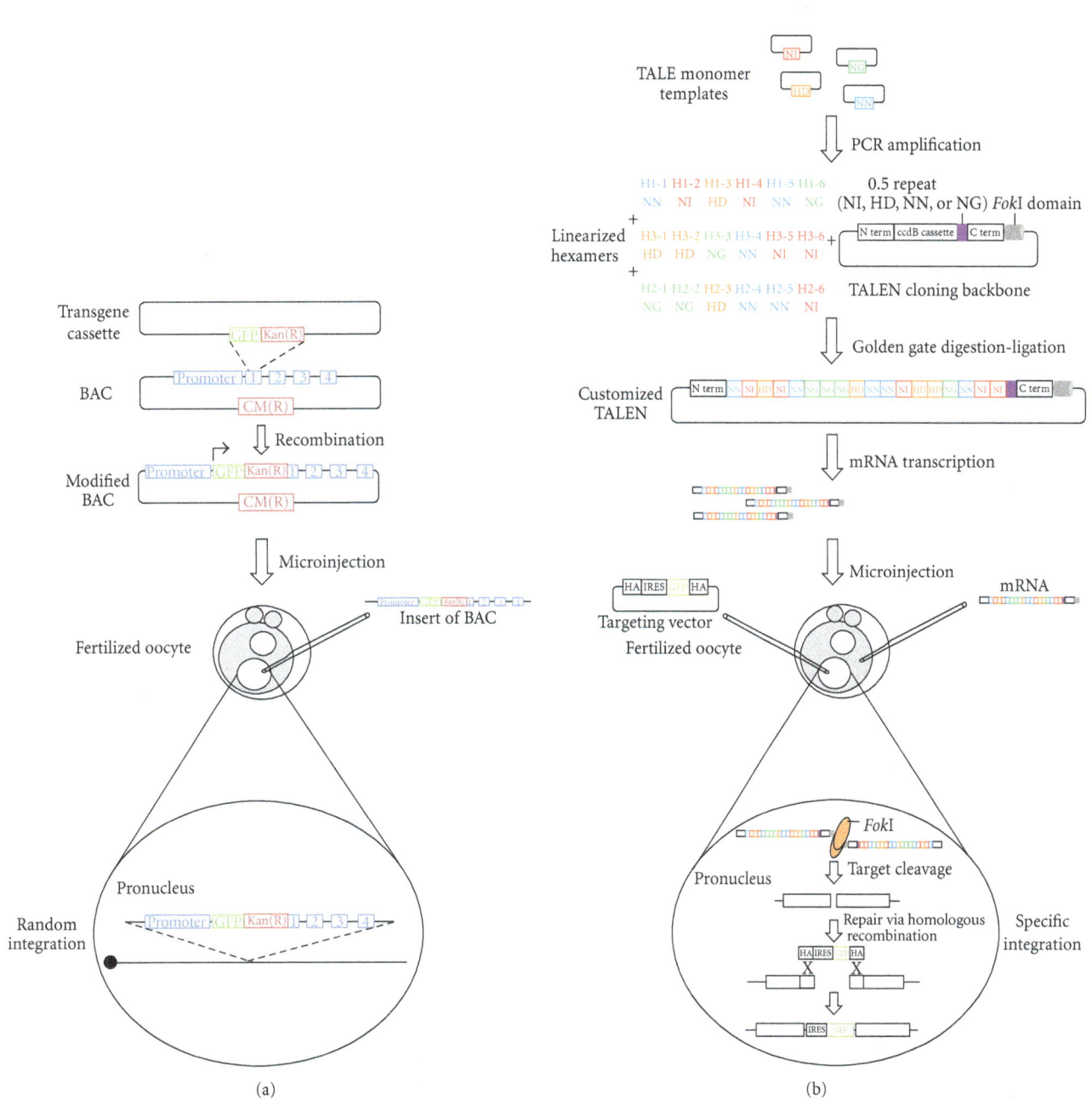

Figure 1: Genomic modifications through BACs and TALEN. (a) The recombination of the BAC and a transgene (GFP and the antibiotic resistance gene kanamycin) leads to a modified BAC which is then randomly integrated into the DNA after microinjection into the male pronucleus. (b) The customized TALEN, specific to the sequence of interest, is constructed by PCR and digestion-ligation of hexamers that are cloned into the TALEN backbone. Subsequently, the TALEN is transcribed into RNA followed by the microinjection of the RNA into the cytoplasm of the zygote simultaneous to the injection of the GFP containing targeting vector into the pronucleus. This results in specific integration of the GFP via homologous recombination generating a knock-in locus.

forward approach ensured that BAC transgenesis remained a viable alternative to the more complex and time-consuming targeted transgene insertion via embryonic stem cell technology [2].

Despite its increasing popularity the use of BAC transgenesis for the generation of transgenic animals is still accompanied by several problems. BAC transgenes are generated by nonspecific integration into the target genome; therefore a variable number of copies can be inserted into an unknown locus in the genome of the target organism. Moreover, because the generation of BAC transgenic founders is less efficient than conventional transgenes, it usually results in less founder lines being generated in the course of a similar microinjection effort. Furthermore, the construction of a BAC transgene can be extremely time consuming; while a BAC transgene construct may be assembled within 2 weeks, in some instances it takes several months. The construction of the BAC is then followed, depending on the facility, by

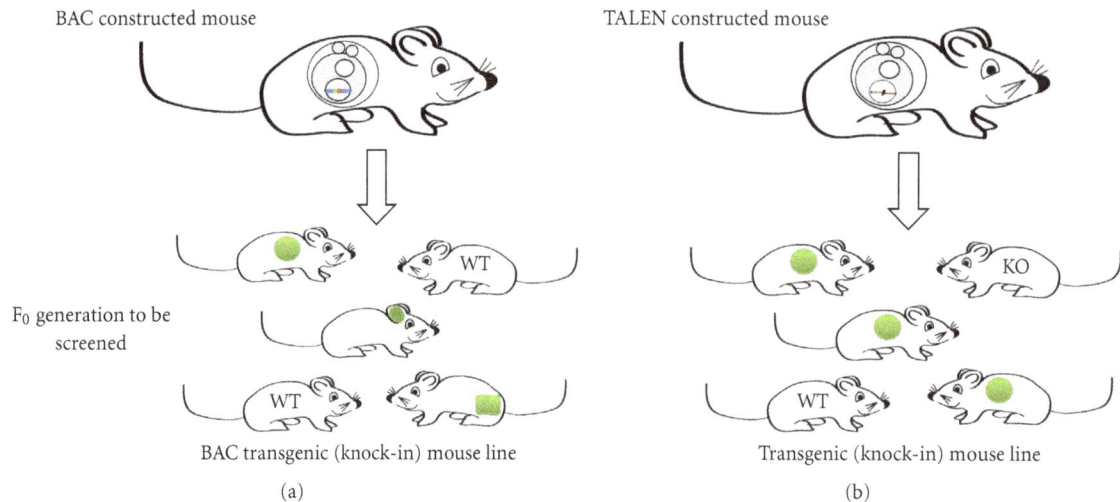

FIGURE 2: Generation of new mouse lines through BAC transgenesis or TALEN-mediated transgene insertion. (a) In consequence of the random integration of the vector during the construction of BAC transgenic mice the founders need to be screened for the appropriate expression pattern. (b) In contrast, TALEN-constructed transgenic mice only need to be checked by molecular biology methods for presence of the appropriate knock-in event. As a side product, knock-out alleles are usually generated as well.

several months of oocyte injection and screening of the founder lines (Figure 2(b)). New promising technologies have been developed which may overcome the problems associated with BAC transgenesis while retaining or even improving on its advantages over conventional transgenesis. These new techniques are based on designer nucleases and their ability to generate double strand breaks (DSBs) at a predetermined sequence in the target genome, even in the oocyte. Such DSBs can be repaired by nonhomologous end-joining (NHEJ) or homologous recombination (HR). While NHEJ results in deletions or insertions that can lead to gene deficiencies, HR can lead to the faithful integration of a coinjected targeting vector. At the time of writing, two classes of nucleases, ZFNs [3–5] and TALENs [6, 7], show the most promise of making an impact on transgenic technology in the mouse and beyond. Both assemble on the target DNA as heterodimers and each feature a DNA-binding domain designed for a specific sequence and a *Fok*I domain containing the endonuclease activity.

The DNA-binding domains of ZFNs are assembled through multimerization of Cys_2His_2 motifs, each motif recognizing three nucleotides of the DNA. ZFNs are commonly designed with 3- to 4-zincfinger domains [8], sometimes even up to 6 fingers [9], which are then fused to the *Fok*I domain. The ZFNs dimerize at the target sequence and cleave the double-stranded DNA thus forming DSBs. It is thought that increasing the number of ZF domains leads to an increase in the specificity and decreases the toxicity of the ZFNs. Constructs containing ZFNs encoding for a specific target can either be obtained from a commercial source such as Sigma-Aldrich or constructed in-house. The most common methods of in house construction are through open-source protocols such as Oligomerized Pool Engineering (OPEN) and Context-Dependent Assembly (CoDA) arrays. OPEN utilizes an archive of zinc finger pools

specific to the sequence of interest. Multifinger arrays are then generated from these pools and arrays which have high affinity and specificity are identified via a bacterial two-hybrid system [10]. As with OPEN, CoDA is based on a library and involves cloning the 3 fingers of the ZFNs out of a large archive of 319 N terminal (F1) and 344 C terminal (F3) fingers designed to work in combination with 18 fixed second fingers (F2) [11]. Compared to OPEN, this method is easier and cheaper for laboratories to adopt as no special techniques or proprietary reagents are required. CoDA does however have some drawbacks; the identity of the second finger is limited to the 18 fixed F2 units and the combined affinity or specificity of the 3 fingers in combination is not taken into account. Therefore despite the increased technical expertise required, at present OPEN may still be the most reliable choice to achieve the high specificity needed for certain therapeutic applications. Unfortunately at present both OPEN and CoDA are restricted to 3 fingers per heterodimer, thus limiting the possibility for increased specificity through inclusion of additional fingers.

TALEN nucleases are similar to ZFNs as they rely on the *Fok*I domain for their nuclease activity. However, the DNA-binding domain is based on the TAL-like effectors originally derived from *Xanthomonas* plant pathogens [12]. The TAL DNA-binding domain consists of a number of repeats where one repeat recognizes one base pair of the target DNA. Each individual repeat is 33-35-amino-acids long and all repeats are essentially identical with the exception of two key amino acid in positions 12 and 13. It is these "repeat variable diresidues" (RVDs) which determine the repeat's sequence specificity. In short, the RVDs NI, HD, NN, and NG recognize A, C, G or A, and T, respectively. Since the individual repeats appear to be modular and function in a context-independent manner, it is possible to design a TALEN protein that will bind and cleave virtually any target sequence. As with the ZFNs two TALEN monomers are necessary to dimerize a *Fok*I domain at the target sequence for generating a DSB [7, 13]. Online tools are available to custom-design TALEN targeted to a specific sequence. These allow the rapid identification of a suitable target sequence and its respective TALEN pair. Several strategies for efficient generation of TALEN pairs have been published including strategies involving "Golden Gate cloning" that include the assembling of repeat modules to an intermediary array with up to 10 repeats followed by the ligation of intermediary arrays into a backbone [7] or a PCR-based protocol where the monomers of a plasmid library are amplified to first form 3 hexamers which are then ligated to one 18 mer. This protocol enables the construction of customised TALEN within few days [14] (Figure 1(b)). Several other TALEN construction systems have been published [15–17] and TALENs can also be obtained commercially.

The method of construction of TALENs and ZFNs may be different but the functional outcome of both methodologies is the same: they are efficient and reliable methods for the *in vitro* and *in vivo* production of gene deficiencies, gain of function transgenics and perhaps most promisingly new therapeutic applications. But how does this relate to BAC transgenesis? In effect, DN technology allows the insertion of reporter genes such as GFP or Cre into the gene of interest by mouse zygote injection. The ZFNs and TALENs determine the exact position for integration of the transgene via homologous recombination. To achieve this, transgene vectors containing homologous sequences, flanking the DSB as well as the transgene, are constructed and coinjected with the DN expression vectors (or their mRNA) into the zygotes. This results in integration similar to that achieved by time-consuming embryonic stem (ES) cell culture, yet obtained with the speed of conventional transgenesis and without using selection genes such as neomycin resistance. Moreover, the founder lines of transgenic animals generated through BAC technology have to be functionally screened thoroughly due to the expression variegation as result of the randomness of the integration and an appropriate line has to be identified. In the case of DN-assisted HR all founder lines are carrying identical gene modifications and they can all be used for functional confirmation and analysis (Figure 2). While the efficiency of TALENs in mouse zygotes still needs to be evaluated, ZFNs have already been shown to allow efficient integration of DNA into a target site such as the frequently used gt(ROSA)26Sor locus [5, 18]. Thus, it may be envisioned that in the future, BAC transgenes will not be used to drive transgenesis but instead functional transgenes will be directly integrated into endogenous loci. As with insertion transgenes involving ES cell technology, this may be achieved by either integration (or mutation) of the start ATG, leading in most cases to a functional knockout allele, or by placing the transgene in the 5′ untranslated region and using an internal ribosomal entry site (IRES) or 2A element thus maintaining the open reading frame of the "hijacked" gene. Moreover placement of open reading frames under the control of loxP-flanked STOP cassettes or even exogenous promoters through placement of the respective constructs in the target locations may also be envisaged. As a side product these approaches would in many cases also yield deficiency alleles after faulty NHEJ repair.

Though it may appear that the advent of DNs will make BAC transgenesis all but obsolete, applications for BAC transgenes remain. Detailed functional analysis of human genetic elements in model organisms such as the mouse is still likely to be carried out using BAC transgenesis. Also, BACs may still serve as donors for homology regions. Additionally, although this paper highlights some of the future possibilities of genomic manipulation through DNs, many obstacles both small and large still remain. These include the determination of offsite effects (mutations) and what outcomes they may have. Furthermore a gold standard protocol for identification of target sequences and DNs must still be validated *in vivo* and other considerations such as the optimum length of homology regions for integration at DSBs in oocytes, and whether the targeting vector should be supplied in linear or circular form must still be decided. Only then a full financial and scientific comparison of BAC technology versus insertion through use of DNs will be possible.

Taken together, this paper has discussed some of the advantages and disadvantages of BAC technology in the context of the newly evolving DN technology which allows targeted insertions in the oocyte. We believe this is a revolutionary time in transgenic technologies and that the use of DN technology could make not only ES cell culture techniques but also conventional transgenesis, including BAC transgenesis all but obsolete.

References

[1] H. Shizuya, B. Birren, U. J. Kim et al., "Cloning and stable maintenance of 300-kilobase-pair fragments of human DNA in *Escherichia coli* using an F-factor-based vector," *Proceedings of the National Academy of Sciences of the United States of America*, vol. 89, no. 18, pp. 8794–8797, 1992.

[2] T. Johansson, I. Broll, T. Frenz et al., "Building a zoo of mice for genetic analyses: a comprehensive protocol for the rapid generation of BAC transgenic mice," *Genesis*, vol. 48, no. 4, pp. 264–280, 2010.

[3] H. J. Lee, E. Kim, and J. S. Kim, "Targeted chromosomal deletions in human cells using zinc finger nucleases," *Genome Research*, vol. 20, no. 1, pp. 81–89, 2010.

[4] C. Öllü, K. Pars, T. I. Cornu et al., "Autonomous zinc-finger nuclease pairs for targeted chromosomal deletion," *Nucleic Acids Research*, vol. 38, no. 22, pp. 8269–8276, 2010.

[5] M. Meyer, M. H. De Angelis, W. Wurst, and R. Kühn, "Gene targeting by homologous recombination in mouse zygotes mediated by zinc-finger nucleases," *Proceedings of the National Academy of Sciences of the United States of America*, vol. 107, no. 34, pp. 15022–15026, 2010.

[6] F. Zhang, L. Cong, S. Lodato, S. Kosuri, G. M. Church, and P. Arlotta, "Efficient construction of sequence-specific TAL effectors for modulating mammalian transcription," *Nature Biotechnology*, vol. 29, no. 2, pp. 149–154, 2011.

[7] T. Cermak, E. L. Doyle, M. Christian et al., "Efficient design and assembly of custom TALEN and other TAL effector-based constructs for DNA targeting," *Nucleic Acids Research*, vol. 39, no. 12, article e82, 2011.

[8] J. Wu, K. Kandavelou, and S. Chandrasegaran, "Custom-designed zinc finger nucleases: what is next?" *Cellular and Molecular Life Sciences*, vol. 64, no. 22, pp. 2933–2944, 2007.

[9] D. J. Segal, J. W. Crotty, M. S. Bhakta, C. F. Barbas 3rd, and N. C. Horton, "Structure of Aart, a designed six-finger zinc finger peptide, bound to DNA," *Journal of Molecular Biology*, vol. 363, no. 2, pp. 405–421, 2006.

[10] M. L. Maeder, S. Thibodeau-Beganny, J. D. Sander, D. F. Voytas, and J. K. Joung, "Oligomerized pool engineering (OPEN): an "open-source" protocol for making customized zinc-finger arrays," *Nature Protocols*, vol. 4, no. 10, pp. 1471–1501, 2009.

[11] J. D. Sander, E. J. Dahlborg, M. J. Goodwin et al., "Selection-free zinc-finger-nuclease engineering by context-dependent assembly (CoDA)," *Nature Methods*, vol. 8, no. 1, pp. 67–69, 2011.

[12] J. Boch, H. Scholze, S. Schornack et al., "Breaking the code of DNA binding specificity of TAL-type III effectors," *Science*, vol. 326, no. 5959, pp. 1509–1512, 2009.

[13] J. C. Miller, S. Tan, G. Qiao et al., "A TALE nuclease architecture for efficient genome editing," *Nature Biotechnology*, vol. 29, no. 2, pp. 143–148, 2011.

[14] N. E. Sanjana, L. Cong, Y. Zhou, M. M. Cunniff, G. Feng, and F. Zhang, "A transcription activator-like effector toolbox for genome engineering," *Nature Protocols*, vol. 7, no. 1, pp. 171–192, 2012.

[15] J. D. Sander, L. Cade, C. Khayter et al., "Targeted gene disruption in somatic zebrafish cells using engineered TALENs," *Nature Biotechnology*, vol. 29, no. 8, pp. 697–698, 2011.

[16] T. Li, S. Huang, X. Zhao et al., "Modularly assembled designer TAL effector nucleases for targeted gene knockout and gene replacement in eukaryotes," *Nucleic Acids Research*, vol. 39, no. 14, pp. 6315–6325, 2011.

[17] D. Reyon, S. Q. Tsai, C. Khgayter, J. A. Foden, J. D. Sander, and J. K. Joung, "FLASH assembly of TALENs for high-throughput genome editing," *Nature Biotechnology*, vol. 30, no. 5, pp. 460–465, 2012.

[18] M. Hermann, M. L. Maeder, K. Rector et al., "Evaluation of OPEN zinc finger nucleases for direct gene targeting of the ROSA26 locus in mouse embryos," *PLoS ONE*. In press.

11

Genotyping of *CYP2C9* and *VKORC1* in the Arabic Population of Al-Ahsa, Saudi Arabia

Abdullah M. Alzahrani,[1] **Georgia Ragia,**[2] **Hamza Hanieh,**[1] **and Vangelis G. Manolopoulos**[2]

[1] *Biological Sciences Department, College of Science, King Faisal University, Hofouf 31982, Saudi Arabia*
[2] *Laboratory of Pharmacology, Medical School, Democritus University of Thrace, Alexandroupolis 68100, Greece*

Correspondence should be addressed to Abdullah M. Alzahrani; aalzahra@kfu.edu.sa

Academic Editor: M. Ilyas Kamboh

Polymorphisms in the genes encoding CYP2C9 enzyme and VKORC1 reductase significantly influence the dose variability of coumarinic oral anticoagulants (COAs). Substantial inter- and intraethnic variability exists in the frequencies of *CYP2C9**2 and *3 and *VKORC1* −1639A alleles. However, the prevalence of *CYP2C9* and *VKORC1* genetic variants is less characterized in Arab populations. A total of 131 healthy adult subjects from the Al-Ahsa region of Saudi Arabia were genotyped for the *CYP2C9**2 and *3 and *VKORC1* −1639G>A polymorphisms by PCR-RFLP method. The frequencies of the *CYP2C9**2 and *3 and *VKORC1* −1639A alleles were 13.3%, 2.3%, and 42.4%, respectively, with no subjects carrying 2 defective alleles. The frequencies of the *CYP2C9**3 and *VKORC1* −1639A alleles were significantly lower than those reported in different Arabian populations. None of the subjects with the *VKORC1* −1639AA genotype were carriers of *CYP2C9**1/*3 genotypes that lead to sensitivity to COAs therapy. The low frequency of the *CYP2C9**3 allele combined with the absence of subjects carrying 2 defective *CYP2C9* alleles suggests that, in this specific population, pharmacogenetic COAs dosing may mostly rely upon *VKORC1* genotyping.

1. Introduction

Pharmacogenomics is the first step towards personalized medicine and is a promising field of investigation that may explain some of the interindividual variations in responses to various classes of drugs [1, 2]. Pharmacogenomics can be applied to all fields of medicine. In particular, recent advances in genotype-phenotype associations in cardiology point towards the application of pharmacogenomics to oral coumarinic anticoagulants (COAs), including warfarin, acenocoumarol, and phenprocoumon, in routine clinical practice [3].

Among other factors, interindividual COAs dose variability is significantly influenced by variations in the genes encoding two enzymes: cytochrome P450 2C9 (CYP2C9), the enzyme that metabolizes COAs, and vitamin K epoxide reductase (VKORC1), the pharmacologic target of these drugs [3].

Polymorphisms in the *CYP2C9* gene seriously affect the enzymatic activity of the encoded CYP2C9 protein. Based on phenotype, populations can be divided into extensive (EM), intermediate (IM), and poor metabolizers (PM), and more than 35 different allelic variants have been identified in the *CYP2C9* gene [4]. Among these alleles, the *CYP2C9**2 (rs1799853) and *3 (rs1057910) variants, which reduce CYP2C9 enzymatic activity, allow for the prediction of more than 85% of PMs [5]. In addition, it has recently been shown that *VKORC1* gene polymorphisms also affect COAs dosing requirements [6]. The *VKORC1* −1639G>A polymorphism (rs9923231) is located in the promoter of the *VKORC1* gene and results in reduced promoter activity and lower mRNA levels, which lead to lower levels of synthesized protein and eventually the reduced production of active clotting factors in subjects with the AA genotype [6].

Whereas the *CYP2C9**2 and *3 alleles affect coumarin pharmacokinetics, the *VKORC1* −1639G>A polymorphism affects the pharmacodynamic response to coumarins [3]. It has been reported that polymorphisms in the *CYP2C9* and *VKORC1* genes together account for 35%–50% of the variability in COAs dose requirements for initiation and maintenance [7]. Moreover, carriers of the *CYP2C9**2 or *3 alleles and the *VKORC1* –1639G>A polymorphism are at higher risk for bleeding and require lower mean daily doses [3]. These associations between genotype and COAs response led the U.S. Food and Drug Administration (FDA) to release a warning in the warfarin insert to indicate the range of expected therapeutic warfarin dosages based on *CYP2C9* and *VKORC1* genotypes [8]. Furthermore, the clinical feasibility of incorporating *CYP2C9* and *VKORC1* genotyping-based COAs dosing regimens into routine clinical practice is being tested in large prospective clinical trials [9–11].

Substantial inter- and intraethnic variability in the frequencies of the *CYP2C9* and *VKORC1* alleles has been reported [5, 12]. The *CYP2C9**2 allele is absent in East Asian populations, whereas its frequency in African-Americans and Ethiopians has been estimated to be as low as 3.2% [5]. By contrast, a higher frequency of the *CYP2C9**2 allele (5%–19%) has been reported in Caucasians [5, 13]. Furthermore, the frequency of the *CYP2C9**3 allele is significantly lower in Asian populations (as low as 3.3%, compared to 4%–16% in Caucasians) [5, 13]. In Arabian populations, intraethnic variability in the frequency of the *CYP2C9**2 and *3 alleles has been reported, ranging from 7% to 21% and 3% to 9%, respectively [14–18]. In the case of the *VKORC1* –1639G>A polymorphism, interethnic variability in –1639AA frequency has also been reported [19]. Moreover, the –1639AA genotype, which is highly correlated with COAs sensitivity, is more common in Asian (frequency 80%) than Caucasian or African populations (estimated frequency 16%–25%) [19, 20]. Whereas the frequency of the –1639A COAs sensitivity allele ranges between 52% and 56% in the Arab populations studied to date [18, 21], no investigation of the *VKORC1* –1639G>A polymorphism in Saudi Arabians has been reported.

Few data are available concerning the prevalence of the *CYP2C9**2 and *3 alleles and the *VKORC1* –1639G>A polymorphism in distant populations of Saudi Arabia where social and in some areas religious beliefs favor consanguineous marriages and therefore limit genetic flow. In the Al-Ahsa region, which is part of the eastern province of Saudi Arabia, the rate of consanguineous marriage has been reported to be 59.1%, with marriages between first-degree relatives at 40% [22]. The reported inter- and intraethnic variability in the frequencies of the *CYP2C9**2 and *3 and *VKORC1* –1639A alleles was our rationale for studying the incidence of these variant alleles in the Al-Ahsa population. The results of this investigation will be critical for the coming era, in which genotype-guided dosing algorithms will be increasingly utilized to guide the prescription of COAs [9–11].

The aim of the present study was to investigate the frequency of the *CYP2C9**2 and *3 alleles and the *VKORC1* –1639G>A polymorphism as well as the number and percentages of individuals with genotypes predictive of COAs response in a representative sample of the population of Al-Ahsa, Saudi Arabia. We also sought to compare the data obtained with existing published data for other populations residing in a wider area of the Middle East.

2. Materials and Methods

2.1. Subjects. All participants were of Arabian origin and residents of the Al-Ahsa urban area, which is located on the east coast of Saudi Arabia. The study protocol was approved by the Ethics Committee of King Faisal University, Hofouf, Saudi Arabia. All of the study participants were nonrelative volunteers and provided informed consent. A total of 131 healthy adult subjects (70 males and 61 females) were genotyped to determine the frequencies of the *CYP2C9**2 and *3 alleles and the *VKORC1* –1639G>A polymorphism. The mean (±SD) age of the subjects was 25 (±7) years (range: 19–52 years). The majority of the subjects (57%) reported to originate from a consanguineous marriage.

2.2. Genotyping. Genomic DNA was extracted from peripheral blood leukocytes using the QIAamp DNA Blood Mini Kit (Qiagen, Germany) according to the manufacturer's instructions. All subjects were genotyped for the *CYP2C9**2 and *3 alleles and the *VKORC1* –1639G>A polymorphism using PCR-restriction fragment length polymorphism (RFLP) protocols, as previously described [13, 20]. PCR amplifications were performed in duplicate by two independent researchers in an MJ Research PTC-200 thermocycler (Watertown, MA, USA). To ensure the accuracy of the results, an internal positive control was utilized for each polymorphism (rare allele) in each PCR-RFLP run.

2.3. Statistics. Data were analyzed using the Statistical Package for Social Sciences (SPSS) version 17.0 and are presented as the medians with 95% confidence intervals. Departure from the Hardy-Weinberg equilibrium was estimated using an exact 2-sided probability test using the formula provided by Weir [23]. Allele frequencies were compared to other ethnic population utilizing the two-tailed Fisher's exact test [24].

3. Results

The distributions of genotypes and alleles of *CYP2C9**2 and *3 and *VKORC1* –1639G>A polymorphisms in the studied population are shown in Table 1.

For the *CYP2C9**2 and *3 alleles, 90 subjects (68.7%) were genotyped as *CYP2C9**1/*1, 35 subjects (26.7%) as *CYP2C9**1/*2, and 6 subjects (4.6%) as *CYP2C9**1/*3. The frequency of the *CYP2C9**2 and *3 alleles was estimated at 13.3% and 2.3%, respectively. None of the subjects were found to be homozygous or combined heterozygous for the *CYP2C9**2 and *3 alleles (*CYP2C9**2/*2, *CYP2C9**2/*3, and *CYP2C9**3/*3 genotypes). The genotype-derived PM phenotype was absent in the study population, whereas 68.7% of the population was predicted to be EM and 31.3% to be IM.

For the *VKORC1* –1639G>A polymorphism, 49 subjects (37.4%) were genotyped as GG, 52 subjects (39.7%) as GA, and 30 subjects (22.9%) as AA. The frequency of the A allele was 42.7% (112 alleles). Consistent results for each genotype call were obtained by two researchers who performed the genotyping independently.

TABLE 1: Frequencies of the *CYP2C9* *2 and *3 and *VKORC1* −1639G>A genotypes and alleles in a sample of the Saudi Arabian population (*n* = 131).

Genotypes and alleles	No. of individuals, relative frequency, and 95% confidence intervals	
	n (%)	95% CI
CYP2C9 genotype		
CYP2C9 *1/*1	90 (68.7)	60.4–76.2
CYP2C9 *1/*2	35 (26.7)	19.7–34.7
CYP2C9 *1/*3	6 (4.6)	1.9–9.2
CYP2C9 *2/*2	0	—
CYP2C9 *2/*3	0	—
CYP2C9 *3/*3	0	—
CYP2C9 allele		
CYP2C9 *2	35 (13.3)	9.7–17.9
CYP2C9 *3	6 (2.3)	1.0–4.7
VKORC1 genotype		
GG	49 (37.4)	29.5–45.9
GA	52 (39.7)	31.6–48.2
AA	30 (22.9)	16.4–30.6
VKORC1 allele		
G	150 (57.3)	51.2–63.1
A	112 (42.7)	36.9–48.8

TABLE 2: Combination of *CYP2C9* and *VKORC1* genotypes in a sample of the Saudi Arabian population (*n* = 131).

CYP2C9 genotype	VKORC1 genotype			Total
	GG, *n* (%)	GA, *n* (%)	AA, *n* (%)	
CYP2C9 *1/*1	35 (26.7)	30 (22.9)	25 (19.1)	90 (68.7)
CYP2C9 *1/*2	11 (8.4)	19 (14.5)	5 (3.8)	35 (26.7)
CYP2C9 *1/*3	3 (2.3)	3 (2.3)	—	6 (4.6)
Total	49 (37.4)	52 (39.7)	30 (22.9)	131 (100)

Table 2 lists all of the combinations of the variant genotypes identified in the present cohort study. Of 49 subjects with the *VKORC1* −1639GG genotype, 35 subjects were genotyped as *CYP2C9* *1/*1, 11 subjects as *CYP2C9* *1/*2, and 3 subjects as *CYP2C9* *1/*3. Among the 52 subjects with the *VKORC1* −1639GA genotype, 30 subjects were *CYP2C9* *1/*1, 19 subjects were *CYP2C9* *1/*2, and 3 subjects were *CYP2C9* *1/*3. Among the 30 subjects with the *VKORC1* −1639AA genotype, 25 subjects were *CYP2C9* *1/*1, 5 subjects were *CYP2C9* *1/*2, and none carried the *CYP2C9* *1/*3 genotype.

According to the range of *CYP2C9*- and *VKORC1*-based warfarin dosages suggested by the U.S. FDA [8], 58% of our population (those carrying the genotypes *VKORC1* −1639GG or 1639GA and *CYP2C9* *1/*1, or *VKORC1* −1639GG and *CYP2C9* *1/*2) would require higher dosages (5–7 mg). In contrast, the remainder of the studied subjects (42%; those carrying the genotypes *VKORC1* −1639GG and *CYP2C9* *1/*3, *VKORC1* −1639GA and *CYP2C9* *1/*2 or

CYP2C9 *1/*3, or *VKORC1* −1639AA and *CYP2C9* *1/*1 or *CYP2C9* *1/*2) may require intermediate dosages (3-4 mg). Due to the lack of *CYP2C9* *2/*2 or *CYP2C9* *3/*3 genotypes and the lack of the combined genotype *VKORC1* −1639AA and *CYP2C9* *1/*3, none of the studied subjects belonged to the sensitive group that would require low warfarin dosages (range of 0.5–2 mg).

We also analyzed the data for potential gender differences. However, there were no significant differences in the genotype frequencies of the *CYP2C9* *2 and *3 alleles and the *VKORC1* −1639G>A polymorphism in our study group of 131 Saudi Arabian subjects (70 male and 61 female) (data not shown).

Finally, to investigate possible differences between individuals who originated from consanguineous marriages and those who did not, we analyzed the distribution of the *CYP2C9* *2 and *3 alleles and the *VKORC1* −1639G>A polymorphism according to this factor. However, the frequencies of the genotypes and alleles studied did not differ with respect to origin from a consanguineous marriage (data not shown).

4. Discussion

The pharmacogenomics of COAs is one field that is most ready to apply genotype-guided dosing in clinical practice. It has been estimated that polymorphisms in the *CYP2C9* and *VKORC1* genes together account for 35%–50% of the variability in COAs initiation and maintenance dosage requirements [20, 40]. Thus, efforts are focused on incorporating this knowledge into the dosing regimens currently used, and genotype-based algorithms are currently being tested in large randomized trials to validate the accuracy, safety, and cost effectiveness of incorporating *CYP2C9* and *VKORC1* genotypes into the optimization of anticoagulant therapy [3, 9].

In the era of developing and testing genotype-guided COAs dosing algorithms, there remain populations in which the frequency of the major *CYP2C9* and *VKORC1* polymorphisms has not been assessed. Thus, potential differences in the prevalence of *CYP2C9* and *VKORC1* genetic variants in different populations may lead to adjustments of genotype-based COAs dosing algorithms or may serve as motivation to identify novel genetic variants that influence COAs therapeutic responses. Towards this goal, the current study reported the frequency of the *CYP2C9* *2 and *3 alleles and *VKORC1* −1639G>A polymorphism in a distant population residing on the east coast of Saudi Arabia. Some of the *CYP2C9* variants that lead to decreased enzymatic activity, such as *CYP2C9* *5, *6, and *11, were not included in this study because they have not been observed in Caucasian or Middle Eastern populations, in contrast to their higher frequency in African populations. Although consanguineous marriage is traditionally favored among Saudis, genetic inflow in Al-Ahsa is further limited due to social and religious beliefs, which could have led to the differences in the prevalence of the *CYP2C9* and *VKORC1* genotypes and alleles studied.

In general, the genetic characteristics of Arabian populations are not well characterized. In the case of *CYP2C9*,

TABLE 3: Prevalence of CYP2C9*2 and *3 in different ethnic groups.

Population	Allele frequencies of CYP2C9					Ref.
	N	*2	(P)	*3	(P)	
Middle East Arab						
Saudi (Al-Ahsa)	131	0.133		0.023		Current
Saudi (Riyadh)	192	0.117	0.73	0.091	0.03	[14]
Egyptian	247	0.120	0.87	0.060	0.13	[16]
Jordanian	263	0.135	1.0	0.068	0.09	[25]
Lebanese	161	0.112	0.72	0.096	0.03	[18]
Omani	189	0.074	0.18	0.029	0.72	[17]
Caucasian						
American	100	0.080	0.29	0.060	0.19	[26]
Croatian	200	0.165	0.53	0.095	0.02	[27]
German	118	0.140	0.86	0.050	0.32	[28]
Greek	283	0.129	1.0	0.081	0.03	[13]
Italian	157	0.112	0.86	0.092	0.03	[29]
Turkish	499	0.106	0.54	0.100	0.006	[30]
Belgian	121	0.10	0.56	0.074	0.081	[31]
Asian						
Japanese	218	0	<0.0001	0.021	0.72	[32]
Korean	574	0	<0.0001	0.011	0.22	[33]
Chinese (Mongolian)	280	0	<0.0001	0.03	1.0	[34]
Vietnamese	157	0	<0.0001	0.022	1.0	[35]
Malaysian (Malay)	202	0.019	0.0003	0.024	1.0	[36]
African						
Beninese	111	0	<0.0001	0	<0.0001	[31]
Ethiopian	150	0.04	0.02	0.02	1.0	[29]
Ghanaian	204	0	<0.0001	0	0.06	[37]
Iranian	200	0.13	1.0	0	0.06	[38]
African-American	490	0.011	<0.0001	0.018	0.48	[39]

there are scattered reports on CYP2C9 frequencies in Arabs in general and Saudis in particular, whereas to the best of our knowledge, this is the first report of the frequency of the VKORC1 –1639G>A polymorphism in a population in Saudi Arabia. Among Arabs in general, there exists great intraethnic variability in the frequencies of the CYP2C9*2 and *3 alleles. In Saudi Arabia, Mirghani et al. [14] reported that the frequencies of the CYP2C9*2 and *3 alleles among Saudis residing in Riyadh were similar to those in Caucasian populations (11.7% and 9.1%, resp.) [14]. Unpredictably, the frequency of CYP2C9*3 was significantly different from the frequency of *3 in our study group (Table 3). In the Omani population, CYP2C9*2 and *3 allele frequencies are markedly lower and have been estimated at 7.4% and 2.9%, respectively [17]. The later was the closest Among Arab populations to the frequency of CYP2C9*3 in our study subjects (2.3%). In the Egyptian population, the frequencies of the CYP2C9*2 and *3 alleles have been estimated to be 12% and 6%, respectively [16]. In Lebanese individuals, the frequencies of CYP2C9*2 and *3 have been reported to be 11.2% and 9.6%, respectively [18].

In our study population, the CYP2C9*2 allele frequency (13.4%) was similar to that reported for other Arabian and Caucasian populations (Table 3). However, we found a significantly reduced frequency of the CYP2C9*3 allele (2.3%). In addition, genotypes predicting the CYP2C9 PM phenotype (i.e., CYP2C9*2/*2, CYP2C9*2/*3, and CYP2C9*3/*3) were absent in the subjects studied. One possible explanation for this finding is that our study population comprised residents of Al-Ahsa, where the population can be divided based on religious beliefs into two main populations (Sunni and Shiaah) between which intermarriage rarely occurs. Moreover, within each of the two populations, social customs may further limit intermarriage between nonrelatives, further resulting in decreased genetic inflow.

Although we did not find statistically significant differences in the distribution of CYP2C9*2 and *3 alleles and VKORC1 –1639G>A polymorphisms in our study group according to presence or absence of consanguineous marriage, different frequencies in variants predicting low CYP2C9 enzymatic activity should be expected in similar populations elsewhere in Saudi Arabia.

Regarding the VKORC1 –1639G>A polymorphism, we found that the frequencies of genotypes and alleles were similar to those reported in Caucasian populations. The VKORC1 –1639AA genotype and VKORC1 –1639A allele frequencies were estimated to be 22.9% and 42.7%, respectively, which suggests that approximately 23% of the studied population is sensitive to COAs and would require lower dosages of COAs. However, this finding should not be generalized to all Saudi Arabians prior to assessing the relative genotype frequencies in populations residing in different regions with different social backgrounds. Indeed, we found that the frequency of the *VKORC1* −1639A allele was lower than that reported for other Arabian populations (i.e., 52.4% in the Lebanese population) [18].

The relatively low frequencies of the *CYP2C9*∗3 allele and *CYP2C9* genotype-derived PMs indicate that, in this population, COAs dosage adjustments and responses may depend more on *VKORC1* gene polymorphisms. This finding is of utmost importance for personalizing COAs therapy in the Al-Ahsa region. The presence of other rare alleles that cause reduced CYP2C9 activity and could potentially interfere with COAs response requires further investigation. In addition to COAs metabolism, normal to slightly decreased metabolism of other drugs that are CYP2C9 substrates would be expected in the studied population in addition to the incidence of adverse effects due to diminished metabolism, as is the case with antidiabetic drugs metabolized by CYP2C9 [41].

In conclusion, we report that the frequency of the *CYP2C9*∗3 allele varies substantially among Saudi Arabian populations. In light of the frequency of these genetic variants, the *VKORC1* −1639G>A polymorphism may be the major determinant of COAs pharmacogenomics in the studied population. Overall, it appears that some *CYP2C9* genotypes known to be associated with sensitivity to COAs are less common in the studied population, particularly *CYP2C9*∗3. Investigation of the frequencies of other *CYP2C9* alleles in this population as well as other Saudi populations, especially those variants known to be associated with decreased levels of the CYP2C9 enzyme, such as *CYP2C9*∗5, ∗8, ∗11, ∗13–18, ∗30, and ∗33, is recommended. To apply COA pharmacogenomics in clinical practice in Saudi Arabia, we need to understand the frequencies of genetic variants in various Saudi Arabian populations in order to facilitate clinical decision making and improve patient management.

Conflict of Interests

The authors declare no competing financial or other conflict of interests.

Acknowledgment

This work was supported by the Deanship of Scientific Research, King Faisal University (Grant no. 130148).

References

[1] V. G. Manolopoulos, "Pharmacogenomics and adverse drug reactions in diagnostic and clinical practice," *Clinical Chemistry and Laboratory Medicine*, vol. 45, no. 7, pp. 801–814, 2007.

[2] V. G. Manolopoulos, G. Ragia, and A. Tavridou, "Pharmacogenomics of oral antidiabetic medications: current data and pharmacoepigenomic perspective," *Pharmacogenomics*, vol. 12, no. 8, pp. 1161–1191, 2011.

[3] V. G. Manolopoulos, G. Ragia, and A. Tavridou, "Pharmacogenetics of coumarinic oral anticoagulants," *Pharmacogenomics*, vol. 11, no. 4, pp. 493–496, 2010.

[4] http://www.cypalleles.ki.se/cyp2c9.htm.

[5] H. G. Xie, H. C. Prasad, R. B. Kim, and C. M. Stein, "CYP2C9 allelic variants: ethnic distribution and functional significance," *Advanced Drug Delivery Reviews*, vol. 54, no. 10, pp. 1257–1270, 2002.

[6] H. Y. Yuan, J. J. Chen, M. T. M. Lee et al., "A novel functional VKORC1 promoter polymorphism is associated with inter-individual and inter-ethnic differences in warfarin sensitivity," *Human Molecular Genetics*, vol. 14, no. 13, pp. 1745–1751, 2005.

[7] L. Bodin, C. Verstuyft, D. A. Tregouet et al., "Cytochrome P450 2C9 (CYP2C9) and vitamin K epoxide reductase (VKORC1) genotypes as determinants of acenocoumarol sensitivity," *Blood*, vol. 106, no. 1, pp. 135–140, 2005.

[8] http://www.fda.gov/downloads/NewsEvents/Newsroom/MediaTranscripts/ucm123583.pdf.

[9] R. M. F. Van Schie, M. Wadelius, F. Kamali et al., "Genotype-guided dosing of coumarin derivatives: the European pharmacogenetics of anticoagulant therapy (EU-PACT) trial design," *Pharmacogenomics*, vol. 10, no. 10, pp. 1687–1695, 2009.

[10] B. French, J. Joo, N. L. Geller et al., "Statistical design of personalized medicine interventions: the Clarification of Optimal Anticoagulation through Genetics (COAG) trial," *Trials*, vol. 11, article 108, 2010.

[11] E. J. Do, P. Lenzini, C. S. Eby et al., "Genetics informatics trial (GIFT) of warfarin to prevent deep vein thrombosis (DVT): rationale and study design," *Pharmacogenomics Journal*, 2011.

[12] H. Schelleman, N. A. Limdi, and S. E. Kimmel, "Ethnic differences in warfarin maintenance dose requirement and its relationship with genetics," *Pharmacogenomics*, vol. 9, no. 9, pp. 1331–1346, 2008.

[13] K. Arvanitidis, G. Ragia, M. Iordanidou et al., "Genetic polymorphisms of drug-metabolizing enzymes CYP2D6, CYP2C9, CYP2C19 and CYP3A5 in the Greek population," *Fundamental and Clinical Pharmacology*, vol. 21, no. 4, pp. 419–426, 2007.

[14] R. A. Mirghani, G. Chowdhary, and G. Elghazali, "Distribution of the major Cytochrome P450 (CYP) 2C9 genetic variants in a Saudi population," *Basic and Clinical Pharmacology and Toxicology*, vol. 109, no. 2, pp. 111–114, 2011.

[15] J. N. Saour, A. W. Shereen, B. J. Saour, and L. A. Mammo, "CYP2C9 polymorphism studies in the Saudi population," *Saudi Medical Journal*, vol. 32, no. 4, pp. 347–352, 2011.

[16] S. I. Hamdy, M. Hiratsuka, K. Narahara et al., "Allele and genotype frequencies of polymorphic cytochromes P450 (CYP2C9, CYP2C19, CYP2E1) and dihydropyrimidine dehydrogenase (DPYD) in the Egyptian population," *British Journal of Clinical Pharmacology*, vol. 53, no. 6, pp. 596–603, 2002.

[17] M. O. Tanira, M. K. Al-Mukhaini, A. T. Al-Hinai, K. A. Al Balushi, and I. S. Ahmed, "Frequency of CYP2C9 genotypes among Omani patients receiving warfarin and its correlation with warfarin dose," *Community Genetics*, vol. 10, no. 1, pp. 32–37, 2007.

[18] I. Djaffar-Jureidini, N. Chamseddine, S. Keleshian, R. Naoufal, L. Zahed, and N. Hakime, "Pharmacogenetics of coumarin dosing: prevalence of CYP2C9 and VKORC1 polymorphisms in the Lebanese population," *Genetic Testing and Molecular Biomarkers*, vol. 15, no. 11, pp. 827–830, 2011.

[19] L. Yang, W. Ge, F. Yu, and H. Zhu, "Impact of VKORC1 gene polymorphism on interindividual and interethnic warfarin dosage requirement—a systematic review and meta analysis," *Thrombosis Research*, vol. 125, no. 4, pp. e159–e166, 2010.

[20] A. Tavridou, I. Petridis, M. Vasileiadis et al., "Association of VKORC1 -1639 G>A polymorphism with carotid intima-media thickness in type 2 diabetes mellitus," *Diabetes Research and Clinical Practice*, vol. 94, no. 2, pp. 236–241, 2011.

[21] E. Efrati, H. Elkin, E. Sprecher, and N. Krivoy, "Distribution of CYP2C9 and VKORC1 risk alleles for warfarin sensitivity and resistance in the Israeli population," *Current Drug Safety*, vol. 5, no. 3, pp. 190–193, 2010.

[22] M. A. F. El-Hazmi, A. R. Al-Swailem, A. S. Warsy, A. M. Al-Swailem, R. Sulaimani, and A. A. Al-Meshari, "Consanguinity among the Saudi Arabian population," *Journal of Medical Genetics*, vol. 32, no. 8, pp. 623–626, 1995.

[23] B. Weir, "Disequilibrium," in *Genetic Data Analysis II*, M. Sunderland, Ed., pp. 91–139, Sinaur Associates, 1996.

[24] http://www.graphpad.com/quickcalcs/contingency1.cfm.

[25] A. M. Yousef, N. R. Bulatova, W. Newman et al., "Allele and genotype frequencies of the polymorphic cytochrome P450 genes (CYP1A1, CYP3A4, CYP3A5, CYP2C9 and CYP2C19) in the Jordanian population," *Molecular Biology Reports*, vol. 39, no. 10, pp. 9423–9433, 2012.

[26] T. H. Sullivan-Klose, B. I. Ghanayem, D. A. Bell et al., "The role of the CYP2C9-Leu359 allelic variant in the tolbutamide polymorphism," *Pharmacogenetics*, vol. 6, no. 4, pp. 341–349, 1996.

[27] N. Božina, P. Granić, Z. Lalić, I. Tramišak, M. Lovrić, and A. Stavljenić-Rukavina, "Genetic polymorphisms of cytochromes P450: CYP2C9, CYP2C19, and CYP2D6 in Croatian population," *Croatian Medical Journal*, vol. 44, no. 4, pp. 425–428, 2003.

[28] M. Burian, S. Grösch, I. Tegeder, and G. Geisslinger, "Validation of a new fluorogenic real-time PCR assay for detection of CYP2C9 allelic variants and CYP2C9 allelic distribution in a German population," *British Journal of Clinical Pharmacology*, vol. 54, no. 5, pp. 518–521, 2002.

[29] M. G. Scordo, E. Aklillu, U. Yasar, M. L. Dahl, E. Spina, and M. Ingelman-Sundberg, "Genetic polymorphism of cytochrome P450 2C9 in a Caucasian and a black African population," *British Journal of Clinical Pharmacology*, vol. 52, no. 4, pp. 447–450, 2001.

[30] A. S. Aynacioglu, J. Brockmöller, S. Bauer et al., "Frequency of cytochrome P450 CYP2C9 variants in a Turkish population and functional relevance for phenytoin," *British Journal of Clinical Pharmacology*, vol. 48, no. 3, pp. 409–415, 1999.

[31] A. C. Allabi, J. L. Gala, J. P. Desager, M. Heusters.preute, and Y. Horsmans, "Genetic polymorphisms of CYP2C9 and CYP2C19 in the Beninese and Belgian populations," *British Journal of Clinical Pharmacology*, vol. 56, no. 6, pp. 653–657, 2003.

[32] K. Nasu, T. Kubota, and T. Ishizaki, "Genetic analysis of CYP2C9 polymorphism in a Japanese population," *Pharmacogenetics*, vol. 7, no. 5, pp. 405–409, 1997.

[33] Y. R. Yoon, J. H. Shon, M. K. Kim et al., "Frequency of cytochrome P450 2C9 mutant alleles in a Korean population," *British Journal of Clinical Pharmacology*, vol. 51, no. 3, pp. 277–280, 2001.

[34] J. Zuo, D. Xia, L. Jia, and T. Guo, "Genetic polymorphisms of drug-metabolizing phase I enzymes CYP3A4, CYP2C9, CYP2C19 and CYP2D6 in Han, Uighur, Hui and Mongolian Chinese populations," *Pharmazie*, vol. 67, no. 7, pp. 639–644, 2012.

[35] S. S. Lee, K. M. Kim, H. Thi-Le, S. S. Yea, I. J. Cha, and J. G. Shin, "Genetic polymorphism of CYP2C9 in a Vietnamese Kinh population," *Therapeutic Drug Monitoring*, vol. 27, no. 2, pp. 208–210, 2005.

[36] Z. Zainuddin, L. K. Teh, A. W. M. Suhaimi, and R. Ismail, "Malaysian Indians are genetically similar to Caucasians: CYP2C9 polymorphism," *Journal of Clinical Pharmacy and Therapeutics*, vol. 31, no. 2, pp. 187–191, 2006.

[37] W. Kudzi, A. N. O. Dodoo, and J. J. Mills, "Characterisation of CYP2C8, CYP2C9 and CYP2C19 polymorphisms in a Ghanaian population," *BMC Medical Genetics*, vol. 10, article 124, 2009.

[38] N. Zand, N. Tajik, A. S. Moghaddam, and I. Milanian, "Genetic polymorphisms of cytochrome P450 enzymes 2C9 and 2C19 in a healthy Iranian population," *Clinical and Experimental Pharmacology and Physiology*, vol. 34, no. 1-2, pp. 102–105, 2007.

[39] N. A. Limdi, J. A. Goldstein, J. A. Blaisdell, T. M. Beasley, C. A. Rivers, and R. T. Acton, "Influence of CYP2C9 genotype on warfarin dose among African-Americans and European-Americans," *Personalized Medicine*, vol. 4, no. 2, pp. 157–169, 2007.

[40] M. Wadelius, L. Y. Chen, J. D. Lindh et al., "The largest prospective warfarin-treated cohort supports genetic forecasting," *Blood*, vol. 113, no. 4, pp. 784–792, 2009.

[41] G. Ragia, I. Petridis, A. Tavridou, D. Christakidis, and V. G. Manolopoulos, "Presence of CYP2C9*3 allele increases risk for hypoglycemia in Type 2 diabetic patients treated with sulfonylureas," *Pharmacogenomics*, vol. 10, no. 11, pp. 1781–1787, 2009.

12

Development of Pineapple Microsatellite Markers and Germplasm Genetic Diversity Analysis

Suping Feng,[1] **Helin Tong,**[2] **You Chen,**[1] **Jingyi Wang,**[2] **Yeyuan Chen,**[3] **Guangming Sun,**[4] **Junhu He,**[3] **and Yaoting Wu**[1]

[1] *Bioscience and Biotechnology College, Qiongzhou University, Sanya 572200, China*
[2] *Institute of Tropical Bioscience and Biotechnology, Chinese Academy of Tropical Agricultural Science, Haikou 571101, China*
[3] *Institute of Tropical Crop Variety Resources, Chinese Academy of Tropical Agricultural Sciences, Danzhou 571737, China*
[4] *South Subtropical Crops Research Institute, Chinese Academy of Tropical Agricultural Sciences, Zhanjiang 524091, China*

Correspondence should be addressed to Yaoting Wu; wuyaoting@tsinghua.org.cn

Academic Editor: Momiao Xiong

Two methods were used to develop pineapple microsatellite markers. Genomic library-based SSR development: using selectively amplified microsatellite assay, 86 sequences were generated from pineapple genomic library. 91 (96.8%) of the 94 Simple Sequence Repeat (SSR) loci were dinucleotide repeats (39 AC/GT repeats and 52 GA/TC repeats, accounting for 42.9% and 57.1%, resp.), and the other three were mononucleotide repeats. Thirty-six pairs of SSR primers were designed; 24 of them generated clear bands of expected sizes, and 13 of them showed polymorphism. EST-based SSR development: 5659 pineapple EST sequences obtained from NCBI were analyzed; among 1397 nonredundant EST sequences, 843 were found containing 1110 SSR loci (217 of them contained more than one SSR locus). Frequency of SSRs in pineapple EST sequences is 1SSR/3.73 kb, and 44 types were found. Mononucleotide, dinucleotide, and trinucleotide repeats dominate, accounting for 95.6% in total. AG/CT and AGC/GCT were the dominant type of dinucleotide and trinucleotide repeats, accounting for 83.5% and 24.1%, respectively. Thirty pairs of primers were designed for each of randomly selected 30 sequences; 26 of them generated clear and reproducible bands, and 22 of them showed polymorphism. Eighteen pairs of primers obtained by the one or the other of the two methods above that showed polymorphism were selected to carry out germplasm genetic diversity analysis for 48 breeds of pineapple; similarity coefficients of these breeds were between 0.59 and 1.00, and they can be divided into four groups accordingly. Amplification products of five SSR markers were extracted and sequenced, corresponding repeat loci were found and locus mutations are mainly in copy number of repeats and base mutations in the flanking region.

1. Introduction

Pineapple (*Ananas comosus* (L.) Merr.), belonging to Bromeliaceae, ananas, is a perennial evergreen herbaceous fruit tree that produces one of the most famous four tropical fruits beside banana, coconut, and mango. During cultivation and propagation, due to the different naming habits of the propagators and local cultivators, homonym and synonym are very common, nomenclature of pineapple was in chaos, and breeds vary greatly within major groups, which not only hinders rational use of pineapple germplasm resources, but also impedes breeding of better pineapple strains.

Molecular marker technology, such as RFLP, RAPD, and AFLP, has been reported to be used in pineapple germplasm analysis; for example, Duval et al. [1] used RFLP marker in research on germplasm diversity of pineapple. De Fátima Ruas et al. [2] analyzed 18 germplasms of pineapple using RAPD marker and concluded that the cultivated germplasms in this study had a similarity coefficient lower than 0.85. Duval et al. [3] determined pineapple chloroplast DNA polymorphism using RFLP analysis. Kato et al. [4] analyzed intraspecific DNA polymorphism of pineapple using AFLP assay. Popluechai et al. [5] assessed genetic diversity of nine germplasms of pineapple and divided them into three

TABLE 1: Pineapple materials included in the study.

Name	Remarks	Seientifiename	Remarks
Sarawak		OK	
Tainong-6	Yellow Mauritius × Cayenne	Unknown	
Tainong-20		Creanme Pine	
Tainong-17	Cayenne♀ × Rough♂	China Local 2	
Xuli-Tainong		Common Rough	
Japan		Natal Queen	
2000sh 1		Queensland Cayenne	
2000sh 2		Riply Queen	
Tainong-19	Cayenne♀ × Rough♂	MacGregor	
Indonesia Cayenne		Jin	
HB		Maroochy	
Comte de Paris 1		Fresh Premium	
Comte de Paris 2		Perolera	
Boli 1	Cayenne♀ × Queen♂	Alexandria	
Thailand THR		Kallara Local	
ST		Smooth Cayenne 1	
Tainong-18		Pattavia	
China Local 1		Nanglae	
Siyuetian		Smooth cayenne 2	
Hawaii 1		New Puket	
Red Spanish		Phuket	
Hongpi		Smooth Cayenne 3	
Tainong-16	Cayenne♀ × Rough♂	Hwaaii 2	
Boli 2	Cayenne♀ × Queen♂	Tainong-4	Cayenne♀ × Queen♂

Chinese Academy of Tropic Agricultural Science, Danzhou.

groups based on a 0.77 similarity coefficient. Wöhrmann and Weising [6] developed EST-SSR markers to carry out cross-amplification study within the pineapple bromeliad species, genus, and subfamily. Their results have shown that most genetic markers had low polymorphism, especially when the subjects are closely related. The recently developed microsatellite marker attracts a lot of interests and is being widely used due to its comparatively high polymorphism and genome specificity [7].

SSR markers can be detected by PCR amplification using specific primers which can be developed mainly by classical library screening [8], microsatellite enriching [9, 10], 5′-anchoring PCR technology [11], sequence tagged microsatellite profiling (STMP) [12], selectively amplified microsatellite (SAM) [13], and bioinformatics methods [6, 14, 15]. Among these methods, SAM can generate SSR markers generating multilocus SSR fingerprints, which requires only one pair of primers and has high efficiency in developing informative SSRs. In this study, we designed SSR primers using SAM or bioinformatics method. Those highly informative and reproducible SSR primers were used to carry out germplasm diversity analysis for 48 breeds of pineapples, so as to reveal the genetic relationship among them, provide reference for improvement of the current chaotic situation of pineapple nomenclature, and reveal the regularity of mutation of pineapple SSR loci through amplification, extraction, and sequencing of SSR loci.

2. Materials and Methods

2.1. Materials. The Tainong 17 pineapple was used to develop SSR markers; materials for genetic diversity analysis were obtained from Institute of Tropical Crop Variety Resources and South Subtropical Crops Research Institute, Chinese Academy of Tropical Agricultural Sciences (Table 1). DNA was extracted using a modified CTAB method [16]. *E. coli* strain DH5α for transformation was kept by our laboratory.

2.2. Development of Genomic SSR Markers. Genomic library was constructed in reference to the SAM method [13]. *Pst*I (15 U/μL, 0.3 μL), *Mse*I (10 U/μL, 0.5 μL), 10x NEB buffer II (5 μL), and BSA (10 μg/μL, 0.5 μL) were added to 1 μg of genomic DNA. Reaction was allowed at 37°C for 1 h and terminated by incubation at 65°C for 10 min. 5 pmol *Pst*I adaptor and 50 pmol *Mse*I adaptor were then added and incubated at 45°C for 5 min; then T4 DNA ligase (0.5 U), dATP (100 mM, 1.8 μL), and sufficient reaction buffer were added to reach a total volume of 30 μL, and the system was incubated at 16°C for 12 h for ligation. Product of SAM-PCR was separated using denaturing polyacrylamide gel electrophoresis. Based on Hayden and Sharp's [13] work, we increased the number of adaptors, sequences of adaptors, and primers used in this study are shown in Table 2. Target sequences were extracted, cloned, and sequenced, then screened for SSR sequences using the Microsatellite software

TABLE 2: Sequences of the adapters and primers.

Name of primers	Sequences of primers (5'-3')
PstI adapter	Sense strand: CTC GGA AGC CTC AGT CCC AGA CTG CGT ACA TGC A-OH
	Antisense strand: phos-TGT ACG CAG TCT GGG ACT GAG GCT TCC GAG A-OH
MseI adapter	Sense strand: GAG CAA GGC TCT CAC AAG GAC GAC CGA CGA G-OH
	Antisense strand: phos-TAC TCG TCG GTC GTC CTT GTG AGA GCC TTG CT-OH.
MseI suppressed amplification primer	GAG CAA GGC TCT CAC A
PstI suppression amplification primer	CTC GGA AGC CTC AGT C
MseI preamplification primer	GAC GAC CGA CGA GTA AC
PstI preamplification primer	AGA CTG CGT ACA TGC AGG A
PstI SAM primers	Adapter 1: AGA CTG CGT ACA TGC AGG ACC
	Adapter 2: AGA CTG CGT ACA TGC AGG ACG
	Adapter 3: AGA CTG CGT ACA TGC AGG AGC
	Adapter 4: AGA CTG CGT ACA TGC AGG A CT
	Adapter 5: AGA CTG CGT ACA TGC AGG A TC
	Adapter 6: AGA CTG CGT ACA TGC AGG A CA
	Adapter 7: AGA CTG CGT ACA TGC AGG A AC
	Adapter 8: AGA CTG CGT ACA TGC AGG A GT
	Adapter 9: AGA CTG CGT ACA TGC AGG A TG
	Adapter 10: AGA CTG CGT ACA TGC AGG A GA
	Adapter 11: AGA CTG CGT ACA TGC AGG A AG
	Adapter 12: AGA CTG CGT ACA TGC AGG A AT
5'-Anchored SSR primers	PAC: a: KKR YRY YAC ACA CAC ACA C
	b: KKY RYR YCA CAC ACA CAC A
	PCT: a: KKV RVR VCT CTC TCT CTC T
	b: KKR VRV RTC TCT CTC TCT C

K = G/T, R = G/A, Y = T/C, V = G/C/A, H = A/C/T for 5'-anchored SSR primer sequences.

(MISA) (http://pgrc.ipk-gatersleben.de/misa/), the criteria for SSR screening were as follows: mononucleotide must be repeated for 10 or more times, dinucleotide and trinucleotide be repeated for six or more times, and ≥4 nucleotide units be repeated for five or more times. Complicated SSRs that are interrupted by no more than 100 bases were also included. Dinucleotide repeats such as AT/TA, CT/AG were regarded as the same type.

Cluster analysis was carried out using stackPACK v 2.2 program [17]. Primers were designed using RIMER5.0 [18] and the main parameters were GC content, 40%–60%, annealing temperature, 48–60°C; anticipated product length, 100–300 bp. The primers were synthesized by Invitrogen.

20 μL PCR reaction system consists of 10x PCR buffer, MgCl$_2$ (1.5 mM), dNTPs (0.25 mM), forward/backward primers (5 pmol each), DNA template (20 ng), and Taq DNA polymerase (0.15 U); reaction program was predenaturation at 94°C for 3 min, 30 cycles of denaturation-annealing-elongation (94°C for 30 sec, 55°C for 45 sec, and 72°C for 1 min), and a final elongation at 72°C for 7 min. PCR product was separated using 8% nondenaturing polyacrylamide gel electrophoresis and visualized by silver staining.

2.3. Development of EST-SSR Markers. ESTs were obtained from the dbEST database of NCBI (http://www.ncbi.nlm.nih.gov/projects/dbEST) registered before February 2012.

ESTs that had PolyT or polyA (≥5 repeats) within 50 bp downstream of 5'-end or upstream of 3'-end or shorter than 100 bp were excluded using the EST-trimmer software (http://pgrc.ipk-gatersleben.de/misa/download/est_trimmer.pl); for ESTs that were longer than 700 bp, only the first 700 bp at the 5'-end were kept. Then, SSRs were screened using the MISA software. Screening criteria were the same as genome-based development.

Cluster analysis was carried out using stackPACK. Design and synthesis of primers were the same as genome-based development.

2.4. Genetic Diversity Analysis and of Locus Mutation Detection. Eighteen pairs of these newly developed primers that were highly informative and reproducible were selected to carry out genetic diversity analysis for 48 breeds of pineapple. After silver staining, electrophoresis bands were recorded using the Banscan software, for the same migration distance, positive band was recorded as "1," negative band as "0," and failure of amplification as "9." Genetic distance matrix was calculated using NTSYSpc ver 2.1 software (http://www.exetersoftware.com/), evolutionary tree was constructed using the Unweighted Pair Group Method with Arithmetic Mean (UPGMAM) method, primer polymorphism informativeness was calculated using the formula

TABLE 3: Different units of anchored primer and adapter primer.

	PAC	PCT	Total
Adapter 1	2	6	8[b]
Adapter 2	3	5	8[b]
Adapter 3	3	4	7[b]
Adapter 4	1	3	4[b]
Adapter 5	5	6	11[b]
Adapter 6	3	3	6[b]
Adapter 7	3	3	6[b]
Adapter 8	2	4	6[b]
Adapter 9	1	4	5[b]
Adapter 10	4	4	8[b]
Adapter 11	5	3	8[b]
Adapter 12	3	6	9[b]
Total	35[a]	51[a]	86[c]

[a]Number of sequences derived from the units of different *PstI* adapter and two 5′-anchor primers.

[b]Number of sequences derived from the units of twelve *PstI* adapters and different 5′-anchor primers.

[c]Number of sequences derived from the units of twelve *PstI* adapters and two 5′-anchor primers.

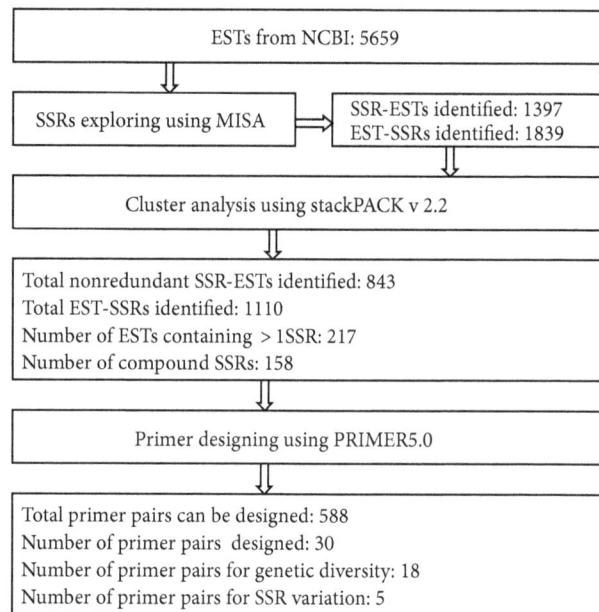

FIGURE 1: Scheme used for data exploring and development of EST-SSRs markers.

PIC $= 1 - \sum (P_i)^2$, wherein P_i stands for the frequency of *i*th locus in all alleles [21].

Repeat types, (CA) *n*, (GCAGGA) *n*, (AG) *n*, (TCGCAG) *n*, and (TCT) *n* primers were used to amplify 10 samples that included bands corresponding to all the previous five repeat types; the bands were then recovered, sequenced, and subjected to SSR locus mutation analysis. The ClustalX software was used to compare original sequence and sequencing results.

3. Results

3.1. Development of Genomic SSR Markers. Products of SAM PCR were separated using denaturing polyacrylamide gel, and 200–750 bp bands were recovered after silver staining. A total of 99 bands were cloned and sequenced, 86 of them contained SSR loci. Numbers of bands obtained by combination of different anchoring primers and adaptor primers were shown in Table 3. Clustering analysis revealed 68 single sequences, and eight groups of repeated sequences; that is, a total of 76 sequences can be used for primer designing. *Pst*I SAM primer in combination with 5′ anchoring primer PAC/PGT developed 44 sequences and *Pst*I SAM primer in combination with 5′ anchoring primer PCT/PGA developed 55 sequences, indicating that CT/AG is more abundant than AC/GT in pineapple genome (Table 3). All sequences were screened for SSR loci using MISA; 52 GA/CT repeat loci were found and 39 AC/GT repeat loci were found, which is in accordance with the result developed by different anchoring primers. Three mononucleotide repeats were found and no tri- or more nucleotide repeat locus was found.

Thirty pairs of primers flanking the SSR locus were designed for each of the 36 SSR-containing DNA sequences;

24 of them generated clear, reproducible bands of expected size, and 13 of them showed polymorphism when amplifying the selected samples.

3.2. Development of EST-SSR Markers. Fifty-six hundred and fifty-nine EST sequences with a total length of 4,141.084 kb were downloaded from NCBI database. MISA was used to analyze these sequences and 1397 EST sequences containing 1839 microsatellite loci were developed (Figure 1). Frequency of SSR-containing sequences among all sequences was 24.68% (one SSR locus every 4.05 ESTs) or one microsatellite locus every 2.25 kb.

Eight hundred and forty-three nonredundant SSR-containing EST sequences were obtained after cluster analysis on the 1397 EST sequences using stackPACK v 2.2. 620 of them were single sequence and 223 were redundant groups. 1110 SSR loci were identified with MISA, and 217 of these sequences contained more than one SSR locus. Of the 1110 SSR loci, 952 were simple SSRs, and 158 were complicated.

Frequencies of nonredundant EST-SSRs in pineapple ESTs were 1SSR/3.73 kb; most of them were small repeating units; taken away mononucleotide repeats, there were 381 (34.3%) dinucleotide repeats, mostly AG/CT accounting for 83.5%, followed by AT/AT accounting for 10.2%, AC/GT accounted for 6.0%, and CG/CG appeared only once; 158 (14.2%) trinucleotide repeats were found, mostly AGC/GCT (24.1%) and AAG/CTT (20.9%); 23 (2.1%) 4-nucleotide repeats were found, mostly AAAG/CTTT, 12 (1.1%) 5-nucleotide repeats were found, 33.3% of them were AAAAG/CTTTT; and 14 (1.3%) AAAAAG/CTTTTT 6-nucleotide repeats were found. In total, 44 types of SSRs were found (Table 4).

Of the 1110 SSR loci identified, taken away the 522 mononucleotide repeats, the other 588 EST-SSRs can be used

TABLE 4: Frequency and distribution of SSRs in the analysed nonredundant 1110 pineapple ESTs.

Repeats motif	Number of repeat units											Total repeats
	5	6	7	8	9	10	11	12	14	15	>16	
A/T	—	—	—	—	—	149	87	67	35	26	100	511
C/G	—	—	—	—	—	5	1	1	2			11
AC/GT	—	8	4	1	1	2	2		1		3	23
AG/CT	—	54	34	40	19	25	19	16	19	14	60	318
AT/AT	—	17	4	9	5	3		1				39
CG/CG	—		1									1
AAC/GTT	—	2	1	1	2		1		1			8
AAG/CTT	—	14	8	7		2	1					33
AAT/ATT	—	3	6	1	2	2						14
ACC/GGT	—	1	2	1		1						5
ACG/CTG	—	6	6	2	2	1	1	1				19
ACT/ATG	—	3	2	1				2				8
AGC/CGT	—	10	6	5	6	2	8					38
AGG/CCT	—	14	8	2								24
AGT/ATC	—		2									2
CCG/CGG	—	3	1		2	1						7
AAAC/GTTT		1										1
AAAG/CTTT	5	2	1									8
AAAT/ATTT	1			1								2
AATC/AGTT	1											1
ACGT/ATGC		1										1
ACTG/ACTG					1							1
AGAT/ATCT	1											1
AGCC/CGGT		1										1
AGCG/CGCT	2											2
AGCT/ATCG	1											1
AGGT/ATCC				4								4
AAAAG/CTTTT	3	1										4
AAAAT/ATTTT	2	1										3
AACAC/GTGTT	1											1
AAGAG/CTCTT	1											1
AATCG/AGCTT	1											1
ACCAT/ATGGT	1											1
AGCCG/CGGCT	1											1
AAAAAG/CTTTTT	3											3
AACCCT/ATTGGG	1											1
AACTAC/ATGTTG	1											1
AAGAGG/CCTTCT		1										1
AAGGAG/CCTCTT	2											2
AAGGCG/CCGCTT	2											2
AATCCC/AGGGTT	1											1
ACGGCG/CCGCTG		1										1
AGCAGG/CCTCGT	1											1
AGCGTC/AGTCGC			1									1

for primer designing. Thirty pairs of primers were designed for each of 30 randomly selected EST-SSRs; 26 of them generated clear, reproducible bands, and 22 of them showed polymorphism.

3.3. Genetic Diversity Analysis and Detection of Mutation Locus. Eighteen pairs of highly informative primers developed by EST-SSR or from genomic library were selected to carry out PCR amplification and genetic diversity analysis for

TABLE 5: Details of the SSR primers for genetic diversity analysis.

Prime Pairs	Sequence ID	SSR motif	Forward primer (5′-3′)	Reverse primer (5′-3′)	Expected product size (bp)
Bp-01	AC 1.3	(CA)8	TCACACACACACACACAAAAAC	ATGGATTGCGCTGAGCTG	119
Bp-08	AC 6.3	(TG)8	ATGATGCCAGTGGAGTGTTC	ACACACACACACTTTTCTCATTG	152
EP-02	DT339694	(GAA)7	CGTGCCGCATAAATCAT	TATCTCCTCGCTCCTCTTG	116
EP-05	DT339172	(CCAT)8···(AT)7	CAGCCAATAACAACCTCAAG	TCCATACACACAGTACGTCG	263
EP-06	DT339094	(CTTTTT)5	CGACTCGAGGATTACATTACG	GAGCACAAAGAACCACACAG	270
EP-09	DT338799	(TC)19	CCGAGGAAGAAGAAGAGGT	GGTCCACAGTTGTTTCAGTT	160
EP-10	DT338783	(GAT)7	GACCTTTATCCATCGCATC	CCATCAAACGTGAAATCTTG	266
EP-11	DT338752	(CAGGAG)5	AGCGAGATAGCAGAGATAGG	TAGAGCGATGTTCGGATG	180
EP-12	DT338506	(AG)6···(AG)6	TTAACACATGCACGGAGTAC	CTAAGAGACAACCCAGGAAG	236
EP-13	DT338494	(CCAT)8···(AT)8	GCCAATAACAACCTCAAGC	TCCATACACACAGTACGTCG	263
EP-15	DT338171	(GCAGTC)7	ACCTACAAGTGGTACGTCG	GGAGCAAGGAGTTATTCAG	242
EP-16	DT338171	(TC)6···(AG)18	TAGTGAGTCAGGAGGAGAATG	CAAATAAACGGAGCGGAT	212
EP-20	DT337383	(TCT)8	TAATCGGGTGGAGTAAGG	GCTCACATAGGCCAATATG	155
EP-23	DT337096	(TC)20	ATGGTGGTTCACTTATCAGC	AGACATTCAAAGCGGAGAG	126
EP-24	DT337054	(CT)10	GCTGCTCTTGCTGCCAT	AAGCCATAGGACCACCAC	166
EP-26	DT336292	(AT)8	GAAGCGCAGGTTCGTAAT	ACAGAAGTAGAGGAAAGCAGC	227
EP-27	DT336032	(TCT)6	ATACTCTGCTGCTGTGAACG	TTGCACTCCTCTTTGCTAAC	155
EP-29	CO731867	(AGC)9	GCGAGCCTGTTAGACTTTGT	ACGATCTCAGCTGGACCTT	213

TABLE 6: Comparisons of different search criteria and software for SSR development.

	References		
	Wöhrmann and Weising [6]	Ong et al. [19]	In this study
Software to assemble ESTs	Geneious 5.0 software	SeqMan software	StackPACK v2.2 program
Software to search SSRs	SciROKO	SynaRex tool	MISA
Search criteria	Mono- ≥ 15, di- ≥ 7, tri-, tetra-, penta-, hexa- ≥ 5	Di- ≥ 8, tri- ≥ 6, tetra- ≥ 5	Mono- ≥ 10, di-, tri- ≥ 6, tetra-, penta-, hexa- ≥ 5
No. of ESTs analysised	5659 Moyle et al. [20]	5931 Moyle et al. [20], Ong et al. [19]	5659 Moyle et al. [20]
No. of SSRs identified	581	416	588
No. of SSR motif types	42	5	44
Frequency of SSRs	1/4.1 KB	Not mentioned	1/3.73 KB
SSRs for primerdesign	537	133	588
No. of di-	240	203	381
No. of tri-	251	213	158
No. of (CG)n	2	0	1

Mono-, di-, tri-, tetra-, penta- and hexa- represent mononucleotide repeats, dinucleotide repeats, trinucleotide repeats, tetranucleotide repeats and hexanucleotide repeats, respectively.

48 breeds of pineapples (Table 5). The results showed that these 48 germplasms of pineapples had similarity coefficients between 0.59 and 1.0. Based on a similarity coefficient of 0.66, they were divided into four groups: Group 1 containing Sarawak, Tainong-6, Tainong-18, Tainong-19, Comte de Paris 1, Comte de Paris 2, Thailand THR, Kallara local, China Local 1, China Local 2, Phuket, Fresh Premium, New Phuket, Boli 1, Boli 2, Natal Queen, Seiyuetian, OK, Tainong-16, Tainong-4, Xuli Tainong, MacGregor, Common Rough,

Alexandria and Ripley Queen; Group 2 containing Tainong-20, 2000sh 1, Indonesia cayenne, Hongpi, Unknown, Hawaii 1, Smooth cayenne 1, Smooth cayenne 2, Creanme pine, Smooth cayenne 3, Pattavia, Nanglae, Japan, HB, Maroochy, Tainong-17, Perolera, Hawaii 2, ST, and Queensland Cayenne; Group 3 containing 2000sh 2 and Jin; and Group 4 containing only Red Spanish (Figure 2).

Five pairs of primers were used for recovery and sequencing, of which four were developed from EST-SSR, and the

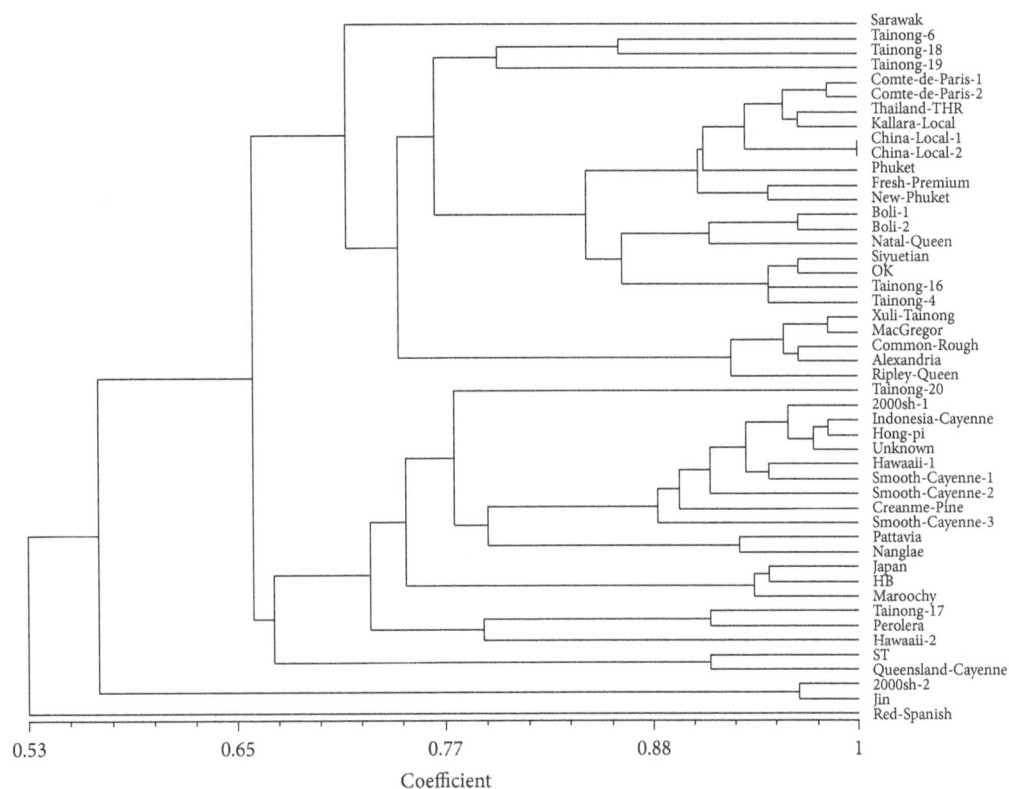

FIGURE 2: Dendrogram of pineapple varieties based on 18 SSRs primer pairs.

other one was developed from genomic library. After PCR amplification and sequencing, these five pairs of primers generated 73 sequences, different SSR markers generated corresponding sequences after amplification and sequencing. Through comparison by ClustalX software, insertions, deletions, transversions, and conversions of these SSR loci and flanking sequences were revealed (Figure 3).

4. Discussion

4.1. Efficiency of SAM Method to Develop SSR Markers. This study used the SAM method invented by Hayden and Sharp [13] to develop positive clones from pineapple genome for sequencing. Ninety nine clones were sequenced and 86 of them contained 94 SSR loci. Thirty-six of these sequences were selected, and 36 pairs of primers flanking the SSR loci were designed, one for each, and 24 of them generated clear and reproducible bands of expected size, and 13 of them showed polymorphism when amplifying selected samples. 86.9% of all sequenced clones were positive, and frequency of SSR marker showing polymorphism was 13.1%, which is lower than [22] results for rubber trees (24.6%) and Wang et al. [23] for banana (19.5%). This may be due to variations between different materials; although SSRs are widely distributed in eukaryotic genomes, their content, type, and copy number vary between different materials. Even within the same species, there would also be variances. Another modification was that primers were designed on repeating sequences of microsatellite, and only a portion of flanking sequence was

used instead of the whole initial 5' anchoring primer. This may have elevated reproducibility of the primers but may lower their polymorphism. Comparatively, the SAM method is much more efficient in developing SSRs than conventional constructing and screening from genomic library of small inserts or STMS method. For example, Ujino et al. [8] acquired only three positive SSR-containing sequences out of 6000 clones (0.05%) using conventional method, and Rajora et al. [24] developed 71 positive clones out of 4028 (1.8%) using STMS method.

4.2. SSR Sequence Analysis. All sequences were screened for repeat loci using MISA, AC/GT repeats accounted for 41.5%, GA/TC repeats accounted for 55.3%, which was in consistence with results acquired by different anchoring primers. Mononucleotide A/T repeats occurred three times (3.2%) no ≥ 3 nucleotide repeats were found. Relatively fewer repeat types were obtained in comparison with Rivera et al. [25], Viruel and Hormaza [26]. Such phenomenon can be explained by the following facts: first, the choice of length and type of the additional 3' bases of preamplification primers reduced SSR productivity at the same time of reducing complexity of the template [13]; second, the parameters set for repeat screening also have certain effects; for example, we have set that 6-nucleotide must repeat five or more times and there was no such loci, but if the parameter was changed to four or more times, there would be one CAAACA/TGTTTG repeat; third, choice of probes may influence frequency of corresponding repeats; for example, Rajora et al. [24] used

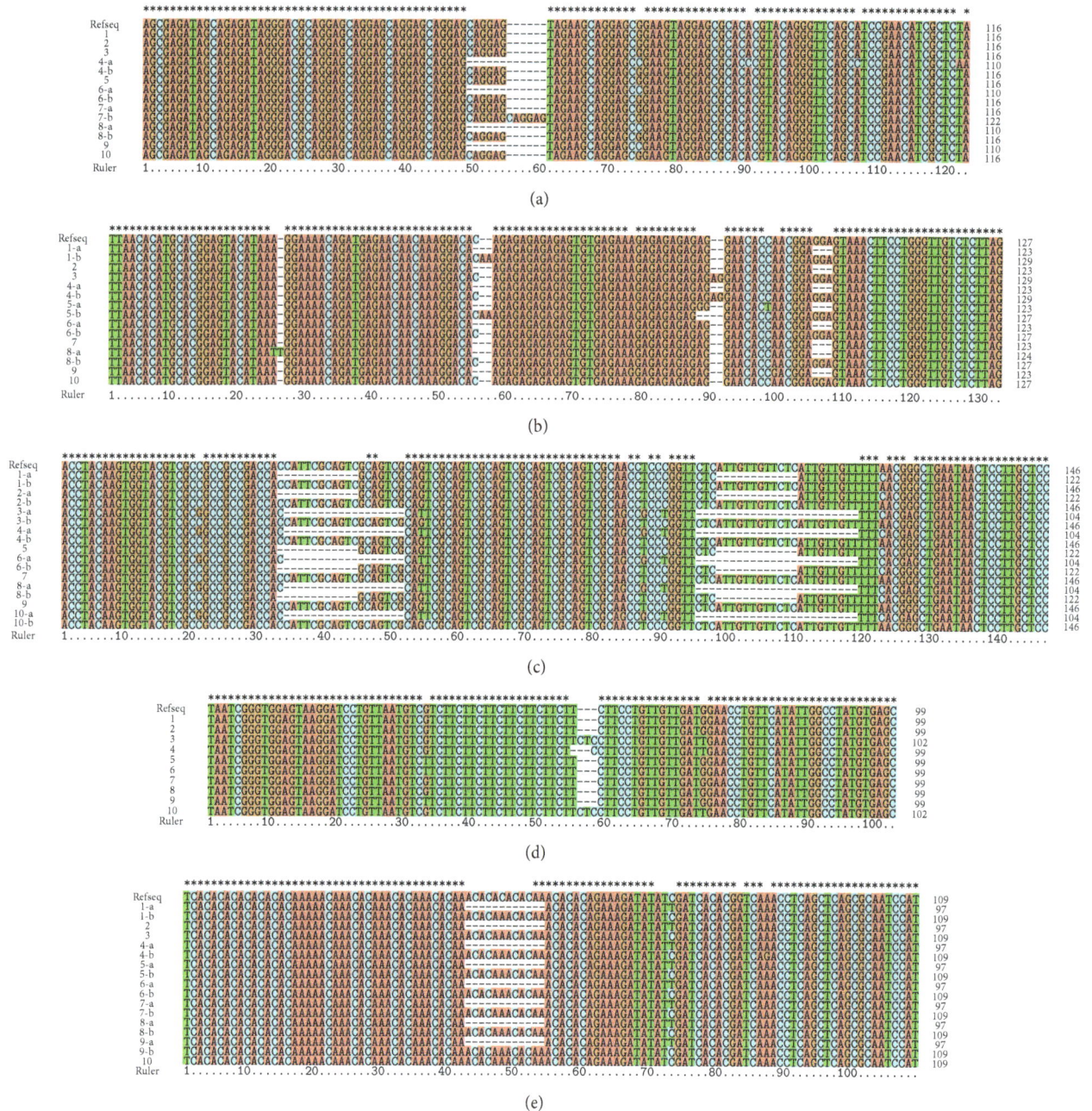

(a)

(b)

(c)

(d)

(e)

FIGURE 3: Sequences obtained using five SSRs markers amplifying across *Ananas comosus*. (a), (b), (c), (d), and (e) represent the marker EP-11, EP-12, EP-15, EP-20, and Bp-01, respectively. RefESTs in (a), (b), (c), and (d) represent the accession numbers: DT338752.1, DT338506.1, DT338171.1, and DT337383.1 in NCBI database, respectively. (e) represents AC1.3. The suffixes "a" and "b" represent the allele numbers. 1–10 represent: 1—Tainong-17, 2—Japan, 3—Comte de Paris, 4—Tainong-18, 5—Red Spanish, 6—MacGregor, 7—Jin, 8—Alexandria, 9—Pattavia; and 10—Phuket.

oligonucleotide sequences corresponding to different SSRs and the resulted SSRs had similar repeat units to the probes.

4.3. Comparison of the Development of Genomic SSR and EST-SSR. EST-SSR marker has unique advantages [27], including being able to detect polymorphism of expressing regions of the genome, high versatility, and relatively low development cost. Thus, it is of great value in genetic mapping, diversity

of genetic resources, discovery, and positioning of functional genes, researches on origin of species, evolution, and genomic comparison [28].

Wöhrmann and Weising [6] screened NCBI database for SSRs, setting the criteria as no less than 15 times for mononucleotide repeats; no less than seven times for dinucleotide repeats, and no less than five times for 3–6 nucleotide repeats. Forty-two types SSRs were revealed from 5659 ESTs; one

SSR occurred every 4.1 kb on average. Trinucleotide repeats was the most common, followed by dinucleotide repeats. Ong et al. [19] also developed SSR markers from 5931 ESTs using SynaRex tool. To ensure comparability between EST-based and genome-based SSR development, different analytic software and the same criteria for SSR screening were used in this study. For EST-based SSR marker development, due to differences in size of database, criteria for SSR screening, and tools for SSR development, the distribution, frequency, and abundance of SSRs also vary (Table 6), as concluded by Varshney et al. [28].

In this study, the rate of polymorphic SSR marker was 13.1% for SAM-developed SSRs, and 2.1% for EST-developed SSRs, showing a higher efficiency of SSR development by SAM method than genomic library-based method which was in consistence with results reported for soybean [29, 30] and rubber trees [22, 31].

4.4. Genetic Diversity Analysis for Pineapple Germplasms Using SSR Markers.

In the present study, 48 germplasms of pineapple were divided into four subgroups, instead of three (Caine, Queen, and Spain) by conventional morphological classification. It can be observed from the cluster analysis, of the 25 germplasms in the first group, Kallara local, Phuket, New Phuket, Natal Queen, MacGregor, Common Rough, Alexandria and Riply Queen, and so forth, belong to the Queen group, and the others such as Tainong-6, Tainong-18, Tainong-19, Boli 1, Boli 2, Tainong-16 and Tainong-4 were the hybridizations of Caine and Queen, and so forth. The germplasms in the second group are morphologically divided into the Caine group, while Indonesia cayenne, Hawaii 1, Smooth cayenne 1, Smooth cayenne 2, Smooth cayenne 3, Pattavia, Nanglae, Hawaii 2, and Queensland Cayenne, and so forth, are hybridizations of Caine and Queen; the fourth group only contains the Red Spanish germplasm of the Spain group.

Some taxonomists regarded Perolera as a new breed, and in this study, it was clustered into the second group, which is closely related to the Caine group. The Sarawak germplasm was thought to belong to the Caine group, but it was actually clustered into the Queen group. This phenomenon may be due to non-unified classification standards for pineapple that leads to different classification; the name for the germplasm's confusion, for exemple, one germplasm has multiple names or a single name used by multiple germplasms because of frequent regional and international exchange of pineapple germplasms; the internal limitation of morphological classification that characteristics of a germplasm is easily affected by environmental conditions. During cultivation and propagation, due to different naming habits of the propagators and local cultivators, homonym and synonym are very common, nomenclature of pineapple was in chaos, and germplasms vary greatly within major groups. In addition, SSR reveals not only genetic variations at the DNA level, but also differences in genotype between germplasms. Genome DNA contains not only structural genes, but also some silence genes which had yet no clear function, and the perceptible phenotype is the results of functional gene expression under influence of both internal and external environment. So, difference in DNA structure may not necessarily lead to differences in morphology.

4.5. Analysis of SSR Mutation.

Mutation of SSR mainly came from base changes in the flanking sequence and repeat region. In this study, we found no insertion or deletion mutations at the EP-11 or EP-20 loci, at the EP-12 locus, Alexandria-a, Tainong-17-b, and Red Spanish-b had "T" and "AA" insertion, respectively; flanking sequence of Bp and EP had deletions (Figure 3).

Flanking sequence of corn had insertion mutations [32]. Gutierrez et al. [33] found in their research on *M. truncatula* that sequence variation was mainly due to variation in copy number of repeats of the SSR region, as well as insertion, deletion, and nucleotide substitution mutations. Symonds and Lloyd [34] pointed out that interruptions in the SSR region shortened the SSR sequence; in this study, it was observed that nucleotide substitution resulted in decrease in copy number of repeats; a single long repeat sequence was divided into several smaller repeat sequences or became shorter. For example, the Golden pineapple had a CAGGAG insertion at the EP-11 "b" locus, increasing repeat number; the "T" of EP-20, Tainong-18 sequence was replaced with "C," leading to decrease in TCT repeats; Red Spanish-b had its "A" replaced with "G" at the EP-12 locus and was thus divided into smaller repeating units.

Conflict of Interests

The authors declare that there is no conflict of interest with MicroSAtellite software (MISA) (http://pgrc.ipk-gatersleben .de/misa/), EST-trimmer software (http://pgrc.ipk-gatersleben.de/misa/download/est_trimmer.pl), and other commercial identities in this paper.

Authors' Contribution

Suping Feng and Helin Tong contributed equally to this research.

Acknowledgment

The present work was supported by Main Programme of National Science Infrastructure Platform (2005DKA21000-5-43).

References

[1] M. F. Duval, J. L. Noyer, X. Perrier, C. D'Eeckenbrugge, and P. Hamon, "Molecular diversity in pineapple assessed by RFLP markers," *Theoretical and Applied Genetics*, vol. 102, no. 1, pp. 83–90, 2001.

[2] C. De Fátima Ruas, P. M. Ruas, and J. R. S. Cabral, "Assessment of genetic relatedness of the genera *Ananas* and *Pseudananas* confirmed by RAPD markers," *Euphytica*, vol. 119, no. 3, pp. 245–252, 2001.

[3] M. F. Duval, G. S. C. Buso, F. R. Ferreira et al., "Relationships in *Ananas* and other related genera using chloroplast DNA restriction site variation," *Genome*, vol. 46, no. 6, pp. 990–1004, 2003.

[4] C. Y. Kato, C. Nagai, P. H. Moore et al., "Intra-specific DNA polymorphism in pineapple (*Ananas comosus* (L.) Merr.) assessed by AFLP markers," *Genetic Resources and Crop Evolution*, vol. 51, no. 8, pp. 815–825, 2005.

[5] S. Popluechai, S. Onto, and P. D. Eungwanichayapant, "Relationships between some Thai cultivars of pineapple (*Ananas comosus*) revealed by RAPD analysis," *Songklanakarin Journal of Science and Technology*, vol. 29, no. 6, pp. 1491–1497, 2007.

[6] T. Wöhrmann and K. Weising, "In silico mining for simple sequence repeat loci in a pineapple expressed sequence tag database and cross-species amplification of EST-SSR markers across Bromeliaceae," *Theoretical and Applied Genetics*, vol. 123, no. 4, pp. 635–647, 2011.

[7] M. J. Kinsuat and S. V. Kumar, "Polymorphic microsatellite and cryptic simple repeat sequence markers in pineapples (*Ananas comosus* var. *comosus*)," *Molecular Ecology Notes*, vol. 7, no. 6, pp. 1032–1035, 2007.

[8] T. Ujino, T. Kawahara, Y. Tsumura, T. Nagamitsu, H. Yoshimaru, and W. Ratnam, "Development and polymorphism of simple sequence repeat DNA markers for *Shorea curtisii* and other Dipterocarpaceae species," *Heredity*, vol. 81, no. 4, pp. 422–428, 1998.

[9] H.-B. Huang, Y.-Q. Song, M. Hsei et al., "Development and characterization of genetic mapping resources for the turkey (*Meleagris gallopavo*)," *Journal of Heredity*, vol. 90, no. 1, pp. 240–242, 1999.

[10] S. N. Nayak, H. Zhu, N. Varghese et al., "Integration of novel SSR and gene-based SNP marker loci in the chickpea genetic map and establishment of new anchor points with Medicago truncatula genome," *Theoretical and Applied Genetics*, vol. 120, no. 7, pp. 1415–1441, 2010.

[11] P. J. Fisher, R. C. Gardner, and T. E. Richardson, "Single locus microsatellites isolated using 5′ anchored PCR," *Nucleic Acids Research*, vol. 24, no. 21, pp. 4369–4371, 1996.

[12] M. J. Hayden and P. J. Sharp, "Sequence-tagged microsatellite profiling (STMP): a rapid technique for developing SSR markers," *Nucleic Acids Research*, vol. 29, no. 8, article e43, 2001.

[13] M. J. Hayden and P. J. Sharp, "Targeted development of informative microsatellite (SSR) markers," *Nucleic Acids Research*, vol. 29, no. 8, article e44, 2001.

[14] L. Ramsay, M. Macaulay, S. Degli Ivanissevich et al., "A simple sequence repeat-based linkage map of Barley," *Genetics*, vol. 156, no. 4, pp. 1997–2005, 2000.

[15] L. F. Gao, J. F. Tang, H. W. Li, and J. Z. Jia, "Analysis of microsatellites in major crops assessed by computational and experimental approaches," *Molecular Breeding*, vol. 12, no. 3, pp. 245–261, 2003.

[16] S. Porebski, L. G. Bailey, and B. R. Baum, "Modification of a CTAB DNA extraction protocol for plants containing high polysaccharide and polyphenol components," *Plant Molecular Biology Reporter*, vol. 15, no. 1, pp. 8–15, 1997.

[17] R. T. Miller, A. G. Christoffels, C. Gopalakrishnan et al., "A comprehensive approach to clustering of expressed human gene sequence: the Sequence Tag Alignment and Consensus Knowledge base," *Genome Research*, vol. 9, no. 11, pp. 1143–1155, 1999.

[18] J. G. Wetmur, "DNA Probes: applications of the principles of nucleic acid hybridization," *Critical Reviews in Biochemistry and Molecular Biology*, vol. 26, pp. 227–259, 1991.

[19] W. D. Ong, C. L. Y. Voo, and S. V. Kumar, "Development of ESTs and data mining of pineapple EST-SSRs," *Molecular Biology Reports*, vol. 39, pp. 5889–5896, 2012.

[20] R. L. Moyle, M. L. Crowe, J. Ripi-Koia, D. J. Fairbairn, and J. R. Botella, "PineappleDB: an online pineapple bioinformatics resource," *BMC Plant Biology*, vol. 5, article 21, 2005.

[21] D. Botstein, R. L. White, M. Skolnick, and R. W. Davis, "Construction of a genetic linkage map in man using restriction fragment length polymorphisms," *American Journal of Human Genetics*, vol. 32, no. 3, pp. 314–331, 1980.

[22] F. Yu, B.-H. Wang, S.-P. Feng, J.-Y. Wang, W.-G. Li, and Y.-T. Wu, "Development, characterization, and cross-species/genera transferability of SSR markers for rubber tree (*Hevea brasiliensis*)," *Plant Cell Reports*, vol. 30, no. 3, pp. 335–344, 2011.

[23] J. Y. Wang, L. S. Zheng, B. Z. Huang, W. L. Liu, and Y. T. Wu, "Development, characterization,and variability analysis of microsatellites from a commerical cultivar of *Musa acuminate*," *Genetic Resources and Crop Evolution*, vol. 57, pp. 553–563, 2010.

[24] O. P. Rajora, M. H. Rahman, S. Dayanandan, and A. Mosseler, "Isolation, characterization, inheritance and linkage of microsatellite DNA markers in white spruce (*Picea glauca*) and their usefulness in other spruce species," *Molecular and General Genetics*, vol. 264, no. 6, pp. 871–882, 2001.

[25] R. Rivera, K. J. Edwards, J. H. A. Barker et al., "Isolation and characterization of polymorphic microsatellites in *Cocos nucifera* L.," *Genome*, vol. 42, no. 4, pp. 668–675, 1999.

[26] M. A. Viruel and J. I. Hormaza, "Development, characterization and variability analysis of microsatellites in lychee (*Litchi chinensis* Sonn., *Sapindaceae*)," *Theoretical and Applied Genetics*, vol. 108, no. 5, pp. 896–902, 2004.

[27] P. K. Gupta, S. Rustgi, S. Sharma, R. Singh, N. Kumar, and H. S. Balyan, "Transferable EST-SSR markers for the study of polymorphism and genetic diversity in bread wheat," *Molecular Genetics and Genomics*, vol. 270, no. 4, pp. 315–323, 2003.

[28] R. K. Varshney, A. Graner, and M. E. Sorrells, "Genic microsatellite markers in plants: features and applications," *Trends in Biotechnology*, vol. 23, no. 1, pp. 48–55, 2005.

[29] J. Rongwen, M. S. Akkaya, A. A. Bhagwat, U. Lavi, and P. B. Cregan, "The use of microsatellite DNA markers for soybean genotype identification," *Theoretical and Applied Genetics*, vol. 90, no. 1, pp. 43–48, 1995.

[30] Q. J. Song, L. F. Marek, R. C. Shoemaker et al., "A new integrated genetic linkage map of the soybean," *Theoretical and Applied Genetics*, vol. 109, no. 1, pp. 122–128, 2004.

[31] S. P. Feng, W. G. Li, H. S. Huang, J. Y. Wang, and Y. T. Wu, "Development, characterization and cross-species/genera transferability of EST-SSR markers for rubber tree (*Hevea brasiliensis*)," *Molecular Breeding*, vol. 23, no. 1, pp. 85–97, 2009.

[32] Y. Matsuoka, S. E. Mitchell, S. Kresovich, M. Goodman, and J. Doebley, "Microsatellites in Zea—variability, patterns of mutations, and use for evolutionary studies," *Theoretical and Applied Genetics*, vol. 104, no. 2-3, pp. 436–450, 2002.

[33] M. V. Gutierrez, M. C. Vaz Patto, T. Huguet, J. I. Cubero, M. T. Moreno, and A. M. Torres, "Cross-species amplification of *Medicago truncatula* microsatellites across three major pulse crops," *Theoretical and Applied Genetics*, vol. 110, no. 7, pp. 1210–1217, 2005.

[34] V. V. Symonds and A. M. Lloyd, "An analysis of microsatellite loci in *Arabidopsis thaliana*: mutational dynamics and application," *Genetics*, vol. 165, no. 3, pp. 1475–1488, 2003.

13

Patterns of Synonymous Codon Usage on Human Metapneumovirus and Its Influencing Factors

Qiao Zhong, Weidong Xu, Yuanjian Wu, and Hongxing Xu

Department of Laboratory Medicine, Suzhou Municipal Hospital Affiliated Nanjing Medical University, 26 Daoqian Street, Jiangsu, Suzhou 215002, China

Correspondence should be addressed to Qiao Zhong, zhongqiao83@yahoo.com.cn

Academic Editor: Sanford I. Bernstein

Human metapneumovirus (HMPV) is an important agent of acute respiratory tract infection in children, while its pathogenicity and molecular evolution are lacking. Herein, we firstly report the synonymous codon usage patterns of HMPV genome. The relative synonymous codon usage (RSCU) values, effective number of codon (ENC) values, nucleotide contents, and correlation analysis were performed among 17 available whole genome of HMPV, including different genotypes. All preferred codons in HMPV are ended with A/U nucleotide and exhibited a great association with its high proportion of these two nucleotides in their genomes. Mutation pressure rather than natural selection is the main influence factor that determines the bias of synonymous codon usage in HMPV. The complementary pattern of codon usage bias between HMPV and human cell was observed, and this phenomenon suggests that host cells might be also act as an important factor to affect the codon usage bias. Moreover, the codon usage biases in each HMPV genotypes are separated into different clades, which suggest that phylogenetic distance might involve in codon usage bias formation as well. These analyses of synonymous codon usage bias in HMPV provide more information for better understanding its evolution and pathogenicity.

1. Introduction

Human metapneumovirus (HMPV) is a negative single-stranded RNA virus of the family Paramyxoviridae and closely related to the avian metapneumovirus (AMPV) subgroup C [1, 2]. HMPV is an important aetiological agent of respiratory tract infection (RTI) in infants, or senior and immunocompromised individuals. This infection caused different symptoms ranging from influenza like syndromes (i.e., fever, cough, and rhinorrhea) to severe lower respiratory tract infection. Previous studies have shown that many children exposed to this virus and also easily to be reinfected as common [3–5]. Therefore, HMPV is becoming as a major concern in child respiratory tract viral infection. However, its pathogenicity is still unclear.

Genome sequencing and comparative analysis provides us a useful approach to analyze the pathogenicity of organisms. Moreover, this analysis might also provide us an approach to understand its evolution history and cell-host interaction. As previously reported HMPV genome is approximately 13 Kb in length, and the gene composition from 3′ terminal to 5′ terminal is N-P-M-F-M2-1/M2-2-SH-G-L [6, 7]. Comparative analysis suggests that its genomic organization is similar to human respiratory syncytial virus (HRSV), which just lacks 2 nonstructural genes, NS1 and NS2. Moreover, HMPV has been demonstrated the existence with two main genetic lineages termed as subtype A and B, which also containing within them the subgroups A1/A2 and B1/B2, respectively [2]. The genetic diversity analysis shows the A2 sublineage exhibits the greatest diversity among all the sublineages of HMPV.

As we all know, there are differences in the frequency of occurrence on synonymous codons in coding DNA, which termed as synonymous codon usage bias. Briefly, there are 64 different codons (61 codons encoding for amino acids plus 3 stop codons) in each organism, but only 20 different translated amino acids. These alternative codons for the same amino acids are termed as synonymous codons. In general, codon usage variation may be the product of natural selection and/or mutation pressure for accurate and efficient translation in various organisms [8–10]. Synonymous codon usage bias on virus can provide us with a better understanding on the evolution profile, gene expression, and virus-host interaction [11–14]. However, there is still lacking about codon usage pattern of HMPV genome and its major influence factors. Herein, we firstly performed the comparative analysis of synonymous codon usage in HMPV genomes and analyzed their influencing factors. This study will provide a new insight to understand the pathogenicity and its evolution history of HMPV.

2. Materials and Methods

2.1. HMPV Genome Sequences. In this study, a total 17 complete HMPV genomes which representing two genotypes were retrieved from NCBI (http://www.ncbi.nlm.nih.gov/) until December, 2011. The serial number (SN), Genbank number, genotype, and other information are listed in Table 1. Moreover, 10 AMPV genomes sequences were retrieved from NCBI database as reference frame (in Supplemental Table 1 supplementary material available online at doi:10.1155/2012/460837).

2.2. Analysis of Codon Usage Pattern. To investigate the characteristics of synonymous codon usage, relative synonymous codon usage (RSCU) values of each complete coding region in 17 HMPV genomes and 10 AMPV genomes were calculated [15]. The RSCU value of each codon for their amino acid was calculated as previously described [16]. A codon with an RSCU value of more than 1.0 has a positive codon usage bias, while a value of less than 1.0 has a negative codon usage bias. When the codon with RSCU values close to 1.0, it means that this codon is chosen equally and randomly. The codon usage data of human cell and bird cell were obtained from the codon usage database online (http://www.kazusa.or.jp/codon/) [17].

The effective number of codons (ENC) is used to measure deviation from expected random codon usage of HMPV and is independent of hypotheses involving natural selection [18]. The ENC values range from 20 to 61. If only one codon is used for each amino acid, this value would be 20, while all of codons are used equally, it will be 61. Moreover, the index of GC3s was used to calculate the fraction of the nucleotides G + C content at the synonymous third codon position (excluding AUG [Met], UGG [Trp], and the termination codons) [19].

2.3. Correspondence Analysis. Multivariate statistical analysis can be used to explore the relationships between variables and samples [20]. In this study, correspondence analysis was used to investigate the major trend in codon usage variation among genomes. In this study, the complete coding region of all 17 HMPV genomes was represented as a 59 dimensional vector, and each dimension corresponds to the RSCU value of one sense codon (excluding Met, Trp, and the termination codons).

Table 1: 17 HMPV genomes sequences used in this study.

SN	Genbank accession	Genotype	Source
1	AB503857.1	A2	Shiga, Japan
2	AF371337.2	A1	Amsterdam, The Netherlands
3	AY297748.1	B2	Quebec, Canada
4	AY297749.1	A2	Quebec, Canada
5	AY525843.1	B1	Amsterdam, The Netherlands
6	DQ843658.1	B1	Beijing, China
7	DQ843659.1	A2	Beijing, China
8	EF535506.1	B2	Taiwan, China
9	FJ168778.1	B2	Amsterdam, The Netherlands
10	FJ168779.1	A2	Amsterdam, The Netherlands
11	GQ153651.1	A2	Guangzhou, China
12	HM197719.1	B2	Rwanda
13	JN184399.1	A1	Nashville, TN, USA
14	JN184400.1	A2	Nashville, TN, USA
15	JN184401.1	B2	Nashville, TN, USA
16	JN184402.1	B2	Nashville, TN, USA
17	NC_004148.2	A2	Quebec, Canada

Abbreviation: SN is serial number.

2.4. Correlation Analysis. Correlation analysis was used to identify the relationship between nucleotide composition and synonymous codon usage pattern [21]. This analysis was implemented based on the Spearman's rank correlation analysis way.

All statistical processes were carried out with statistical software STATA11.5 for windows.

3. Results

3.1. Pattern of Synonymous Codon Usage on HMPV. In order to investigate the synonymous codon usage of HMPV, we calculated various RSCU values of various codons from 17 different strains, including different genotypes. As shown in Table 3, the preferred codons in HMPV are GCA, AGA, AAU, GAU, UGU, CAA, GAA, GGA, CAU, CAU, AUU, UUA, AAA, UUU, CCA, UCA, AGU, ACA, UAU, GUU. Interestingly, all preferred codons in HMPV genomes are ended with A/U, while none of them is ended with G/C. This result suggests that HMPV genome has a great synonymous codon usage bias, and this phenomenon might highly associate with the nucleotide composition in its genomes. Therefore, we analyzed the GC content among HMPV genomes. As shown in Table 2, the G + C content of HMPV genome is 36.91%, which shares similar extent with another RNA virus. There are over 68.57% codons are ended with A/U among HMPV genomes, and 40.87% codons are A3 end, and 27.7% codons are U3 end. This high abundance of A/U nucleotides is consistent with all preferred codons, which are ended with A/U. This phenomenon reflects that nucleotide composition is the main force to affect the codon usage bias in HMPV genome. Moreover, as shown in Table 2, the ENC values among HMPV genomes show a range from 45.127 to 48.28, and its average value of 45.785 and SD value of 0.8458.

TABLE 2: Nucleotide contents and effective number of codons in complete coding region in HMPV genomes.

SN	A%	U%	G%	C%	GC%	A3%	U3%	C3%	G3%	GC3%	*ENC
1	37.51	25.18	19.52	17.8	37.3	41.76	27.7	17.68	11.6	29.28	45.741
2	37.49	25.58	19.55	17.4	36.93	41.67	28.6	16.83	11.52	28.35	45.148
3	37.78	25.27	19.63	17.3	36.95	36.09	21.5	15.18	17.51	32.69	47.608
4	37.4	25.5	19.61	17.5	37.09	41.55	27.9	17.44	11.76	29.2	45.667
5	37.58	25.41	19.76	17.3	37.02	41.32	28.1	16.71	12.47	29.18	45.529
6	37.65	25.64	19.75	17	36.71	41.54	28.5	16.25	12.35	28.6	45.662
7	37.33	25.66	19.59	17.4	37.01	36.36	23.1	16.45	15.7	32.16	48.28
8	37.53	26.27	19.88	16.3	36.2	41.52	29.3	15.5	12.34	27.84	45.127
9	37.61	25.69	19.72	17	36.7	41.48	28.8	16.05	12.36	28.41	45.468
10	37.46	25.58	19.51	17.5	36.96	41.62	28.2	17.26	11.58	28.85	45.707
11	37.22	25.83	19.76	17.2	36.95	41.16	28.9	16.56	12.07	28.63	45.379
12	37.73	25.68	19.7	16.9	36.59	41.55	28.9	15.85	12.42	28.27	45.28
13	37.42	25.72	19.59	17.3	36.86	41.47	28.9	16.61	11.67	28.28	45.274
14	37.4	25.47	19.61	17.5	37.13	41.38	28.1	17.46	11.77	29.24	45.58
15	37.47	25.51	19.87	17.2	37.02	41.16	28.4	16.52	12.64	29.16	45.442
16	37.61	25.43	19.78	17.2	36.96	41.56	28.4	16.32	12.45	28.77	45.778
17	37.4	25.5	19.61	17.5	37.09	41.55	27.9	17.44	11.76	29.2	45.667
Mean	37.51	25.58	19.67	17.2	36.91	40.87	27.7	16.6	12.59	29.18	45.785
SD	0.144	0.241	0.114	0.33	0.25	1.755	2.1	0.713	1.589	1.296	0.8458

Abbreviation: SN is serial number; *ENC is effective number of codons.

The stable ENC values suggest that their genomic compositions are much conserved among HMPV genomes.

3.2. Nucleotide Contents of HMPV Genomes.

Natural selection and mutation pressure have been considered to be two key factors which have effect on codon usage patterns of organisms [22]. In order to investigate whether mutation pressure or natural selection as a determinative factor for codon usage mutation in HMPV, we calculated correlation relation between A%, U%, G%, C%, GC% and A3%, U3%, G3%, C3%, GC3%. As shown in Table 4, there exhibits a very complex correlation map observed in nucleotide compositions. In detail, U3% has a significant positive correlation with U% ($r = 0.7821$, $P < 0.01$), while shared negative correlation with C% ($r = -0.7153$, $P < 0.01$) and GC% ($r = -0.7474$, $P < 0.01$). C3% has positive correlation with C% ($r = 0.8331$, $P < 0.01$) and GC% ($r = 0.7880$, $P < 0.01$), and negative correlation with A% ($r = -0.6199$, $P < 0.01$) and G% ($r = -0.5846$, $P < 0.05$). GC3% has significant positive correlation with C% ($r = 0.7178$, $P < 0.01$) and GC% ($r = 0.8052$, $P < 0.01$), but has negative correlation with U% ($r = -0.7342$, $P < 0.01$). Interestingly, the highest third end nucleotide A3% has no correlation with any nucleotides. Interestingly, the GC3% shows positive correlation with C%, while shows negative correlation with G%. We calculated the GC3 skew by using formula as CG3 skew = (C3 − G3)/(C3 + G3) [22]. The GC3 skew of HMPV range s from 0.371 to −0.592, which reveals that GC composition involved in the codon usage bias. These data suggest that the nucleotide constraint might play an important role in influencing synonymous codon usage bias.

Furthermore, the correlation analysis of first two principle axes (f_1' and f_2') in HMPV and its nucleotide contents were performed (Table 5). Apparently, the first principle axis (f_1') has a significantly negative correlation with U3%, and negative correlation with GC3%. This result suggests that nucleotide U3% and GC3% are the major factor influencing the synonymous codon usage pattern in HMPV genome. Moreover, we observed the second principle axis (f_2') shared a significant positive correlation with G3%, and negative correlation with C3% and GC3%. Therefore, compositional constraint is a major factor which is involved in shaping the pattern of synonymous codon usage bias in HMPV genome.

3.3. Phylogenetic Distant Effect on Synonymous Codon Usage.

To investigate the effect of different HPMV genotypes on synonymous codon usage, we analyzed the codon usage bias of different genotypes with correspondence analysis. From the correspondence analysis, the first dimension variable f_1' and the second dimension variable f_2' can reflect 43.27% and 33.38% of total mutation, respectively. As the plot of correspondence analysis shown (Figure 1), each genotype is mainly separated and clustered into two clades. This phenomenon implied that phylogenetic relationship has some extent effect on codon usage bias. However, the sublineage of each genotype did not exhibit any significant difference among them. But this might be due to the limited number of HPMV genomes available in current study. Therefore, the phylogenetic distant might effect on the variation of synonymous codon usage in HMPV, and this difference might reflect on their biological effect, such as viral replication, virulence, and so forth.

TABLE 3: Synonymous codon usages of complete coding region of HMPV, human cell, AMPV, and bird cell.

AA	Codon	[a]RSCU	[b]RSCU	[c]RSCU	[d]RSCU
Ala	GCU	1.36	1.08	1.02	**1.42**
	GCC	0.55	**1.6**	0.77	1.33
	GCA	**1.89**	0.92	**1.99**	0.98
	GCG	0.2	0.44	0.22	0.27
Arg	CGU	0.12	0.48	0.15	0.00
	CGC	0.13	1.08	0.14	0.55
	CGA	0.31	0.66	0.34	0.27
	CGG	0.23	1.2	0.41	1.09
	AGA	**3.58**	**1.26**	**2.86**	2.18
	AGG	1.63	**1.26**	2.10	1.91
Asn	AAU	**1.13**	0.94	**1.08**	1.41
	AAC	0.87	**1.06**	0.92	0.59
Asp	GAU	**1.26**	0.92	**1.12**	0.88
	GAC	0.74	**1.08**	0.89	1.12
Cys	UGU	**1.04**	0.92	**1.11**	0.67
	UGC	0.96	**1.08**	0.89	1.33
Gln	CAA	**1.27**	0.54	**1.19**	0.48
	CAG	0.73	**1.46**	0.81	1.52
Glu	GAA	**1.26**	0.84	**1.11**	0.75
	GAG	0.74	**1.16**	0.89	1.25
Gly	GGU	1.08	0.64	0.89	0.43
	GGC	0.73	**1.36**	0.68	1.19
	GGA	**1.49**	1	**1.43**	1.08
	GGG	0.71	1	1.01	1.30
His	CAU	**1.13**	0.84	1.00	0.35
	CAC	0.87	**1.16**	1.00	1.65
Ile	AUU	**1.2**	1.08	1.02	1.78
	AUC	1	**1.41**	0.90	1.03
	AUA	0.8	0.51	**1.08**	0.19
Leu	UUA	**1.44**	0.48	1.19	0.42
	UUG	1.08	0.78	**1.36**	0.85
	CUU	1.14	0.78	0.83	0.56
	CUC	0.64	1.2	0.57	0.99
	CUA	0.85	0.42	0.92	0.28
	CUG	0.84	**2.4**	1.13	2.89
Lys	AAA	**1.21**	0.86	**1.12**	0.93
	AAG	0.79	**1.14**	0.88	1.07
Phe	UUU	**1.32**	0.92	**1.27**	1.33
	UUC	0.68	**1.08**	0.73	0.67
Pro	CCU	1.15	1.16	1.12	1.22
	CCC	0.82	**1.28**	0.74	1.22
	CCA	**1.8**	1.12	**1.78**	1.47
	CCG	0.23	0.44	0.36	0.08
Ser	UCU	0.97	1.14	0.93	0.71
	UCC	0.63	1.32	0.72	1.15
	UCA	**1.46**	0.9	**1.37**	1.06
	UCG	0.13	0.3	0.18	0.26
	AGU	**1.46**	0.9	**1.41**	0.44
	AGC	1.36	**1.44**	1.30	**2.38**

TABLE 3: Continued.

AA	Codon	[a]RSCU	[b]RSCU	[c]RSCU	[d]RSCU
Thr	ACU	1.17	1	0.98	0.16
	ACC	0.92	**1.44**	0.83	1.28
	ACA	**1.75**	1.12	**1.91**	**1.76**
	ACG	0.16	0.44	0.27	0.80
Tyr	UAU	**1.34**	0.88	**1.14**	0.95
	UAC	0.66	**1.12**	0.86	**1.05**
Val	GUU	**1.46**	0.72	1.17	0.57
	GUC	0.73	0.96	0.72	0.91
	GUA	0.93	0.48	0.87	0.80
	GUG	0.88	**1.84**	**1.24**	**1.71**

The number in this table is shown as relative synonymous codon usage values (RSCU) and [a]RSCU values represent values in HMPV, [b]RSCU values represent values in human host cell, [c]RSCU values represent values in AMPV, and [d]RSCU values represent values in bird cell. All preferred codons are shown in bold. AA is abbreviation of amino acid.

TABLE 4: Summary of correlation analysis between A%, U%, C%, G%, and GC% and A3, U3, C3, G3, and GC3 in all HMPV samples.

	A3%	U3%	C3%	G3%	GC3%
A%	0.0959[NS]	0.1144[NS]	−0.6199**	0.3788[NS]	−0.2595[NS]
U%	−0.1596[NS]	0.7821**	−0.4150[NS]	−0.0933[NS]	−0.7342**
C%	0.2417[NS]	−0.7153**	0.8331**	−0.4699[NS]	0.7178**
G%	−0.4012[NS]	0.4234[NS]	−0.5846*	0.6129**	−0.3138[NS]
GC%	0.0774[NS]	−0.7474**	0.7880**	−0.1328[NS]	0.8052**

The number in the table represents as correlation coefficient r value, which is calculated from each correlation analysis. Abbreviation: NS represents as nonsignificant ($P > 0.05$), *represents $P < 0.05$, **represents $P < 0.01$.

TABLE 5: Summary of correlation between the first two principle axes and nucleotide contents in samples.

Base compositions	f_1' (43.27%)	f_2' (33.38%)
A3%	$r = -0.2247^{NS}$	$r = -0.2491^{NS}$
U3%	$r = -0.8091^{**}$	$r = 0.4454^{NS}$
C3%	$r = 0.3106^{NS}$	$r = -0.8233^{**}$
G3%	$r = 0.1449^{NS}$	$r = 0.6393^{**}$
GC3%	$r = 0.8447^{**}$	$r = -0.5043^{*}$

The number in the table represents as correlation coefficient r value, which is calculated from each correlation analysis. Abbreviation: NS represents as nonsignificant ($P > 0.05$), *represents $P < 0.05$, **represents $P < 0.01$.

3.4. Relationship between Codon Usage Pattern of HMPV and Its Host. From the ENC-GC3% plot analysis (Figure 2), the plots of each HMPV genomes are all under the expected curve, none of them shows above the curve. This result implied that mutation pressure is the major factor influencing the codon usage [19]. Moreover, there are still some other factors that can effect on the codon usage bias of HMPV. In the current study, we compared the patterns of codon usage in HMPV and human host. As shown in Table 3, the pattern of synonymous codon usage in HMPV shows a complementary profile, which shows in human cell. In detail,

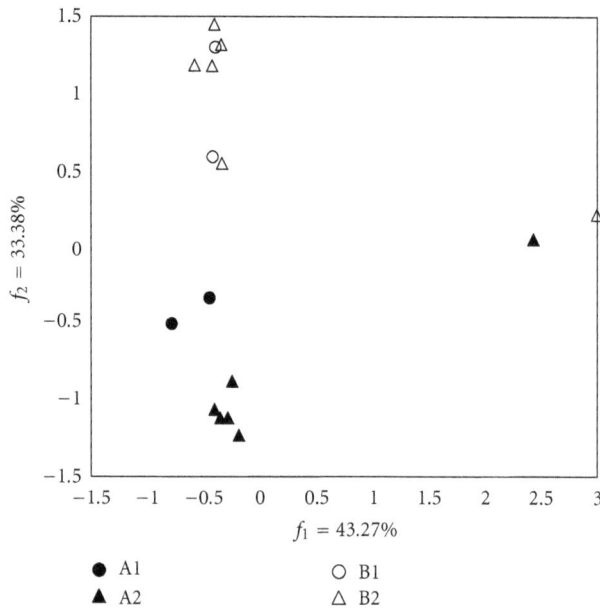

FIGURE 1: The plot of value of the first and second axis of each complete coding region in COA. The first axis (f_1') accounts for 43.27% of the total variation, and the second axis (f_2') accounts for 33.38% of the total variation. Each HMPV complete coding region was divided by genotype.

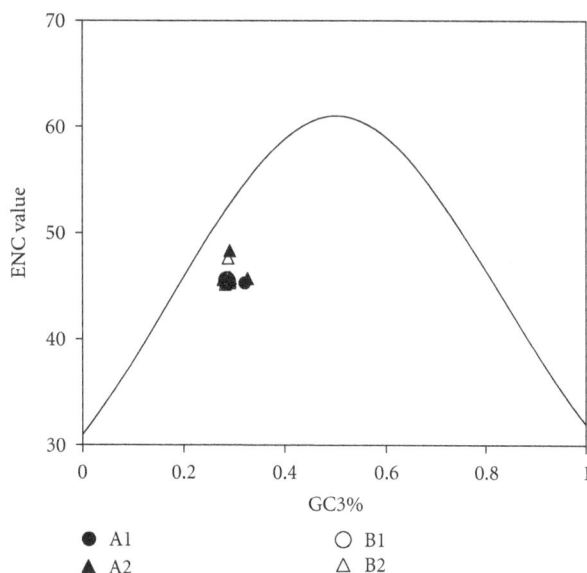

FIGURE 2: Effective number of codons used in each complete coding region plotted against the GC3. The continuous curve plots the relationship between GC3 and ENC in the absence of selection. All of plots lie below the expected curve.

HMPV and human host cell shared only 1 preferred codon (AGA), which encoded for Arginine, while there are 17 different preferred codons between them. As a reference frame, AMPVs were enrolled in this study, and it also shows a complementary pattern with its host, bird cell. The comparative analysis among HMPV and human host cell, HMPV

and AMPV host bird cell, and HMPV and AMPV were analyzed. To compare the complementary ability of HMPV with bird cell, there are more overlays (6 preferred codons overlay) than HMPV with its human host cell (only 1 pre-ferred codons overlay). This result shows human cell has much more complementary pattern with HMPV than bird cells. This might be more suitable for HMPV survive and persist infect in human host environment. This result also suggests that host factor plays an important role in codon usage bias in HMPV. This complementary trend will benefit for virus replication instead of competitive with its host and it might help us to understand the mechanism of HMPV persistent infection. Interestingly, HMPVs are shares with more than 15 preferred codons to AMPV, which might be due to a close phylogenetic distance.

4. Discussion

Synonymous codon usage analysis can reveal much about virus genome. To understand the extent and causes of codon usage bias is essential for studying the viral evolution, particularly the interaction between viruses and host immune response. In this study, we analyzed the codon usage bias in HMPV, and its influencing factors. As we know, the variation and evolution of virus generally happened in the changes of nucleotide composition [23]. Therefore, the nucleotide composition bias is the main force to influence the synonymous codon usage patterns. In this study, several evidences can support this statement in HMPV genomes. First of all, in HMPV, all preferred codons are ended in A/U nucleotide, which occupied the majority of nucleotide composition in HMPV genome. This phenomenon confirmed that nucleotide composition was the main force in shaping the pattern of codon usage. Secondly, ENC was used to quantify the codon usage bias, which is one of the best overall estimators of absolute synonymous codon usage bias [18]. In this study, we observed ENC of these genomes fluctuated from 45.13 to 48.28 with a mean 45.78±0.85. This ENC value of HMPV is consistent with other previously reported RNA virus in the same family Paramyxoviridae, ranging from 43.8 to 55.1, that is, Measles virus 55.1, Mumps virus 54.3, Para-influenza-3 virus 43.8, and Respiratory syncytial virus 44.3 [8, 24, 25]. Moreover, the ENC of HMPV is more close to respiratory infection agents, RSV and parainfluenza-3, which reveals that the similar extent of codon usage bias among viruses might have similar infection syndrome. This observation helps us to address an interesting assumption that synonymous codon usage bias of virus might associate with its pathogenicity.

Mutation pressure and natural selection are generally treated as the main factors that account for codon usage bias in different organisms [22]. ENC-GC plot was considered as a part of the general strategy to investigate patterns of synonymous codon usage [9, 18, 26]. Herein, all the plots are laid below the expected curve, suggesting that codon usage bias in all these 17 HMPV genomes was principally influenced by mutation bias, which consistent with that mutation pressure rather than natural selection is the most important determinant of the codon usage in human RNA virus [8, 27–31]. This observation can be explained as the mutation rates in RNA viruses much higher than those in DNA viruses.

Mutation pressure is the main force in shaping synonymous codon usage bias of RNA virus. However, based on correspondence analysis, we observed an interesting phenomenon that codon usage bias in HMPV showed distinct differences among different phylogenetic types. It might suggest that codon usage bias plays an important role in HMPV evolution history. This similar phenomenon also observed in several other viruses, this might reflect that phylogenetic difference is a common influencing factor in shaping codon usage bias [25, 27, 30–34]. This difference could potentially affect the viral protein expression rate or its replication manner. Therefore, we hypothesize that the difference of codon usage bias might influence its virulence in different genotypes.

In this study, we also observed that HMPV showed a complementary trend with human cells by comparing the codon usage. This complementary will be benefit for the survive of virus, which can keep replication by using the nonpreferred codons in the host cell without competition, and this could be one of the mechanisms of virus persistent infection in the human environment. Moreover, this pattern might also be caused by the longitude selection and evolution between human hosts with virus. Therefore, this characteristic is important for HMPV keeping balanced with their host on the codon usage side, and also for understanding the cell-host interaction and viral evolution.

5. Conclusions

In summary, we firstly reported the synonymous codon usage pattern in HMPV genomes and revealed that mutation pressure is the main force in shaping its codon usage bias. Phylogenetic difference and host factors are also discussed, and this information can provide better understanding on the molecular evolution and its pathogenicity of HMPV.

Acknowledgment

This work was supported by the Grant from Nature & Science project in Suzhou (SYSD2010136).

References

[1] S. Broor, P. Bharaj, and H. S. Chahar, "Human metapneumovirus: a new respiratory pathogen," *Journal of Biosciences*, vol. 33, no. 4, pp. 483–493, 2008.

[2] F. Feuillet, B. Lina, M. Rosa-Calatrava, and G. Boivin, "Ten years of human metapneumovirus research," *Journal of Clinical Virology*, vol. 53, no. 2, pp. 97–105, 2012.

[3] Y. Abed and G. Boivin, "Human metapneumovirus infection in immunocompromised child," *Emerging Infectious Diseases*, vol. 14, no. 5, pp. 854–856, 2008.

[4] M. Arabpour, A. Samarbafzadeh, M. Makvandi et al., "The highest prevalence of human metapneumovirus in Ahwaz children accompanied by acute respiratory infections," *Indian Journal of Medical Microbiology*, vol. 26, no. 2, pp. 123–126, 2008.

[5] M. J. Carr, A. Waters, F. Fenwick, G. L. Toms, W. W. Hall, and E. O'Kelly, "Molecular epidemiology of human metapneumovirus in Ireland," *Journal of Medical Virology*, vol. 80, no. 3, pp. 510–516, 2008.

[6] R. Piyaratna, S. J. Tollefson, and J. V. Williams, "Genomic analysis of four human metapneumovirus prototypes," *Virus Research*, vol. 160, no. 1-2, pp. 200–205, 2011.

[7] B. G. van den Hoogen, T. M. Bestebroer, A. D. M. E. Osterhaus, and R. A. M. Fouchier, "Analysis of the genomic sequence of a human metapneumovirus," *Virology*, vol. 295, no. 1, pp. 119–132, 2002.

[8] G. M. Jenkins and E. C. Holmes, "The extent of codon usage bias in human RNA viruses and its evolutionary origin," *Virus Research*, vol. 92, no. 1, pp. 1–7, 2003.

[9] G. A. Palidwor, T. J. Perkins, and X. Xia, "A general model of Codon bias due to GC mutational bias," *PLoS ONE*, vol. 5, no. 10, Article ID e13431, 2010.

[10] P. M. Sharp, L. R. Emery, and K. Zeng, "Forces that influence the evolution of codon bias," *Philosophical Transactions of the Royal Society B*, vol. 365, no. 1544, pp. 1203–1212, 2010.

[11] T. Warnecke and L. D. Hurst, "Evidence for a trade-off between translational efficiency and splicing regulation in determining synonymous codon usage in *Drosophila melanogaster*," *Molecular Biology and Evolution*, vol. 24, no. 12, pp. 2755–2762, 2007.

[12] N. Stoletzki and A. Eyre-Walker, "Synonymous codon usage in *Escherichia coli*: selection for translational accuracy," *Molecular Biology and Evolution*, vol. 24, no. 2, pp. 374–381, 2007.

[13] A. O. Urrutia and L. D. Hurst, "Codon usage bias covaries with expression breadth and the rate of synonymous evolution in humans, but this is not evidence for selection," *Genetics*, vol. 159, no. 3, pp. 1191–1199, 2001.

[14] Y. Kim, "Effect of strong directional selection on weakly selected mutations at linked sites: implication for synonymous codon usage," *Molecular Biology and Evolution*, vol. 21, no. 2, pp. 286–294, 2004.

[15] S. G. E. Andersson and P. M. Sharp, "Codon usage and base composition in *Rickettsia prowazekii*," *Journal of Molecular Evolution*, vol. 42, no. 5, pp. 525–536, 1996.

[16] M. Bulmer, "The selection-mutation-drift theory of synonymous codon usage," *Genetics*, vol. 129, no. 3, pp. 897–907, 1991.

[17] Y. Nakamura, T. Gojobori, and T. Ikemura, "Codon usage tabulated from international DNA sequence databases: status for the year 2000," *Nucleic Acids Research*, vol. 28, no. 1, article 292, 2000.

[18] A. Fuglsang, "The effective number of codons for individual amino acids: some codons are more optimal than others," *Gene*, vol. 320, no. 1-2, pp. 185–190, 2003.

[19] E. Elhaik, G. Landan, and D. Graur, "Can GC content at third-codon positions be used as a proxy for isochore composition?" *Molecular Biology and Evolution*, vol. 26, no. 8, pp. 1829–1833, 2009.

[20] M. V. Boland and R. F. Murphy, "Multivariate analysis," in *Current Protocols in Cytometry*, chapter 10, p. unit 108, 2001.

[21] C. B. Bagwell, "Data analysis through modeling," in *Current Protocols in Cytometry*, chapter 10, p. unit 107, 2001.

[22] T. V. Tatarinova, N. N. Alexandrov, J. B. Bouck, and K. A. Feldmann, "GC3 biology in corn, rice, sorghum and other grasses," *BMC Genomics*, vol. 11, no. 1, article 308, 2010.

[23] G. Sablok, K. C. Nayak, F. Vazquez, and T. V. Tatarinova, "Synonymous codon usage, GC3, and evolutionary patterns across plastomes of three pooid model species: emerging grass

genome models for monocots," *Molecular Biotechnology*, vol. 49, no. 2, pp. 116–128, 2011.

[24] T. Zhou, W. Gu, J. Ma, X. Sun, and Z. Lu, "Analysis of synonymous codon usage in H5N1 virus and other influenza A viruses," *BioSystems*, vol. 81, no. 1, pp. 77–86, 2005.

[25] W. Q. Liu, J. Zhang, Y. Q. Zhang et al., "Compare the differences of synonymous codon usage between the two species within cardiovirus," *Virology Journal*, vol. 8, article 325, 2011.

[26] J. M. Comeron and M. Aguadé, "An evaluation of measures of synonymous codon usage bias," *Journal of Molecular Evolution*, vol. 47, no. 3, pp. 268–274, 1998.

[27] X. S. Liu, Y. G. Zhang, Y. Z. Fang, and Y. L. Wang, "Patterns and influencing factor of synonymous codon usage in porcine circovirus," *Virology Journal*, vol. 9, article 68, 2012.

[28] M. R. Ma, X. Q. Ha, H. Ling et al., "The characteristics of the synonymous codon usage in hepatitis B virus and the effects of host on the virus in codon usage pattern," *Virology Journal*, vol. 8, Article ID Article number544, 2011.

[29] Y. S. Liu, J. H. Zhou, H. T. Chen et al., "The characteristics of the synonymous codon usage in enterovirus 71 virus and the effects of host on the virus in codon usage pattern," *Infection, Genetics and Evolution*, vol. 11, no. 5, pp. 1168–1173, 2011.

[30] J. H. Zhou, J. Zhang, H. T. Chen, L. N. Ma, and Y. S. Liu, "Analysis of synonymous codon usage in foot-and-mouth disease virus," *Veterinary Research Communications*, vol. 34, no. 4, pp. 393–404, 2010.

[31] P. Tao, L. Dai, M. Luo, F. Tang, P. Tien, and Z. Pan, "Analysis of synonymous codon usage in classical swine fever virus," *Virus Genes*, vol. 38, no. 1, pp. 104–112, 2009.

[32] Y. S. Liu, J. H. Zhou, H. T. Chen et al., "Analysis of synonymous codon usage in porcine reproductive and respiratory syndrome virus," *Infection, Genetics and Evolution*, vol. 10, no. 6, pp. 797–803, 2010.

[33] S. Zhao, Q. Zhang, X. Liu et al., "Analysis of synonymous codon usage in 11 Human Bocavirus isolates," *BioSystems*, vol. 92, no. 3, pp. 207–214, 2008.

[34] P. L. Meintjes and A. G. Rodrigo, "Evolution of relative synonymous codon usage in human immunodeficiency virus type-1," *Journal of Bioinformatics and Computational Biology*, vol. 3, no. 1, pp. 157–168, 2005.

14

Artificial Chromosomes to Explore and to Exploit Biosynthetic Capabilities of Actinomycetes

Rosa Alduina and Giuseppe Gallo

Department of Science and Molecular and Biomolecular Technology, University of Palermo, Viale delle Scienze, 90128 Palermo, Italy

Correspondence should be addressed to Giuseppe Gallo, valeria.alduina@unipa.it

Academic Editor: Jozef Anné

Actinomycetes are an important source of biologically active compounds, like antibiotics, antitumor agents, and immunosuppressors. Genome sequencing is revealing that this class of microorganisms has larger genomes relative to other bacteria and uses a considerable fraction of its coding capacity (5–10%) for the production of mostly cryptic secondary metabolites. To access actinomycetes biosynthetic capabilities or to improve the pharmacokinetic properties and production yields of these chemically complex compounds, genetic manipulation of the producer strains can be performed. Heterologous expression in amenable hosts can be useful to exploit and to explore the genetic potential of actinomycetes and not cultivable but interesting bacteria. Artificial chromosomes that can be stably integrated into the *Streptomyces* genome were constructed and demonstrated to be effective for transferring entire biosynthetic gene clusters from intractable actinomycetes into more suitable hosts. In this paper, the construction of several shuttle *Escherichia coli-Streptomyces* artificial chromosomes is discussed together with old and new strategies applied to improve heterologous production of secondary metabolites.

1. Introduction

Actinomycetes, Gram-positive bacteria, represent an important source of biologically active compounds. In fact, they synthesize one-third of the antibiotics commercially available—that is, the macrolide erythromycin, the glycopeptide vancomycin, and the cyclic lipopeptide daptomycin—and other drugs like neuroprotectants (e.g., meridamycin), and anticancer compounds (e.g., migrastatin or geldanamycin) [1]. Most natural products possess complex structures (Figure 1) that are very difficult and expensive to be chemically synthesized and they are obtained at an industrial scale mainly by fermentation processes from the producer microorganism. However, industrial production requires microbial strains that produce unnaturally high yields of a secondary metabolite. In many cases, industrial fermentation yields more than 10 g/L of product; that is, penicillin production process was steadily improved over the years to allow titres of 70 g/L [2]. To increase production yields, classical strain improvement (CSI) or rational genetic methods are used. CSI consists of sequential random mutagenesis and screening of the mutant strains: after each cycle of CSI a population of improved mutants is identified from which the single best performer is selected. CSI does not give information on genes and molecular mechanisms during the improvement and it is likely that some mutations may be detrimental or neutral in term of production.

Rational methods consist of deleting or overexpressing specific genes that, finally, control biosynthesis of a secondary metabolite [3], relying on the knowledge of gene sequences and functions and on suitable protocols for the genetic manipulation of natural producers. The biosynthetic genes devoted to secondary metabolite biosynthesis are organised in clusters that can occasionally reach 100 kb in size or more [4, 5]. Since suitable protocols for the genetic manipulation of natural producers are not always available and given the size of most gene clusters, artificial chromosome-based vectors that can be maintained in *Streptomyces* have been constructed. Such vectors allow genetic information for secondary metabolite production to be transferred from the

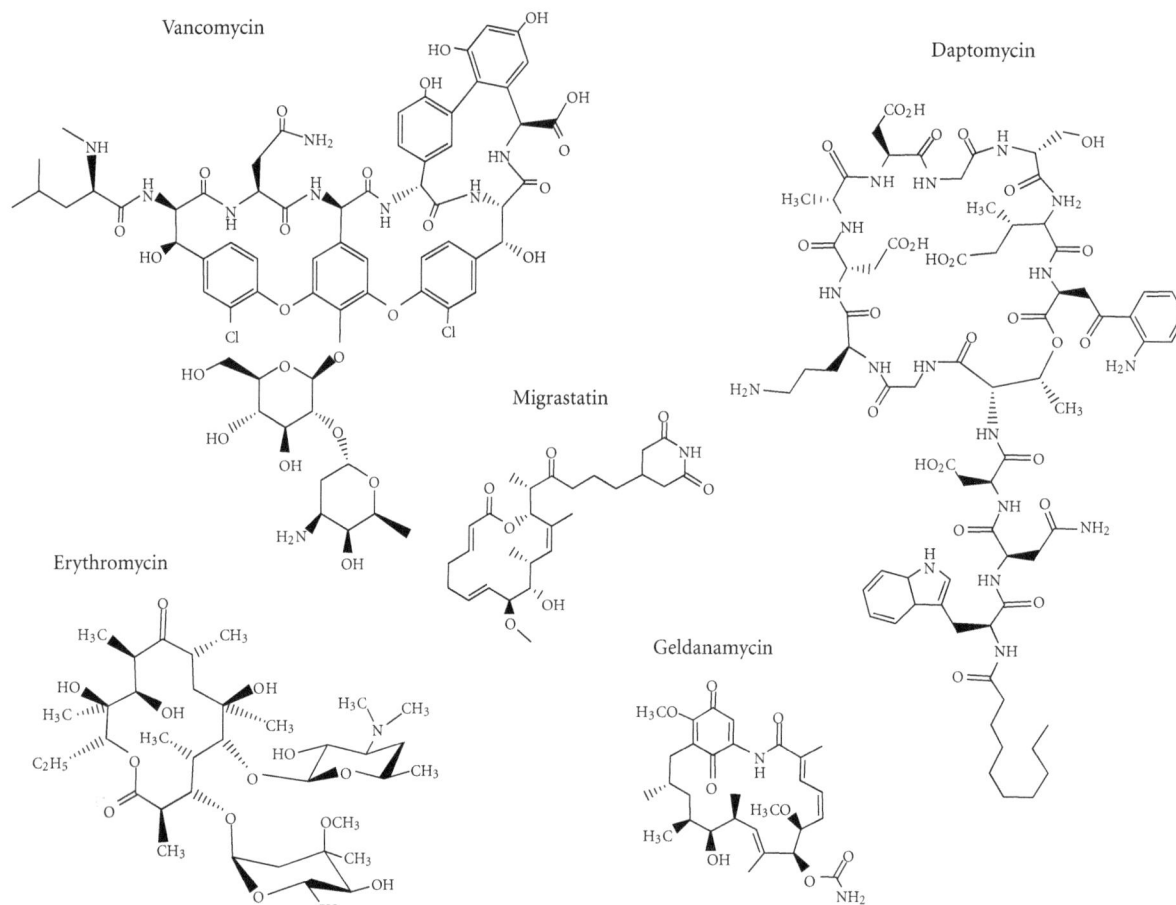

FIGURE 1: Examples of natural compounds produced by actinomycetes. Vancomycin and daptomycin are synthesized by NRPSs from *Amycolatopsis orientalis* and *Streptomyces roseosporus*, respectively; erythromycin, migrastatin, and meridamycin are produced by PKSs in *Saccharopolyspora erythraea, Streptomyces platensis,* and *Streptomyces* sp. *NRRL 30748.*

original producer to a host with well-defined genetics and physiology, where the genes can be expressed. Heterologous expression of large biosynthetic pathways can also be necessary in all cases in which producing bacteria are not cultivable or to examine new metabolites produced by cryptic gene clusters, usually revealed by genome sequencing and mining. This paper deals with artificial chromosome-based vectors and related methods currently available to analyze biosynthetic genes and to express biosynthetic pathways in heterologous systems.

2. Handling Large Actinomycete DNA Fragments: Construction of *E. coli-Streptomyces* Artificial Chromosomes

The first step for studying and analyzing biosynthesis at the genetic level is to obtain gene sequences. The recent genomic technologies allow to sequence entire genomes and thereby to identify biosynthetic gene clusters [6]. However, the majority of natural-product-producing microorganisms are refractory to standard gene manipulation techniques, and some are not cultivable outside their ecological context

[7, 8]. For these reasons, the study and modification of whole pathways has greatly benefited from the possibility to manipulate them in vectors, such as cosmids, and artificial chromosomes. Cosmids show high transformation efficiency, with a loading capacity up to 45 kb. Many cosmid libraries of actinomycetes have been successfully constructed to clone and identify biosynthetic gene clusters [9–14]. However, many biosynthetic gene clusters for natural products are larger than the average insert size of common cosmid vectors. To overcome this size limitation, one strategy can be the subcloning of subsets of the biosynthetic gene cluster into compatible expression plasmids followed by their introduction and coexpression in a suitable host strain [15, 16]. Another option is the reassembly of large natural product pathways on a single transferable vector by using the recombineering (recombination-mediated genetic engineering) technology that employs homologous recombination mediated by phage-based recombination systems. Recombineering is independent of the *E. coli* endogenous homologous recombination functions, of restriction sites and of DNA size. Recombineering using the *E. coli* λ phage Red pathway, which is mediated by its Exo and Beta proteins, was successfully applied to reconstitute phenalinolactone and

anthramycin gene clusters from two independent cosmids [17, 18] in E. coli.

For a single step cloning of inserts larger than 100 kb a more straightforward strategy uses bacterial artificial chromosome (BAC), based on the E. coli F factor, and the P1-derived artificial chromosome (PAC), which can carry large DNA fragments (about 100–300 kb) in E. coli cells [19]. E. coli is not suitable to heterologously express actinomycete genes which are G+C rich [20]. Thus, vectors that can be shuttled between E. coli, where library construction and manipulation can be easily conducted, and Streptomyces, where actinomycete genes can be expressed (Figure 2), were developed [21–23].

So far, three E. coli-Streptomyces artificial chromosomes have been constructed. The features of pPAC-S2 (derived from pCYPAC2 [21]), pSTREPTOBAC V (derived from pBACe3.6 [22]), and pSBAC (derived from pCC1BAC [23]) are briefly reported in Table 1 and maps are given in Figure 3. All of them are replicative in E. coli and integrative in Streptomyces. Integrative vectors are maintained in the host without selection and are more stable, since metabolic burden and unwanted recombination events are reduced.

Integration into actinomycete chromosome is ensured by the attP-int system derived from the temperate phages ΦC31 or ΦBT1, which directs the integration of the vector DNA at the corresponding chromosomal attB sites of the host. The availability of two compatible integrating artificial chromosomes can increase the possibility to further manipulate the host. A detrimental effect on endogenous antibiotic production was reported in some producer strains after the integration of vectors in the ΦC31 attB site. For example, the integration of nonrecombinant vectors of 11 kb and 48 kb caused a 90 and 59% reduced production of the endogenous glycopeptide A47934 in Streptomyces toyocaensis [24-26].

In addition, vectors contain a resistance marker to select recombinant clones. Some antibiotic resistance cassettes, like ampicillin resistance gene, are not useful for selection in Streptomyces and different resistance cassettes can be necessary for selection in the two hosts. As an example, the pPAC-S2 contains kanamycin and thiostrepton resistance cassettes used for selecting E. coli and Streptomyces recombinant clones, respectively. pSTREPTOBAC V and pSBAC contain an apramycin resistance cassette that can be used to select both E. coli and Streptomyces recombinant clones.

To transfer BACs or PACs from E. coli to Streptomyces two main strategies can be adopted. In pSTREPTOBAC V and pSBAC, an oriT in the vector allows intergeneric conjugation between an E. coli strain containing Tra encoding genes (carried in an helper plasmid [7] or into the chromosome [27]) and Streptomyces. When oriT is not present (i.e., p-PAC-S2), a more laborious yet easy and well-established protocol is necessary, which comprises protoplast preparation and PEG-assisted protoplast transformation [7].

3. Actinomycete Genomic Libraries Constructed in Artificial Chromosomes

To date, several libraries have been constructed in BAC or PAC artificial chromosomes [22, 23, 28–32] and relevant features of these libraries are mentioned in Table 2.

Using Sau3AI, libraries of S. coelicolor and Planobispora rosea with an average insert size of 60 kb were constructed in pPAC-S2 [28]. A partial genomic library (two-fold coverage) of S. coelicolor was first constructed to set up the methodologies, and using the same procedure the first PAC library of an intractable actinomycete, P. rosea, with an average insert size of 60 kb and the largest insert of 150 kb was constructed. When the same methodology was applied to Nonomuraea sp. ATCC 39727, it was not successful, since its DNA undergoes degradation during PFGE [29, 37]. The PFGE step was therefore omitted, and a high-quality, high molecular weight genomic library of 2,051 recombinant clones with inserts larger than 30 up to 155 kb with an average insert size of 57 kb was constructed in pPAC-S2.

To construct genomic libraries of S. roseosporus and S. platensis, BamHI digested genomic DNA was ligated to BamHI digested pSTREPTOBAC V. Libraries of 2000 and 2400 clones with an average insert size of 71 and 75 kb were obtained from S. roseosporus and S. platensis [22, 31]. The use of the frequently cutting Sau3AI allows a more random selection of inserts. BamHI, on the other hand, allows to obtain large inserts due to its less frequent restriction pattern. The BamHI-digested S. platensis DNA was fractionated twice by PFGE to reduce the number of small inserts.

Since many genome sequences are nowadays available, more specific approaches can be applied. In the case of Streptomyces sp. NRRL 30748, gene sequence analysis had already revealed that two MfeI sites flanked the 90 kb gene cluster devoted to meridamycin biosynthesis and no MfeI sites were present inside the gene cluster. Thus, genomic DNA was digested with MfeI and resulting DNA fragments of approximately 100 kb were ligated to EcoRI sites of pSBAC [23].

If the biosynthetic gene cluster was cloned in a BAC/PAC-derived vector that does not contain the integration site for Streptomyces, the vector can be provided with an integration cassette and a marker for selection in Streptomyces, as in the case of the library of Streptomyces fradiae NRRL 18160 constructed in pECBAC. After the identification of a clone containing the gene cluster devoted to A54145 biosynthesis, the clone was modified and used to transform streptomycetes by adding the ΦC31 attP-int cassette, the apramycin resistance cassette and oriT [30].

4. Heterologous Expression of Biosynthetic Gene Clusters from Actinomycetes

Heterologous expression of secondary metabolic gene clusters can allow process development and molecular biological manipulation of the biosynthetic pathways in those cases of native producer strains recalcitrant to manipulation. Even if this strategy appears effective and straightforward, choice of an appropriate host is always challenging.

E. coli is not a good host for the expression of actinomycete genes, and heavy manipulation has to be carried out to get biosynthesis of actinomycete products. As an example, to obtain erythromycin biosynthesis, the E. coli BL21 (DE3) strain containing the gene for T7 RNA polymerase was

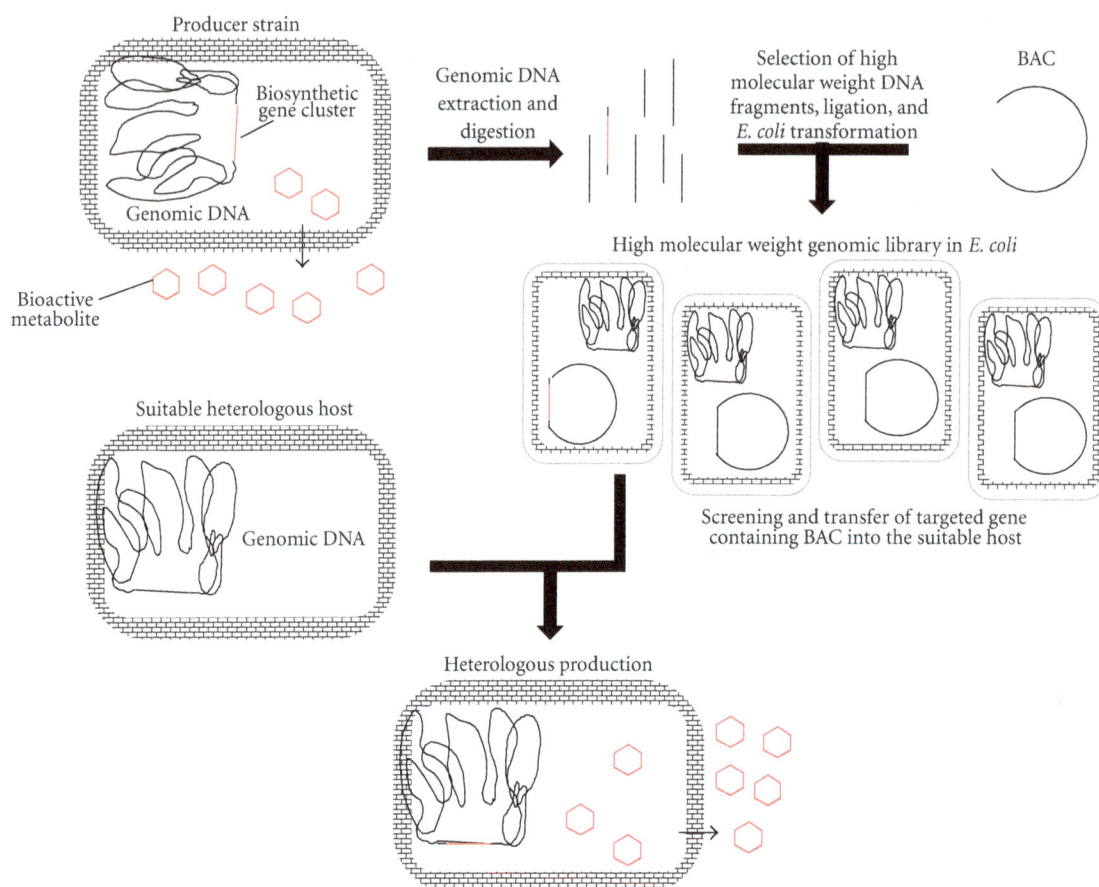

FIGURE 2: Generalized workflow for heterologous production of natural compounds in an amenable host. A high molecular weight genomic library is constructed and screened to identify one or two clones containing the biosynthetic gene cluster of interest. Once the clone is identified and sequenced to confirm the presence of all the genes, the recombinant vector DNA is used to transform the amenable host. Last, the heterologous production can be improved by changing fermentation process or by genetically manipulating the host.

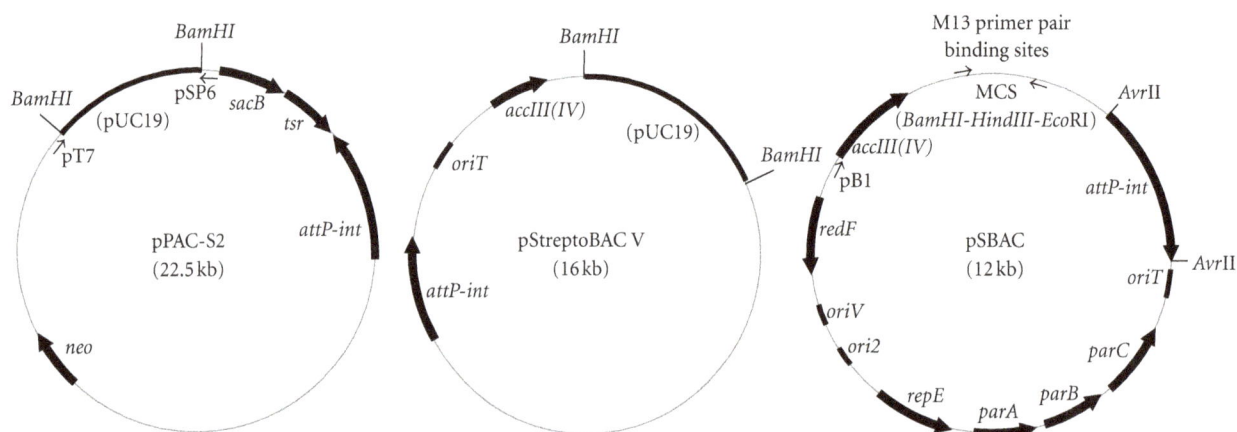

FIGURE 3: Comparison between physical maps of *E. coli-Streptomyces* artificial chromosomes. The vector pPAC-S2 derives from PAC, while pSTREPTOBAC V and pSBAC from BAC. The main features of the vectors are depicted: the resistance genes (*neo, accIII(IV), tsr* conferring resistance to kanamycin, apramycin, thiostrepton), *attP-int* cassettes for the site-specific integration in *Streptomyces,* restriction sites (*Bam*HI, *Hind*III, *Eco*RI), *sacB* (conferring sensitivity to sucrose when expressed), *oriT* and *oriV*, devoted to intergeneric conjugation and inducible high copy number of the vector, respectively.

TABLE 1: Shuttle *E. coli-Streptomyces* artificial chromosomes.

Vector	Derived from	*attP-int* cassette derived from	Resistance to	Reference
pPAC-S2	pCYPAC2	ΦC31	Kanamycin/thiostrepton in *E. coli/Streptomyces*	[21]
pSTREPTOBAC V	pBACe3.6	ΦC31	Apramycin in *E. coli/Streptomyces*	[22]
pSBAC	pCC1BAC	φBT1	Apramycin in *E. coli/Streptomyces*	[23]

TABLE 2: Actinomycetes genomic libraries constructed in artificial chromosomes.

Source	Genome coverage	Average insert size	Genomic DNA digested with	Vector	Reference
Planobispora rosea ATCC 53773	5X	60 kb	*Sau*3AI	pPAC-S2	[28]
Nonomuraea sp. ATCC 3972	5X	57 kb	*Sau*3AI	pPAC-S2	[29]
S. roseosporus NRRL 11379	17X	71 kb	*Bam*HI	pSTREPTOBAC V	[22]
S. platensis NRRL18993	22X	75 kb	*Bam*HI	pSTREPTOBAC V	[31]
Streptomyces sp. NRRL 30748	NR	NR	*Mfe*I	pSBAC*	[23]
S. autolyticus JX-47	50X	150 kb	*Sau*3AI	pIndigoBAC5*	[32]

NR: not reported.
*Not integrative in *Streptomyces*.

transformed with four plasmids: (i) pHZT1 and (ii) pHZT2 containing all *ery* genes organized in two operons under T7 promoter control; (iii) pHZT4 containing *eryK*, the last gene of one of the operons, since it was not expressed at high level; and (iv) plasmid pGro7 containing the genes for the *E. coli* GroEL/ES chaperone system to aid protein folding and/or association of actinomycete gene products. Only the strain carrying the four plasmids produced erythromycin A, while the same *E. coli* strains containing only pHZT1 and pHZT2 did not. This very labour-intensive, time-consuming procedure was specific for erythromycin [38], and a similar strategy could not always be applied for the synthesis of other antibiotic compounds that may require "specific" precursors that *E. coli* cannot provide.

Streptomycetes are naturally the preferred hosts of actinomycete genes. Compared to other actinomycetes, *Streptomyces* is more amenable for strain improvement, grows more rapidly, and generally has the biosynthetic apparatus and primary precursors necessary to support natural product synthesis from exogenous pathways. Among streptomycetes, *S. coelicolor* and *S. lividans* have been widely used as heterologous hosts.

S. coelicolor is considered the model species among actinomycetes and a large array of genetic tools, including its genome sequence, is available to understand and manipulate the organism [7, 39]. Although there are many advantages of using *S. coelicolor* as a host, plasmid-free derivatives of *S. lividans*, a close relative of *S. coelicolor*, have often been used for heterologous production (Table 3) [40, 41]. The primary reason is the absence of a strong restriction system, like in *S. coelicolor*, which heavily restricts methylated DNA [42].

Other streptomycetes have already been used for heterologous expression of secondary metabolite gene clusters, like *Streptomyces albus* G J1074 [35], and other industrial strains, such as *S. avermitilis*. *S. albus* G J1074 is a mutant defective in *Sal*I restriction and modification, readily transformable cloning host that has been used for the expression of several other secondary metabolite gene clusters [35]. The industrial microorganism *S. avermitilis*, producer of the anthelmintic macrolide avermectins, has been already optimized for the efficient supply of primary metabolic precursors and biochemical energy to support multistep biosynthesis [43, 44].

4.1. Optimization of Heterologous Production in Streptomycetes. In some cases heterologous production is either absent or so low that manipulation of the host is necessary. Different strategies, both genetics- and physiology-driven, can be explored in a well-defined host. Fermentation-based approaches can consist of varying carbon or nitrogen sources or different phosphate levels in the medium and in feeding with a biosynthetic precursor that may be limiting during the growth phase. Genetic methods can target both host and cloned genes. Genomic analysis of streptomycetes has revealed that these microorganisms have large linear chromosomes that harbour over 20 gene clusters for secondary metabolites that could compete for precursors or could interfere with chemical analysis. Thus, the deletion of host gene clusters devoted to biosynthesis of other endogenous metabolites can increase precursor supply and, thus, improve heterologous production [33, 34, 36, 45, 46]; in addition, there are point mutations in *rpoB* and *rpsL* genes, coding for RNA polymerase beta-subunit and 30S ribosomal subunit protein S12, respectively, known for increasing antibiotic production [45, 47]. Although the underlying molecular mechanism is not understood, these mutations have a pleiotropic effect on levels of secondary metabolite production that is achieved, at least in part, by elevated transcript levels.

S. coelicolor produces at least four metabolites with antibiotic activity: the Type II polyketide actinorhodin (Act), the NPRS/PKS-derived prodiginines (Red), the NRPS-derived calcium-dependent antibiotic (CDA), and the TypeI polyketide yellow pigment CPK [39]. In the last two years, two groups put their efforts in the engineering of *S. coelicolor*

TABLE 3: Streptomycetes hosts.

Strain	Genotype	Reference
S. lividans TK23	*spc* ⁻ SLP2⁻ SLP3⁻	[7]
S. lividans TK24	*str6* ⁻ SLP2⁻ SLP3⁻	[7]
S. lividans TK64	*pro-2 str-6* SLP2⁻ SLP3⁻	[7]
S. lividans K4-114	*pro-2 str-6*SLP2⁻ SLP3⁻ Δ*act::ermE Streptomyces*	[33]
S. coelicolor M512	Δ*redD* Δ actII-ORF4 SCP1⁻ SCP2⁻	[34]
S. albus G J1074	*ilv-1 sal-2*	[35]
S. avermitilis SUKA4/SUKA5	Δ(*gap1-ptlL*)::*ermE* (Δ3,745,502–3,758,936 nt)	[36]

in order to construct suitable production hosts [45, 46]. In one case four of these endogenous secondary metabolite gene clusters were deleted and point mutations in *rpoB* and *rpsL* genes were introduced [45]. In the more recent case, sequential deletions of all ten polyketide synthase (PKS) and nonribosomal peptide synthetase (NRPS) biosynthetic gene clusters and a 900-kb subtelomeric sequence (total ca. 1.22 Mb, 14% of the genome) from *S. coelicolor* chromosome were performed to construct derivative strains [46]. In addition to lacking antibacterial activity, the engineered strains possess relatively simple extracellular metabolite profiles and they could markedly facilitate the discovery of new compounds by heterologous expression of cloned gene clusters, particularly the numerous cryptic secondary metabolic gene clusters that are prevalent within actinomycete genome sequences [45, 46].

S. lividans has been largely used as heterologous host for its more relaxed restriction/modification system [42], but some strains of *S. lividans* under many fermentation conditions produce actinorhodin that interferes with the detection and purification of other secondary metabolites, so part of the *act* gene cluster was deleted in K4-114, therefore providing a cleaner background for heterologous expression [33].

The industrial microorganism *S. avermitilis* was manipulated by deletion of the telomeric ends of the linear chromosome, which contain nonvital genes, often antibiotic biosynthesis-related [36]. A region of more than 1.4 Mb was deleted stepwise from the 9.02 Mb *S. avermitilis* linear chromosome to generate a series of defined deletion mutants (i.e., SUKA4, SUKA5), that did not produce any of the major endogenous secondary metabolites found in the parent strain.

Also foreign genes can be manipulated, that is, strong promoters can be added to drive transcription of genes or operons, otherwise weakly transcribed.

Here we discuss the successful production of daptomycin (from *dpt* genes), meridamycin (from *mer* genes), and migrastatin (from *mgs* genes), pointing out the strategies applied (Figure 4). Firstly, artificial chromosome libraries from *S. roseosporus* NRRL 11379, *S. platensis* NRRL18993, and *Streptomyces* sp. NRRL 30748 were screened by PCR or hybridization for the presence of *dpt, mgs, mer,* and *lpt* gene clusters, respectively. Once sequenced, the relevant clones were used to transform various *Streptomyces* hosts, where they were stably integrated and maintained.

4.1.1. Production of Daptomycin in S. lividans. The daptomycin gene cluster was transferred in *S. lividans* TK64 and TK23. In both cases, the recombinant strains produced the lipopeptide with a low titre and a high amount of actinorhodin [22, 41]. To reduce the background effect of native metabolites, the *S. lividans* TK23 host was manipulated by inactivating actinorhodin production through replacement of a portion of the gene cluster with a resistance gene, but this did not result in a yield improvement. An improvement in daptomycin biosynthesis was obtained by increasing the levels of K_2HPO_4 during fermentation. It is likely that phosphate, often reported as an inhibitor of antibiotic production, inhibited host metabolite biosynthesis pathways, whose precursors could be necessary for daptomycin biosynthesis. Adjusting the level of phosphate in the medium the yield of the lipopeptide was increased from 20 to 55 mg/L, corresponding to one-third of the yield of the natural producer [22, 41].

4.1.2. Production of Meridamycin in S. lividans. In the case of meridamycin biosynthesis, the transfer of the 90 kb gene cluster to *S. lividans* TK24 and K4-114 did not lead to the detection of any heterologous metabolites, since PKS genes were not properly transcribed, possibly due to the size of operon (3 genes for a 78 kb operon). Indeed, the cloning of a strong promoter (*ermE* promoter) upstream of the *mer* genes coding for the PKS allowed the detection of approximately $100 \, \mu g/L$ of meridamycin in the fermentation extract. To further increase the yield, a fermentation strategy based on feeding with biosynthetic precursors was applied. The supply of ethylmalonyl-CoA, used as an extender unit in meridamycin biosynthesis, is a critical factor that limits the synthesis of some polyketides by heterologous hosts. In fact, when *S. lividans* containing *mer* gene cluster with PKS genes under *ermE* control, was supplemented with diethyl malonate, which is an effective precursor for ethylmalonyl-CoA, the production increased about 2-fold ($240 \, \mu g/L$). On the other hand, feeding with pipecolic acid, another unique precursor for the biosynthesis of meridamycin, did not increase production [23].

4.1.3. Production of Migrastatin in Streptomycetes. To find the most suitable host to express the biosynthetic gene cluster for the production of the polyketide migrastatin of *S. platensis* NRRL18993, five *Streptomyces* hosts were selected: *S. lividans* K4-114, *S. coelicolor* M512, *S. albus G J1074*, and the two

Drug	Daptomycin	Meridamycin	Migrastatin
Native producer drug production yield	*S. roseosporus* NRRL 11379 120 mg/mL	*S. sp.* NRRL 30748	*S. platensis* NRRL18993 58 mg/mL
	↓ *dpt* gene cluster	↓ *mer* gene cluster	↓ *mgr* gene cluster
BAC used for gene cluster transfer	pSTREPTOBAC V	pSBAC	pSTREPTOBAC V
	↓	↓	↓
Heterologous drug production yield	*S. lividans* TK64 and TK23 20 mg/L	*S. lividans* TK24 and K4-114 No production	*S. albus* J1074 46 mg/mL
Production improvement strategy	Fermentation medium / Host genetic manipulation	Cluster genetic manipulation ↓ Fermentation medium / Fermentation medium	Fermentation medium
Improved heterologous drug production yield	55 mg/L / No improvement	240 µg/L / No improvement	186 mg/L

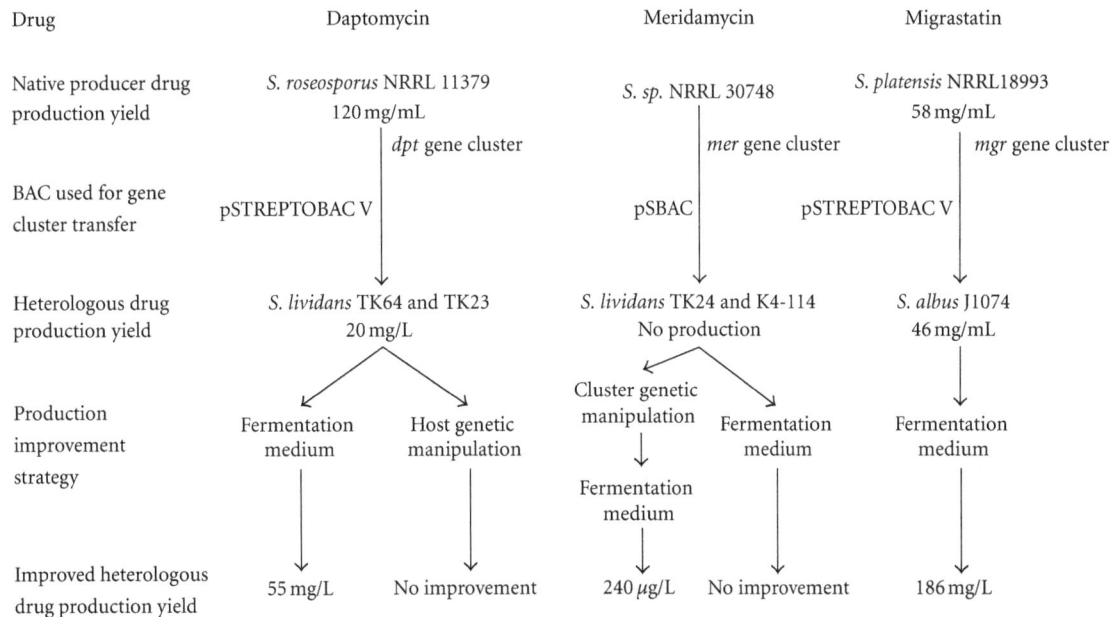

FIGURE 4: Strategies adopted to improve heterologous production of daptomycin, meridamycin, and migrastatin in heterologous hosts. Under "fermentation strategies" fall feeding with biosynthetic precursors or growth in presence of specific nutrients. Host genetic manipulation indicates the inactivation of host biosynthetic pathways or mutations in genes (*rpsL* and *rpoB*) known for increasing secondary metabolite production. Genetic cluster manipulation refers to the cloning of a strong promoter upstream of foreign genes which are weakly transcribed. Details are provided in the text.

S. avermitilis strains, SUKA4 and SUKA5. All the five strains produced the metabolite, but the titres, in two media already optimized for the natural producer, were similar to or less than those obtained with *S. platensis*. The best producers were *S. albus* that produced approximately 46 mg/L in B2 medium (natural producer 58 mg/L), and *S. lividans* K4-114 that produced 15 mg/L (control 17 mg/L) in R2YED medium. Adjusting fermentation conditions, an improvement of the titer (appr. 186 mg/L) was registered in *S. albus*. In addition, feeding with NaHCO₃ that plays a critical role in the biosynthesis of malonyl-CoA, the major precursor for migrastatin biosynthesis, allowed *S. albus* to produce migrastatin at a titer of 213.8 mg/L, about 5-fold higher than the originally [31, 48].

4.1.4. Production of the Lipopeptide A54145 in S. ambofaciens and S. roseosporus. It is noteworthy that also the production of the lipopeptide A54145 from *S. fradiae* was obtained in streptomycetes hosts, but the vector, pECBAC1, was not an *E. coli-Streptomyces* shuttle vector. The recombinant vector, containing the entire *lpt* gene cluster was modified by adding the ΦC31 *attP-int* integration cassette after library construction [30]. The *lpt* gene cluster of *S. fradiae* was transferred into *S. ambofaciens* and *S. roseosporus* strains genetically manipulated in a such way to not being anymore able to produce spiramycin [49] and daptomycin [50], respectively. Interestingly, both sets of heterologous strains were more efficient than wild type *S. fradiae* at producing the lipopeptide, but less productive than the *S. fradiae* high-producer control.

5. Future Perspectives

Artificial chromosomes are useful tools to exploit or to explore the genetic potential of intractable actinomycetes, by creating, for example, large-insert libraries from strains of interest, in order to isolate one or more large gene clusters [22, 23, 28–32], or by reconstructing gene clusters that are already available as a set of smaller fragments cloned in lower capacity vectors [17, 18, 51, 52]. To be transferred in amenable hosts, BAC and PAC are provided with means to be maintained in manipulable streptomycetes, which can synthesize foreign complex molecules, such as antibiotics [21–23]. Sometimes, heterologous expression of actinomycetes biosynthetic gene clusters was successful after changing fermentation conditions, that is, feeding with a biosynthetic precursor, minimizing background endogenous activities or after cloning strong promoters upstream of production genes weakly transcribed. The examples here described can be considered as promising starting points to successfully express heterologously valuable biosynthetic products. A significant improvement in heterologous production will surely come from new technologies, like next generation sequencing and gene stitching and synthesis. Indeed, the genome sequences of both the producing strains and of the hosts can help to find the minimal set of genes necessary to maintain heterologous production in a clean background and to avoid interference between foreign gene products and host that in some cases were reported [53, 54].

A possibility could be to create artificial hosts that maintain as less endogenous genes as possible and that contain the specific genes to provide primary metabolites and all the cofactors necessary to synthesize a specific secondary metabolite. Some studies in this direction have just started with the deletion of large DNA fragments from *S. avermitilis* and *S. coelicolor* chromosomes [36, 46]. More efforts are needed to obtain a universal suitable host of actinomycete gene clusters; although *S. avermitilis* and *S. coelicolor* strains with minimized genomes have been constructed and shown to produce more metabolites than natural producers, further manipulation of their genomes followed by the improvement of the yields will hopefully occur with the help of specific experiments of gene stitching and synthesis. The genome sequencing of *Streptomyces* species (e.g., *S. coelicolor* and *S. avermitilis* [39, 43]) has remarkably revealed that each strain has the genetic information to produce a large number (e.g., 23 and 32, resp.) of secondary metabolites. This implies that as much as 90% of the chemical potential of these organisms remains undiscovered by the conventional screening programs. Genome mining offers a powerful method for tapping into these cryptic natural products [55]. For discovery of new compounds by genome mining, heterologous expression could be employed to examine new products by these new gene clusters.

Heterologous expression through artificial chromosomes may provide the possibility to access gene clusters for the production of bioactive metabolites from the large number of uncultured bacteria present in the environment without the need to actually isolate them [56–61]. Growing many of these organisms at large scale for production purposes is unlikely to be practical but their biosynthetic gene clusters are particularly attractive targets for heterologous expression in a genetically amenable host.

For all these cases, it will be important to develop a suite of streptomycete hosts for high-level heterologous expression of specific types of secondary metabolites, and to facilitate rapid identification and testing of novel products. Given the relatively recent success of heterologous complex natural product biosynthesis, these tools are just beginning to be applied, but they appear to be very promising. Such approaches, in addition to omics global characterization techniques and more traditional process engineering steps, will be of paramount importance for the maximization of natural product outputs from heterologous hosts.

When the new technologies (e.g., gene synthesis) will be cheap and there will be no more novel activities and genes to discover, it is likely that cloning strategy will be abandoned, but at the moment this perspective is quite far.

Acknowledgments

G. Gallo was supported by the European-funded project "LAntibiotic Production: Technology, Optimization and improved Process" (LAPTOP), KBBE-245066. The authors are very grateful to Dr. Luca Dolce for critical reading of and corrections on the paper.

References

[1] A. Raja and P. Prabakarana, "Actinomycetes and drug-an overview," *American Journal of Drug Discovery and Development*, vol. 1, no. 2, pp. 75–84, 2011.

[2] J. L. Adrio and A. L. Demain, "Genetic improvementof processes yielding microbial products," *FEMS Microbiology Reviews*, vol. 30, no. 2, pp. 187–214, 2006.

[3] R. H. Baltz, "Genetic methods and strategies for secondary metabolite yield improvement in actinomycetes," *Antonie van Leeuwenhoek*, vol. 79, no. 3-4, pp. 251–259, 2001.

[4] S. Donadio, P. Monciardini, R. Alduina et al., "Microbial technologies for the discovery of novel bioactive metabolites," *Journal of Biotechnology*, vol. 99, no. 3, pp. 187–198, 2002.

[5] Y. Mast, T. Weber, M. Gölz et al., "Characterization of the "pristinamycin supercluster" of *Streptomyces pristinaespiralis*," *Microbial Biotechnology*, vol. 4, no. 2, pp. 192–206, 2011.

[6] R. Kirby, "Chromosome diversity and similarity within the *Actinomycetales*," *FEMS Microbiology Letters*, vol. 319, no. 1, pp. 1–10, 2011.

[7] T. Kieser, M. J. Bibb, M. J. Buttner et al., *Practical Streptomyces Genetics*, The John Innes Foundation, Norwich, UK; Crowes, London, UK, 2000.

[8] J. Handelsman, "Metagenomics: application of genomics to uncultured microorganisms," *Microbiology and Molecular Biology Reviews*, vol. 68, no. 4, pp. 669–685, 2004.

[9] M. Steffensky, A. Mühlenweg, Z. X. Wang, S. M. Li, and L. Heide, "Identification of the novobiocin biosynthetic gene cluster of *Streptomyces spheroides* NCIB 11891," *Antimicrobial Agents and Chemotherapy*, vol. 44, no. 5, pp. 1214–1222, 2000.

[10] F. Pojer, S. M. Li, and L. Heide, "Molecular cloning and sequence analysis of the clorobiocin biosynthetic gene cluster: new insights into the biosynthesis of aminocoumarin antibiotics," *Microbiology*, vol. 148, no. 12, pp. 3901–3911, 2002.

[11] M. Sosio, S. Stinchi, F. Beltrametti, A. Lazzarini, and S. Donadio, "The gene cluster for the biosynthesis of the glycopeptide antibiotic A40926 by *Nonomuraea* species," *Chemistry and Biology*, vol. 10, no. 6, pp. 541–549, 2003.

[12] M. C. Cone, X. Yin, L. L. Grochowski, M. R. Parker, and T. M. Zabriskie, "The blasticidin S biosynthesis gene cluster from *Streptomyces griseochromogenes*: sequence analysis, organization, and initial characterization," *ChemBioChem*, vol. 4, no. 9, pp. 821–828, 2003.

[13] D. Singh, M. J. Seo, H. J. Kwon et al., "Genetic localization and heterologous expression of validamycin biosynthetic gene cluster isolated from *Streptomyces hygroscopicus* var. *limoneus* KCCM 11405 (IFO 12704)," *Gene*, vol. 376, no. 1-2, pp. 13–23, 2006.

[14] M. Koběrská, J. Kopecký, J. Olšovská et al., "Sequence analysis and heterologous expression of the lincomycin biosynthetic cluster of the type strain *Streptomyces lincolnensis* ATCC 25466," *Folia Microbiologica*, vol. 53, no. 5, pp. 395–401, 2008.

[15] Q. Xue, G. Ashley, C. R. Hutchinson, and D. V. Santi, "A multiplasmid approach to preparing large libraries of polyketides," *Proceedings of the National Academy of Sciences of the United States of America*, vol. 96, no. 21, pp. 11740–11745, 1999.

[16] L. Tang, S. Shah, L. Chung et al., "Cloning and heterologous expression of the epothilone gene cluster," *Science*, vol. 287, no. 5453, pp. 640–642, 2000.

[17] T. M. Binz, S. C. Wenzel, H. J. Schnell, A. Bechthold, and R. Müller, "Heterologous expression and genetic engineering of the phenalinolactone biosynthetic gene cluster by using red/ET recombineering," *ChemBioChem*, vol. 9, no. 3, pp. 447–454, 2008.

[18] Y. Hu, V. V. Phelan, C. M. Farnet, E. Zazopoulos, and B. O. Bachmann, "Reassembly of anthramycin biosynthetic gene cluster by using recombinogenic cassettes," *ChemBioChem*, vol. 9, no. 10, pp. 1603–1608, 2008.

[19] H. Shizuya, B. Birren, U. J. Kim et al., "Cloning and stable maintenance of 300-kilobase-pair fragments of human DNA in *Escherichia coli* using an F-factor-based vector," *Proceedings of the National Academy of Sciences of the United States of America*, vol. 89, no. 18, pp. 8794–8797, 1992.

[20] W. R. Strohl, "Compilation and analysis of DNA sequences associated with apparent streptomycete promoters," *Nucleic Acids Research*, vol. 20, no. 5, pp. 961–974, 1992.

[21] M. Sosio, F. Giusino, C. Cappellano, E. Bossi, A. M. Puglia, and S. Donadio, "Artificial chromosomes for antibiotic-producing actinomycetes," *Nature Biotechnology*, vol. 18, no. 3, pp. 343–345, 2000.

[22] V. Miao, M. F. Coëffet-LeGal, P. Brian et al., "Daptomycin biosynthesis in *Streptomyces roseosporus*: cloning and analysis of the gene cluster and revision of peptide stereochemistry," *Microbiology*, vol. 151, no. 5, pp. 1507–1523, 2005.

[23] H. Liu, H. Jiang, B. Haltli et al., "Rapid cloning and heterologous expression of the meridamycin biosynthetic gene cluster using a versatile *Escherichia coli-Streptomyces* artificial chromosome vector, pSBAC," *Journal of Natural Products*, vol. 72, no. 3, pp. 389–395, 2009.

[24] R. H. Baltz, "Genetic manipulation of antibiotic-producing *Streptomyces*," *Trends in Microbiology*, vol. 6, no. 2, pp. 76–83, 1998.

[25] P. Matsushima, M. C. Broughton, J. R. Turner, and R. H. Baltz, "Conjugal transfer of cosmid DNA from *Escherichia coli* to *Saccharopolyspora spinosa*: effects of chromosomal insertions on macrolide A83543 production," *Gene*, vol. 146, no. 1, pp. 39–45, 1994.

[26] P. Matsushima and R. H. Baltz, "A gene cloning system for '*Streptomyces toyocaensis*,'" *Microbiology*, vol. 142, no. 2, pp. 261–267, 1996.

[27] M. D. McMahon, C. Guan, J. Handelsman, and M. G. Thomas, "Metagenomics in *Streptomyces lividans* reveals host-dependent functional expression," *Applied Environmental Microbiology*, vol. 78, no. 10, pp. 3622–3629, 2012.

[28] R. Alduina, S. De Grazia, L. Dolce et al., "Artificial chromosome libraries of *Streptomyces coelicolor* A3(2) and *Planobispora rosea*," *FEMS Microbiology Letters*, vol. 218, no. 1, pp. 181–186, 2003.

[29] R. Alduina, A. Giardina, G. Gallo et al., "Expression in *Streptomyces lividans* of *Nonomuraea* genes cloned in an artificial chromosome," *Applied Microbiology and Biotechnology*, vol. 68, no. 5, pp. 656–662, 2005.

[30] D. C. Alexander, J. Rock, X. He, P. Brian, V. Miao, and R. H. Baltz, "Development of a genetic system for combinatorial biosynthesis of lipopeptides in *Streptomyces fradiae* and heterologous expression of the A54145 biosynthesis gene cluster," *Applied and Environmental Microbiology*, vol. 76, no. 20, pp. 6877–6887, 2010.

[31] Z. Feng, L. Wang, S. R. Rajski, Z. Xu, M. F. Coeffet-LeGal, and B. Shen, "Engineered production of iso-migrastatin in heterologous *Streptomyces* hosts," *Bioorganic and Medicinal Chemistry*, vol. 17, no. 6, pp. 2147–2153, 2009.

[32] S. Dai, Y. Ouyang, G. Wang, and X. Li, "Streptomyces autolyticus JX-47 large-insert bacterial artificial chromosome library construction and identification of clones covering geldanamycin biosynthesis gene cluster," *Current Microbiology*, vol. 63, no. 1, pp. 68–74, 2011.

[33] R. Ziermann and M. C. Betlach, "Recombinant polyketide synthesis in *Streptomyces*: engineering of improved host strains," *BioTechniques*, vol. 26, no. 1, pp. 106–110, 1999.

[34] B. Floriano and M. Bibb, "afsR is a pleiotropic but conditionally required regulatory gene for antibiotic production in

Streptomyces coelicolor A3(2)," *Molecular Microbiology*, vol. 21, no. 2, pp. 385–396, 1996.

[35] K. F. Chater and L. C. Wilde, "Streptomyces albus G mutants defective in the SalGI restriction-modification system," *Journal of General Microbiology*, vol. 116, no. 2, pp. 323–334, 1980.

[36] M. Komatsu, T. Uchiyama, S. Omura, D. E. Cane, and H. Ikeda, "Genome-minimized *Streptomyces* host for the heterologous expression of secondary metabolism," *Proceedings of the National Academy of Sciences of the United States of America*, vol. 107, no. 6, pp. 2646–2651, 2010.

[37] M. L. Beyazova, B. C. Brodsky, M. C. Shearer, and A. C. Horan, "Preparation of actinomycete DNA for pulsed-field gel electrophoresis," *International Journal of Systematic Bacteriology*, vol. 45, no. 4, pp. 852–854, 1995.

[38] H. Zhang, K. Skalina, M. Jiang, and B. A. Pfeifer, "Improved *E. coli* erythromycin a production through the application of metabolic and bioprocess engineering," *Biotechnology Progress*, vol. 8, no. 1, pp. 292–296, 2012.

[39] S. D. Bentley, K. F. Chater, A. M. Cerdeno-Tarraga et al., "Complete genome sequence of the model actinomycete *Streptomyces coelicolor* A3(2)," *Nature*, vol. 417, pp. 141–147, 2002.

[40] Z. Hu, D. A. Hopwood, and C. Khosla, "Directed transfer of large DNA fragments between *Streptomyces* species," *Applied and Environmental Microbiology*, vol. 66, no. 5, pp. 2274–2277, 2000.

[41] J. Penn, X. Li, A. Whiting et al., "Heterologous production of daptomycin in *Streptomyces lividans*," *Journal of Industrial Microbiology and Biotechnology*, vol. 33, no. 2, pp. 121–128, 2006.

[42] D. J. MacNeil, "Characterization of a unique methyl-specific restriction system in *Streptomyces avermitilis*," *Journal of Bacteriology*, vol. 170, no. 12, pp. 5607–5612, 1988.

[43] H. Ikeda, J. Ishikawa, A. Hanamoto et al., "Complete genome sequence and comparative analysis of the industrial microorganism *Streptomyces avermitilis*," *Nature Biotechnology*, vol. 21, no. 5, pp. 526–531, 2003.

[44] R. H. Baltz, "Streptomyces and Saccharopolyspora hosts for heterologous expression of secondary metabolite gene clusters," *Journal of Industrial Microbiology and Biotechnology*, vol. 37, no. 8, pp. 759–772, 2010.

[45] J. P. Gomez-Escribano and M. J. Bibb, "Engineering *Streptomyces coelicolor* for heterologous expression of secondary metabolite gene clusters," *Microbial Biotechnology*, vol. 4, no. 2, pp. 207–215, 2011.

[46] M. Zhou, X. Jing, P. Xie et al., "Sequential deletion of all the PKS and NRPS biosynthetic gene clusters and a 900-kb subtelomeric sequence of the linear chromosome of *Streptomyces coelicolor*," *FEMS Microbiology Letters*, vol. 333, no. 2, pp. 169–179, 2012.

[47] A. Talà, G. Wang, M. Zemanova, S. Okamoto, K. Ochi, and P. Alifano, "Activation of dormant bacterial genes by *Nonomuraea* sp. strain ATCC 39727 mutant-type RNA polymerase," *Journal of Bacteriology*, vol. 191, no. 3, pp. 805–814, 2009.

[48] D. Yang, X. Zhu, X. Wu et al., "Titer improvement of iso-migrastatin in selected heterologous *Streptomyces* hosts and related analysis of mRNA expression by quantitative RT-PCR," *Applied Microbiology and Biotechnology*, vol. 89, no. 6, pp. 1709–1719, 2011.

[49] M. A. Richardson, S. Kuhstoss, M. L. B. Huber et al., "Cloning of spiramycin biosynthetic genes and their use in constructing *Streptomyces ambofaciens* mutants defective in spiramycin biosynthesis," *Journal of Bacteriology*, vol. 172, no. 7, pp. 3790–3798, 1990.

[50] V. Miao, M. F. Coëffet-Le Gal, K. Nguyen et al., "Genetic engineering in *Streptomyces roseosporus* to produce hybrid lipopeptide antibiotics," *Chemistry and Biology*, vol. 13, no. 3, pp. 269–276, 2006.

[51] M. Sosio, E. Bossi, and S. Donadio, "Assembly of large genomic segments in artificial chromosomes by homologous recombination in *Escherichia coli*," *Nucleic acids research*, vol. 29, no. 7, p. E37, 2001.

[52] O. Perlova, J. Fu, S. Kuhlmann et al., "Reconstitution of the myxothiazol biosynthetic gene cluster by red/ET recombination and heterologous expression in *Myxococcus xanthus*," *Applied and Environmental Microbiology*, vol. 72, no. 12, pp. 7485–7494, 2006.

[53] D. E. Gillespie, S. F. Brady, A. D. Bettermann et al., "Isolation of antibiotics turbomycin A and B from a metagenomic library of soil microbial DNA," *Applied and Environmental Microbiology*, vol. 68, no. 9, pp. 4301–4306, 2002.

[54] A. Giardina, R. Alduina, E. Gottardi, V. Di Caro, R. D. Süssmuth, and A. M. Puglia, "Two heterologously expressed *Planobispora rosea* proteins cooperatively induce *Streptomyces lividans* thiostrepton uptake and storage from the extracellular medium," *Microbial Cell Factories*, vol. 9, article 44, 2010.

[55] R. H. Baltz, "Renaissance in antibacterial discovery from actinomycetes," *Current Opinion in Pharmacology*, vol. 8, no. 5, pp. 557–563, 2008.

[56] O. Béjà, M. T. Suzuki, E. V. Koonin et al., "Construction and analysis of bacterial artificial chromosome libraries from a marine microbial assemblage," *Environmental Microbiology*, vol. 2, no. 5, pp. 516–529, 2000.

[57] A. Martinez, S. J. Kolvek, C. L. T. Yip et al., "Genetically modified bacterial strains and novel bacterial artificial chromosome shuttle vectors for constructing environmental libraries and detecting heterologous natural products in multiple expression hosts," *Applied and Environmental Microbiology*, vol. 70, no. 4, pp. 2452–2463, 2004.

[58] M. R. Rondon, P. R. August, A. D. Bettermann et al., "Cloning the soil metagenome: a strategy for accessing the genetic and functional diversity of uncultured microorganisms," *Applied and Environmental Microbiology*, vol. 66, no. 6, pp. 2541–2547, 2000.

[59] A. E. Berry, C. Chiocchini, T. Selby, M. Sosio, and E. M. H. Wellington, "Isolation of high molecular weight DNA from soil for cloning into BAC vectors," *FEMS Microbiology Letters*, vol. 223, no. 1, pp. 15–20, 2003.

[60] J. Singh, A. Behal, N. Singla et al., "Metagenomics: concept, methodology, ecological inference and recent advances," *Biotechnology Journal*, vol. 4, no. 4, pp. 480–494, 2009.

[61] Y. Ouyang, S. Dai, L. Xie et al., "Isolation of high molecular weight DNA from marine sponge bacteria for BAC library construction," *Marine Biotechnology*, vol. 12, no. 3, pp. 318–325, 2010.

15

Asbestos-Induced Cellular and Molecular Alteration of Immunocompetent Cells and Their Relationship with Chronic Inflammation and Carcinogenesis

Hidenori Matsuzaki,[1] Megumi Maeda,[1,2] Suni Lee,[1] Yasumitsu Nishimura,[1] Naoko Kumagai-Takei,[1] Hiroaki Hayashi,[1,3] Shoko Yamamoto,[1] Tamayo Hatayama,[1] Yoko Kojima,[4] Rika Tabata,[4] Takumi Kishimoto,[4] Junichi Hiratsuka,[5] and Takemi Otsuki[1]

[1] *Department of Hygiene, Kawasaki Medical School, 577 Matsushima, Kurashiki 7010192, Japan*
[2] *Department of Biofunctional Chemistry, Division of Bioscience, Okayama University Graduate School of Natural Science and Technology, 3-1-1 Tsushima-Naka, Okayama 7008530, Japan*
[3] *Department of Dermatology, Kawasaki Medical School, 577 Matsushima, Kurashiki 7010192, Japan*
[4] *Research Center for Asbestos-Related Diseases, Okayama Rosai Hospital, 1-10-25 Chikko-Midorimachi, Minami-Ku, Okayama 7028055, Japan*
[5] *Department of Radiation Oncology, Kawasaki Medical School, 577 Matsushima, Kurashiki 7010192, Japan*

Correspondence should be addressed to Takemi Otsuki, takemi@med.kawasaki-m.ac.jp

Academic Editor: Vassilis Gorgoulis

Asbestos causes lung fibrosis known as asbestosis as well as cancers such as malignant mesothelioma and lung cancer. Asbestos is a mineral silicate containing iron, magnesium, and calcium with a core of SiO_2. The immunological effect of silica, SiO_2, involves the dysregulation of autoimmunity because of the complications of autoimmune diseases found in silicosis. Asbestos can therefore cause alteration of immunocompetent cells to result in a decline of tumor immunity. Additionally, due to its physical characteristics, asbestos fibers remain in the lung, regional lymph nodes, and the pleural cavity, particularly at the opening sites of lymphatic vessels. Asbestos can induce chronic inflammation in these areas due to the production of reactive oxygen/nitrogen species. As a consequence, immunocompetent cells can have their cellular and molecular features altered by chronic and recurrent encounters with asbestos fibers, and there may be modification by the surrounding inflammation, all of which eventually lead to decreased tumor immunity. In this paper, the brief results of our investigation regarding reduction of tumor immunity of immunocompetent cells exposed to asbestos *in vitro* are discussed, as are our findings concerned with an investigation of chronic inflammation and analyses of peripheral blood samples derived from patients with pleural plaque and mesothelioma that have been exposed to asbestos.

1. Introduction

Asbestos causes lung fibrosis known as asbestosis, a few benign pleural diseases such as pleural plaque and effusion, and malignant diseases such as mesothelioma and lung cancers [1–5]. Furthermore, cancers in other organs such as the gastrointestinal tract, larynx, kidney, liver, pancreas, ovary, and hematopoietic systems show ahigh prevalence in asbestos-inhaled people [6–9]. This issue has been tackled as a major medical and social problem throughout the world, especially since asbestos is very useful in various industries for its mineral characteristics such as resistance to heat, cold, chemicals, cheapness, easiness to obtain and weave, and so on [8, 9].

In Japan, the quantity of asbestos produced has increased since the mid-1950s and reached a peak in 1974 (over 350,000z tons) [10–12]. The peak level of usage continued until the late 1980s in construction, car, and other industries. In the summer of 2005, Kubota Corporation, which mainly used asbestos to make water pipes in Amagasaki City, Hyogo Prefecture, Japan, acknowledged the prevalence of asbestos-related diseases among their workers, including several patients living near Kubota's asbestos-handling factory in Amagasaki City [10–12]. Citizens in Japan were suddenly made aware that asbestos causes malignancies in asbestos-handling workers and in residents living near these factories. They were informed that mesothelioma is difficult to diagnose and cure and were angered that workers and neighborhood residents had not been notified that these factories had been handling this silent bomb, asbestos [10–12].

To reduce the anxieties of the Japanese people, epidemiological analyses commenced regarding the Amagasaki area, and clinical and basic research was conducted on the biological effects of asbestos and the early detection of mesothelioma. It is in this context that the authors became involved in the project "Comprehensive Approach on Asbestos-Related Diseases", supported by the "Special Coordination Funds for Promoting Science and Technology" (headed by Dr. Takemi Otsuki, Department of Hygiene, Kawasaki Medical School, Kurashiki, Japan) from 2006 to 2010. In this project, a case and clinical specimen registration system was established. A feasible clinical trial was established and involved a combined trimodality therapy using anticancer chemotherapy with cisplatin and pemetrexed, followed by extrapleural pneumonectomy and postoperative radiation therapy for early-stage mesothelioma patients [13, 14]. Furthermore, early detection procedures were developed using serum or pleural effusions to measure soluble mesothelin-related peptide (SMRP) and other markers such as osteopontin, vascular endothelial growth Factor (VEGF), and angiopoietin-1 as well as procedures for detection of circulating mesothelioma cells and circulating epithelial cells using peripheral blood [15–19].

For the basic research, the project "Comprehensive Approach on Asbestos-Related Diseases" included three subgroups: (1) analyses of cellular and molecular characteristics using mesothelioma cell lines, (2) an investigation of asbestos-induced carcinogenesis using an animal model, and (3) a study of the immunological effects of silica/asbestos.

The basic research project (3) was performed by us, and in this paper, we introduce our viewpoint that asbestos may cause alteration of immunocompetent cells resulting in chronic inflammation as well as tumor development.

2. Immunological Effects of the Mineral Silicate on Asbestos and Silica

Asbestos is a mineral silicate containing iron, magnesium, and calcium, with a core of "Si" and "O" [6, 7]. Individuals exposed to silica and asbestos develop lung fibrosis known as silicosis and asbestosis, respectively [1–5, 20]. Silicosis is a form of occupational lung disease caused by inhalation of crystalline silica dust, and is marked by inflammation and scarring in the form of nodular lesions in the upper lobes of the lungs [20]. However, asbestosis developpsin patients who inhale a relatively high dose of asbestos (compared with patients with pleural lesions such as mesothelioma and plaque). In addition, the pathology of asbestosis differs from that of

silicosis [1–5]. Asbestosis involves the scarring of lung tissue (around the terminal bronchioles and alveolar ducts). There are two types of fibers: amphibole (thin and straight) and serpentine (curved). The former are primarily responsible for human disease as they are able to penetrate deeply into the lungs. When such fibers reach the alveoli in the lung, where oxygen is transferred into the blood, the foreign bodies (asbestos fibers) cause the activation of the lung's local immune system and provoke an inflammatory reaction [1–5]. This inflammatory reaction can be described as chronic rather than acute, slow ongoing activation of the immune system in an attempt to eliminate the foreign fibers. Macrophages phagocytose the fibers and stimulate fibroblasts to deposit connective tissue. Due to the natural resistance of asbestos fibers to digestion, the macrophage dies off, releasing cytokines and attracting further lung macrophages and fibroblastic cells to lay down fibrous tissue, which eventually forms a fibrous mass [21, 22]. The result is interstitial fibrosis. The fibrotic scar tissue causes alveolar walls to thicken, which reduces elasticity and gas diffusion, reducing oxygen transfer to blood as well as the removing of carbon dioxide.

Furthermore, the complications in silicosis and in people who have inhaled asbestos are also different. Patients with silicosis often present with complications involving autoimmune diseases such as rheumatoid arthritis (known as Caplan's syndrome) [23–25], systemic sclerosis [26, 27], systemic lupus erythematosus [28, 29], and antineutrophil cytoplasmic antibody (ANCA)-related vasculitis/nephritis [30–32]. Although these have been considered adjuvant effects of silica, we have assumed that silica may influence circulating immunocompetent cells, particularly T lymphocytes, and have reported that silica can activate CD4 + 25 + FoxP3 (forkhead box P3) + regulatory T cells (Treg) and responder T cells (Tresp) [33–38]. These activations induce overexpression of CD95/Fas in Treg, resulting in early loss and contamination of activated Tresp, which express CD25 as the marker for activation into a peripheral CD4+25+ subpopulation of T cells. These phenomena seem to induce reduced regulation of autoimmunity [37, 38]. Regarding silica-induced fibrosis (lung and skin), the idea proposed previously [27] is that silica affect alveolar macrophages, endothelial cells, and fibroblasts to modify cytokine production and disturbance of collagen synthesis and degradation, subsequently forming fibrosis of lung and skin lesion. In addition, our findings suggested that silica may influence and alter circulating lymphocytes to disturb autoimmune tolerance [37, 38].

On the other hand, a consideration of the complications of asbestos exposure indicates that the development of cancers is the most important aspect [1–5]. As silica can affect immunocompetent cells, asbestos may possess a similar influence on various immune cells, and the results should be the reduction of tumor immunity. As mentioned above, in addition to lung cancer and mesothelioma, there may be a relatively high prevalence of other cancers among asbestos-inhaled patients [6, 7].

We have reported that the natural killer (NK) cell line and freshly isolated NK cells derived from healthy donors (HDs) exposed to asbestos (chrysotile) for a long period (more than half a year *in vitro* for the cell line and approximately two weeks for the fresh NK cells) showed a reduction of cytotoxicity with decreased expression of their activating receptors, such as NKG2D and 2B4 in the cell line and NKp46 in the fresh NK cells [39]. The NK cell line exposed to asbestos also showed suppressed signaling such as extracellular signal-regulated kinas (ERK) 1/2 in the mitogen-activated protein kinase (MAPK) cascade from activating receptors and also producting of granzyme and perforin [40, 41]. At present, the effects of chrysotile on CD8+ cytotoxic T lymphocytes (CTL) are also being analyzed, and findings show reduction of differentiation from naïve CTL to effector/memory CTL and proliferation [42].

3. Chronic Inflammation and Development of Mesothelioma

Inhaled asbestos is usually handled by alveolar macrophages. Recently, the role of a NOD-like receptor family, the pyrin domain containing 3 (NLRP3) (NACHT, LRR, and PYD domains-containing protein 3; Nalp3) inflammasome, has received attention regarding the handling of these foreign bodies as well as various crystalline substances such as uric acid and cholesterol crystalline causing atherosclerosis [43–47]. The following cellular and molecular events then occur. (1) Capture of silica/asbestos by macrophages and entrapment within lysosomes. (2) Activation of NLRP3 inflammasome to cleave procaspase 1 to an active form. (3) Cleavage of prointerleukin (IL)-1β to an active form for release to form fibrotic nodules. (4) Production of reactive oxygen species (ROS) and reactive nitrogen species (RNS) in the macrophages. (5) Induction of cellular and tissue damage due to the production of ROS and RNS. (6) Apoptosis of the alveolar macrophages. (7) Production of various cytokines/chemokines such as IL-1β tumor necrosis factor (TNF)-α, macrophage inflammatory protein (MIP)-1/2, monocyte-chemoattractant protein-1, and IL-8 to cause chronic inflammation and proliferation of collagenic fibers. (8) Release of silica particles and asbestos fibers from alveolar macrophages and the repetition of similar cellular reactions described above by newly recognized nearby macrophages. (9) Transfer of silica particles and (partially cleaved) asbestos fibers to regional lymph nodes. (10) As these cellular and molecular reactions are continuously repeated, pulmonary fibrosis will appear gradually and progressively [48, 49].

As a result of these cellular and molecular events, cleaved asbestos fibers will accumulate in regional lymphonodes, the distal end of the alveolus, and the pleural cavity, particularly at the opening of lymphatic vessels (Figure 1) [1–5]. Circulating and local immunocompetent cells may encounter asbestos fibers repeatedly, recurrently, and continuously.

Asbestos, particularly amphibole in the form of crocidolite and amosite, includes iron and is considered responsible for the production of reactive oxygen and nitrogen species (ROS and RNS) that may cause DNA damage to nearby cells and induce the development of cancers [50–52]. In addition, these ROS and RNS may develop local chronic inflammation. It has been thought that chronic inflammation contributes to a substantial part of environmental carcinogenesis [53–55]. Various infectious diseases and physical, chemical, and immunological factors participate in inflammation-related carcinogenesis [50–55]. For example, hepatocellular carcinoma, cervical cancer of uterine, and bladder cancer are known to be caused by infection of hepatitis C virus, human papilloma virus (HPV), and *Schistosoma haematobium*, respectively. In addition, *Helicobacter pylori* infection causes gastric cancer, extranodal marginal zone lymphoma of mucosa-associated lymphoid tissue type (MALT lymphoma), and diffuse large B-cell lymphoma. Regarding incidences of these diseases, WHO reported about Hepatitis C virus [56]. As many as 2 to 4 million persons may be chronically infected in the United States, 5 to 10 million in Europe, and about 12 million in India, and most do not know that they are infected. About 150,000 new cases occur annually in the USA and in western Europe and about 350,000 in Japan. Of these, about 25% are symptomatic, but 60 to 80% may progress to chronic liver disease, and 20% of these develop cirrhosis. About 5%–7% of patients may ultimately die of the consequences of the infection. In addition, HPV causes cervical cancer which is the second most common cancer in women worldwide. In 2008, there were an estimated 529,000 new cases and 274,00 deaths due to cervical cancer [57]. It was reported that 8-nitroguanine was formed at sites of carcinogenesis in animal models and patients with various cancer-prone infectious and inflammatory diseases, caused by parasites, viruses, and asbestos exposure. In asbestos-exposed mice, 8-nitroguanine was formed in bronchial epithelial cells, and it is noteworthy that crocidolite induced significantly more intense 8-nitroguanine formation than chrysotile, findings that are inconsistent with their respective carcinogenic potentials [58].

From the above-mentioned basic research (2) concerning "investigation of asbestos-induced carcinogenesis using an animal model" in the "Comprehensive Approach on Asbestos-Related Diseases" project, the importance of iron is supposed even in the development of mesothelioma caused by iron-free chrysotile because of its easy binding to hemoglobin and the induced hemolysis [59–61]. Immunocompetent cells may show some alteration as characteristics of chronic inflammation and features involving reduced tumor immunity.

From this viewpoint, even though we do not observe cellular and molecular changes for immunocompetent cells, we reported interesting findings regarding the reduction of tumor immunity in T cells. We then introduce the findings concerning reduction of CXC chemokine receptor (CXCR) 3 and interferon (IFN)-γ with the activated potential expression of IL-6.

4. Chemokine Receptor CXCR3 Expression and Its Relation To Interferon (IFN)-γ and IL-6

Similar to analyses of NK cells, we adopted a human T-cell line, MT-2, as the chronic and continuous exposure model of T cells. Although the altered features of continuously exposed (to chrysotile) sublines (we established six sublines

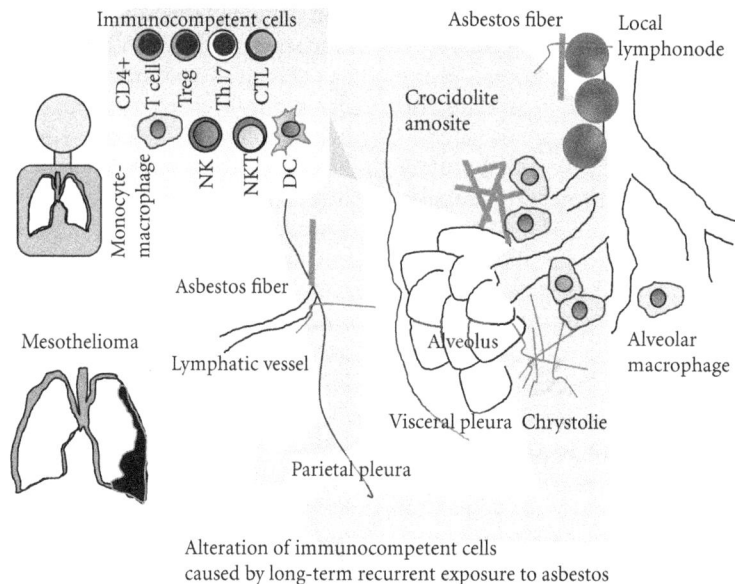

FIGURE 1: Localization of inhaled asbestos fibers in the lung, pleural cavity, and regional lymphonodes. Chronic and recurrent encounters between circulating immunocompetent cells and these moored asbestos fibers causing the long-term alteration of these cells.

independently exposed to chrysotile) of MT-2 were reported previously [62–64], one of the interesting molecular changes regarding tumor immunity is the reduction of CXCR3 expression and IFN-γ production [65–68]. CXCR3 is thought to be important for inflammation, since CXCR3 is known as the receptor for CXCL9, 10, and 11 which induce inflammation. In addition, CXCR3 expressing T cells in the tumor-localized region recruit IFN-γ-producing cells to kill the tumor cells.

Results using the MT-2 cell line model, as shown in the left upper panel of Figure 2, indicated that continuously exposed sublines of MT-2 showed reduced CXCR3 expression on their surface and mRNA expression levels, with reduced production and expression of IFN-γ. Production of the Th1 type CXCR3 ligand CXCL10/IP10 was also significantly reduced in sublines compared with the original line. In addition, another Th1-type chemokine, CCL4/MIP-1β mRNA, was also expressed at low levels in all six sublines compared with the MT-2 original line as previously reported [69, 70]. However, CCR5, the Th1-type receptor for CCL4/MIP-1β, was not reduced significantly through mRNA expression in MT-2Rsts cells. These results indicated that continuous exposure of MT-2 original cells to asbestos altered the expression of Th1-related chemokines (CXCL10/IP10 and CCL4/MIP-1β) and chemokine receptors (CXCR3) [69, 70].

Similar to analyses regarding the effects of asbestos on NK cells, we then tried to determine whether freshly isolated human peripheral CD4+ T cells show a similar alteration *ex vivo* when proliferation is maintained by IL-2-containing medium in the presence of chrysotile as shown in the left lower panel of Figure 2 [69, 70]. After several weeks of co-culture supplemented with IL-2 in the presence or absence of chrysotile, cell surface CXCR3 expression decreased in a dose-dependent manner. Thus, we examined cell surface ex-

pression of CXCR3 and CCR5 in CD4+ T cells derived from six healthy donors, since both receptors are preferentially expressed in Th1/effect or T cells. The expression of CXCR3 was significantly reduced following exposure to 10 μg/mL of chrysotile for 28 days, although this difference seemed to depend on one case in which the expression decreased remarkably [69, 70]. Even if the culture conditions for the CD4+ T cells was limited to a period of around four weeks, four of the six HDs showed a decrease of CXCR3 expression to various degrees, and it might be concluded that asbestos exposure potentiates reduction of CXCR3 expression in CD4+ T cells. These results indicated that CXCR3 expression might be specifically reduced by asbestos exposure. In addition, these experiments revealed decreased IFN-γ expression and production when CD4+ T cells from HDs were cultured with chrysotile for 28 days [69, 70].

Finally, analyses of changes in surface CXCR3 expression on freshly isolated CD4+ T cells from asbestos-exposed patients such as those with pleural plaque (PP) or malignant mesothelioma (MM) were compared with those from HDs. In addition, IFN-γ and IL-6 expression of CD4+ T cells from these patients and HDs was measured with stimulation using anti-CD3/CD28 antibodies with IL-2 [69, 70]. As summarized in the right panel of Figure 2, CXCR3 expression was reduced in CD4+ T cells from asbestos-exposed patients. A comparison of PP and MM patients showed that the expression level of CXCR3 on CD4+ T cells from MM was decreased, although the difference was not statistically significant. Moreover, although IFN-γ expression was only reduced in stimulated CD4+ T cells from MM patients and not in those with PP, IL-6 expression was gradually enhanced in HDs and to a lesser extent in PP, followed by MM. As we reported previously, the plasma level of IL-6 was significantly higher in MM compared to HDs, PP, and silicosis [71].

Asbestos-Induced Cellular and Molecular Alteration of Immunocompetent Cells and Their Relationship with Chronic Inflammation and Carcinogenesis

143

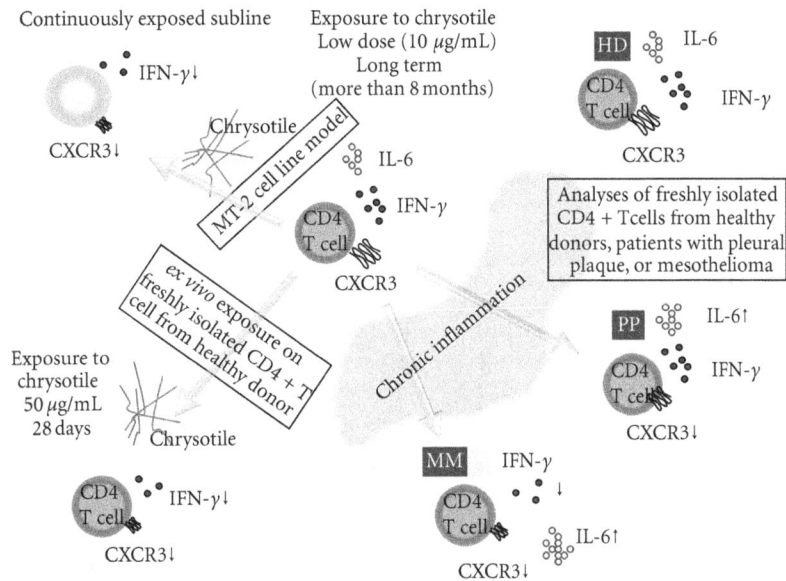

FIGURE 2: Schematic representation of asbestos-induced reduction of expression of a chemokine receptor, CXCR3, and expression and production of IFN-γ with increasing IL-6 using the MT-2 cell line model exposed continuously to a low dose of chrysotile (upper left), an *ex vivo* exposure model using freshly isolated CD4+ T cells from healthy donors (HD) (lower left) as well as analyses of freshly isolated CD4+ T cells from healthy donors and patients with pleural plaque (PP) and malignant mesothelioma (MM) (right).

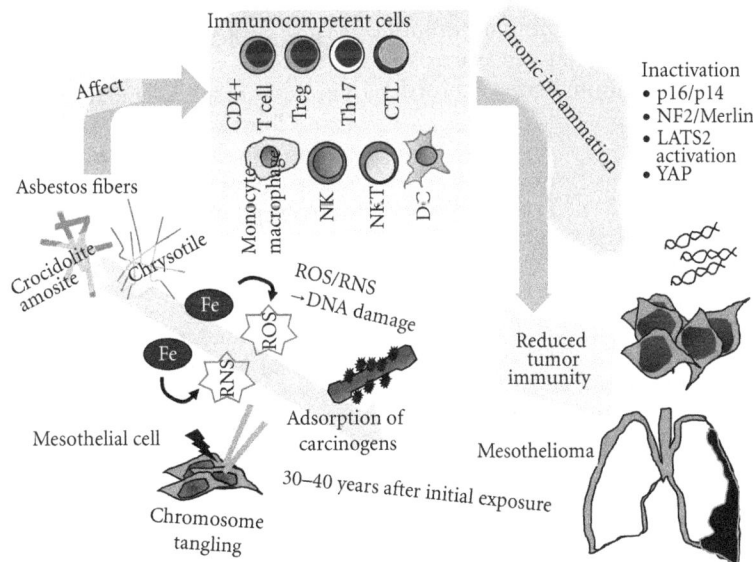

FIGURE 3: Schematic model showing mechanisms of asbestos-induced carcinogenesis and genomic/epigenetic changes found in mesothelioma cells, carcinogenic activities of asbestos fibers, and the relationship of the immunological effects of asbestos in regard to chronic in-flammation and reduced tumor immunity.

Although this may depend on the tumor-producing IL-6 [72–74], our findings indicated that part of the increased IL-6 may be produced by T cells with altered potentials to express cytokines due to continuous exposure to asbestos. Furthermore, IL-6 is an interleukin that acts as a proinflammatory and anti-inflammatory cytokine. It is secreted by T cells and macrophages to stimulate an immune response, for example, during infection and after trauma, especially in the case of burns or other tissue damage leading to inflam-mation [75–78]. For these reasons, it can be assumed that immunocompetent cells possess cellular characteristics of chronic inflammatory alterations during continuous exposure to asbestos, and they then proceed to result in the reduction of tumor immunity as shown in Figure 3. In addition, the supposed difference of immunological effects between silica and asbestos is shown in Figure 4. However, although there is insufficient evidence for all of these sequential modifications of immunocompetent cells,

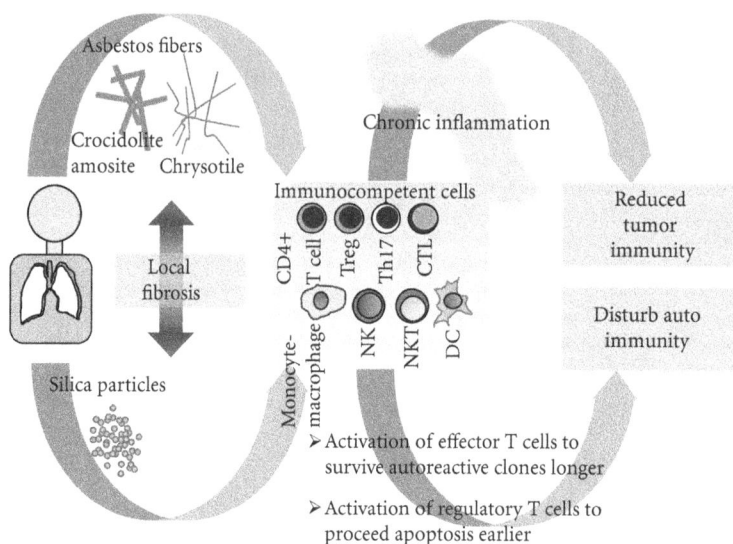

FIGURE 4: Schematic model presenting the differences of immunological effects of silica and asbestos based on our findings. Silica may disturb the autoimmunity, and asbestos may cause reduction of tumor immunity via forming chronic inflammation.

an investigation of their long-term alteration may lead to the development of preventive tools for asbestos-induced malignancies. For example, it may be possible to find some physiologically active substances in the plants or microorganisms to modify or recover the altered function of immunocompetent cells to reconstitute the tumor immunity in asbestos-exposed people, and discriminate iron from bodies exposed to asbesots.

5. Conclusion

As shown in Figure 3, the carcinogenic activity of asbestos encompasses the following phenomena: (1) DNA damage caused by ROS/RNS production due to the iron presentin asbestos fibers, (2) chromosome tangling to result in DNA damage due to the physical features of asbestos fibers, and (3) adsorption of various carcinogensaround the asbestos fibers [59–61]. In addition, the molecular events regarding carcinogenesis found in mesothelioma cells include (1) homogenous deletion of $p16^{INK4a}/p19^{ARF}$ found in more than 90% of cases, (2) inactivation of neurofibromatosis 2 (NF2)/Merlin found in approximately half of the cases, (3) inactivation of the serine/threonine-protein kinase (LATS2) gene in approximately one-third of mesothelioma cell lines and representinga candidate for a novel tumor suppressor in MM, and (4) transcription factor, Yes-associated protein (YAP) involved in the NF2/Merlin-hippo signaling pathway and constitutively dephosphorylated by LATS, and usually acting as an oncogene to bind with the TEAD transcription factor to enhance the cell cycle and resistance to apoptosis [79–81].

Asbestos fibers and the cellular and molecular characteristics of mesothelioma cells may lead to the gradual alteration of immunocompetent cells and subsequent development of chronic inflammation and later reduction of tumor immunity. Investigation of the progression of modification in immunocompetent cells caused by exposure to asbestos may lead to the development of novel methods for the prevention of mesothelioma and other asbestos-related cancers.

Acknowledgments

The authors specially thank Ms. Yoshiko Yamashita, Minako Kato, Tomoko Sueishi, Keiko Kimura, Misao Kuroki, and Naomi Miyahara for their technical help. The experimental results performed by them and presented partly in this paper were supported by Special Funds for Promoting Science and Technology (H18-1-3-3-1), JSPS KAKENHI (22790550, 22700933, 20390178, 20890270, 19689153, 19790431, 19790411, 18390186, 16390175, and 09670500), the Takeda Science Foundation (Tokutei Kenkyu Josei I, 2008), and Kawasaki Medical School Project, Grants (22-B1, 22-A58, 22-A29, 21-401, 21-201, 21-107, 20-411I, 20-210O, 20-109N, 20-402O, and 20-410I).

References

[1] J. E. Craighead and A. R. Gibbs, *Asbestos and Its Diseases*, Oxford University Press, New York, NY, USA, 2008.

[2] R. F. Dodson and S. P. Hammar, *Asbestos: Risk Assessment, Epidemiology, and Health Effects*, CRC Press, Boca Ratton, Fla, USA, 2006.

[3] V. L. Roggli, T. D. Oury, and T. A. Sporn, *Springer Science and Business Media*, Springer Science and Business Media, New York, NY, USA, 2nd edition, 2004.

[4] K. O'Byrne and V. Rusch, *Malignant Pleural Mesothelioma*, Oxford University Press, New York, NY, USA, 2006.

[5] A. Baldi, *Nova Scientific*, Nova Scientific Publishers, New York, NY, USA, 2008.

[6] J. E. Craighead, "Nonthoracic cancers possibly resulting from asbestos exposure," in *Asbestos and Its Diseases*, J. E. Craighead and A. R. Gibbs, Eds., pp. 230–252, Oxford University Press, New York, NY, USA, 2008.

[7] R. Rolston and D. Oury, "Other neoplasia," in *Pathology of Asbestos-Associated Diseases*, V. L. Roggli, T. D. Oury, and T. A. Sporn, Eds., pp. 217–230, Springer Science and Business Media, New York, NY, USA, 2nd edition, 2004.

[8] J. E. Craighead, Al. Gibbs, and F. Pooley, "Mineralogy of asbestos," in *Asbestos and Its Diseases*, J. E. Craighead and A. R. Gibbs, Eds., pp. 23–38, Oxford University Press, New York, NY, USA, 2008.

[9] V. L. Roggli and P. Coin, "Mineralogy of asbestos," in *Pathology of Asbestos-Associated Diseases*, V. L. Roggli, T. D. Oury, and T. A. Sporn, Eds., pp. 1–16, Springer Science and Business Media, New York, NY, USA, 2nd edition, 2004.

[10] N. Kurumatani and S. Kumagai, "Mapping the risk of mesothelioma due to neighborhood asbestos exposure," *American Journal of Respiratory and Critical Care Medicine*, vol. 178, no. 6, pp. 624–629, 2008.

[11] S. Kumagai and N. Kurumatani, "Asbestos fiber concentration in the area surrounding a former asbestos cement plant and excess mesothelioma deaths in residents," *American Journal of Industrial Medicine*, vol. 52, no. 10, pp. 790–798, 2009.

[12] S. Kumagai, N. Kurumatani, T. Tsuda, T. Yorifuji, and E. Suzuki, "Increased risk of lung cancer mortality among residents near an asbestos product manufacturing plant," *International Journal of Occupational and Environmental Health*, vol. 16, no. 3, pp. 268–278, 2010.

[13] T. Yamanaka, F. Tanaka, S. Hasegawa et al., "A feasibility study of induction pemetrexed plus cisplatin followed by extrapleural pneumonectomy and postoperative hemithoracic radiation for malignant pleural mesothelioma," *Japanese Journal of Clinical Oncology*, vol. 39, no. 3, pp. 186–188, 2009.

[14] S. Hasegawa and F. Tanaka, "Malignant mesothelioma: current status and perspective in Japan and the world," *General Thoracic and Cardiovascular Surgery*, vol. 56, no. 7, pp. 317–323, 2008.

[15] R. Ohashi, K. Tajima, F. Takahashi et al., "Osteopontin modulates malignant pleural mesothelioma cell functions in vitro," *Anticancer Research*, vol. 29, no. 6, pp. 2205–2214, 2009.

[16] C. Tabata, N. Hirayama, R. Tabata et al., "A novel clinical role for angiopoietin-1 in malignant pleural mesothelioma," *European Respiratory Journal*, vol. 36, no. 5, pp. 1099–1105, 2010.

[17] A. Yasumitsu, C. Tabata, R. Tabata et al., "Clinical significance of serum vascular endothelial growth factor in malignant pleural mesothelioma," *Journal of Thoracic Oncology*, vol. 5, no. 4, pp. 479–483, 2010.

[18] Y. Okumura, F. Tanaka, K. Yoneda et al., "Circulating tumor cells in pulmonary venous blood of primary lung cancer patients," *Annals of Thoracic Surgery*, vol. 87, no. 6, pp. 1669–1675, 2009.

[19] F. Tanaka, K. Yoneda, N. Kondo et al., "Circulating tumor cell as a diagnostic marker in primary lung cancer," *Clinical Cancer Research*, vol. 15, no. 22, pp. 6980–6986, 2009.

[20] IARC ,International Agency for Research on Cancer, World Health Organization, "Silica, some silicate, coal dust and para-aramid fibrils," *IARC Monographs on the Evaluation of Carcinogenic Risks to Humans*, vol. 68, 1997.

[21] N. H. Heintz and B. T. Mossman, "Molecular responses to asbestos: induction of cell proliferation and apoptosis through modulation of redox-dependent cell signaling pathway," in *Asbestos and Its Diseases*, J. E. Craighead and A. R. Gibbs, Eds., pp. 120–138, Oxford University Press, New York, NY, USA, 2008.

[22] M. A. L. Atkinson, "Molecular and cellular responses to asbestos exposure," in *Asbestos: Risk Assessment, Epidemiology, and Health Effects*, R. F. Dodson and S. P. Hammar, Eds., pp. 39–90, CRC Press, Boca Ratton, Fla, USA, 2006.

[23] A. Caplan, "Rheumatoid disease and pneumoconiosis (Caplan's syndrome)," *Proceedings of the Royal Society of Medicine*, vol. 52, pp. 1111–1113, 1959.

[24] A. Caplan, R. B. Payne, and J. L. Withey, "A broader concept of Caplan's syndrome related to rheumatoid factors," *Thorax*, vol. 17, pp. 205–212, 1962.

[25] J. Schreiber, D. Koschel, J. Kekow, N. Waldburg, A. Goette, and R. Merget, "Rheumatoid pneumoconiosis (Caplan's syndrome)," *European Journal of Internal Medicine*, vol. 21, no. 3, pp. 168–172, 2010.

[26] K. M. Reiser and J. A. Last, "Silicosis and fibrogenesis: fact and artifact," *Toxicology*, vol. 13, no. 1, pp. 51–72, 1979.

[27] U. F. Haustein and U. Anderegg, "Silica induced scleroderma—Clinical and experimental aspects," *Journal of Rheumatology*, vol. 25, no. 10, pp. 1917–1926, 1998.

[28] C. L. Uber and R. A. McReynolds, "Immunotoxicology of silica," *Critical Reviews in Toxicology*, vol. 10, no. 4, pp. 303–320, 1982.

[29] K. Steenland and D. F. Goldsmith, "Silica exposure and autoimmune diseases," *American Journal of Industrial Medicine*, vol. 28, no. 5, pp. 603–608, 1995.

[30] J. W. Cohen Tervaert, C. A. Stegeman, and C. G. M. Kallenberg, "Silicon exposure and vasculitis," *Current Opinion in Rheumatology*, vol. 10, no. 1, pp. 12–17, 1998.

[31] Z. Rihova, D. Maixnerova, E. Jancova et al., "Silica and asbestos exposure in ANCA-associated vasculitis with pulmonary involvement," *Renal Failure*, vol. 27, no. 5, pp. 605–608, 2005.

[32] J. Bartuňková, D. Pelclová, Z. Fenclová et al., "Exposure to silica and risk of ANCA-associated vasculitis," *American Journal of Industrial Medicine*, vol. 49, no. 7, pp. 569–576, 2006.

[33] P. Wu, Y. Miura, F. Hyodoh et al., "Reduced function of CD4+25+ regulatory T cell fraction in silicosis patients," *International Journal of Immunopathology and Pharmacology*, vol. 19, no. 2, pp. 357–368, 2006.

[34] H. Hayashi, M. Maeda, S. Murakami et al., "Soluble interleukin-2 receptor as an indicator of immunological disturbance found in silicosis patients," *International Journal of Immunopathology and Pharmacology*, vol. 22, no. 1, pp. 53–62, 2009.

[35] H. Hayashi, Y. Miura, M. Maeda et al., "Reductive alteration of the regulatory function of the CD4(+)CD25(+) T cell fraction in silicosis patients," *International Journal of Immunopathology and Pharmacology*, vol. 23, no. 4, pp. 1099–1109, 2010.

[36] H. Hayashi, Y. Nishimura, F. Hyodo et al., "Dysregulation of autoimmunity caused by silica exposure: fas-mediated apoptosis in T lymphocytes derived from silicosis patients," in *Autoimmune Disorders: Symptoms, Diagnosis and Treatment*, M. E. Petro, Ed., pp. 293–301, Nova Science Publishers, Hauppauge, NY, USA, 2011.

[37] N. Kumagai, H. Hayashi, M. Maeda et al., "Immunological effects of silica and related dysregulation of autoimmunity," in *Autoimmune Disorder*, C. Mavragani, Ed., vol. 1, InTech Open Access Publisher, Croatia.

[38] T. Otsuki, H. Hayashi, Y. Nishimura et al., "Dysregulation of autoimmunity caused by silica exposure and alteration of Fas-mediated apoptosis in T lymphocytes derived from silicosis patients," *International Journal of Immunopathology and Pharmacology*, vol. 24, no. 1, supplement, pp. 11S–16S, 2011.

[39] Y. Nishimura, Y. Miura, M. Maeda et al., "Impairment in cytotoxicity and expression of NK cell-activating receptors on human NK cells following exposure to asbestos fibers," *International Journal of Immunopathology and Pharmacology*, vol. 22, no. 3, pp. 579–590, 2009.

[40] Y. Nishimura, M. Maeda, N. Kumagai, H. Hayashi, Y. Miura, and T. Otsuki, "Decrease in phosphorylation of ERK following decreased expression of NK cell-activating receptors in human NK cell line exposed to asbestos," *International Journal of Immunopathology and Pharmacology*, vol. 22, no. 4, pp. 879–888, 2009.

[41] Y. Nishimura, N. Kumagai, M. Maeda et al., "Suppressive effect of asbestos on cytotoxicity of human NK cells," *International Journal of Immunopathology and Pharmacology*, vol. 24, no. 1, supplement, pp. 5S–10S, 2011.

[42] M. Maeda, Y. Nishimura, N. Kumagai et al., "Dysregulation of the immune system caused by silica and asbestos," *Journal of Immunotoxicology*, vol. 7, no. 4, pp. 268–278, 2010.

[43] C. Dostert, V. Pétrilli, R. Van Bruggen, C. Steele, B. T. Mossman, and J. Tschopp, "Innate immune activation through Nalp3 inflammasome sensing of asbestos and silica," *Science*, vol. 320, no. 5876, pp. 674–677, 2008.

[44] S. L. Cassel, S. C. Eisenbarth, S. S. Iyer et al., "The Nalp3 inflammasome is essential for the development of silicosis," *Proceedings of the National Academy of Sciences of the United States of America*, vol. 105, no. 26, pp. 9035–9040, 2008.

[45] S. Benko, D. J. Philpott, and S. E. Girardin, "The microbial and danger signals that activate Nod-like receptors," *Cytokine*, vol. 43, no. 3, pp. 368–373, 2008.

[46] F. Martinon and J. Tschopp, "Inflammatory caspases: linking an intracellular innate immune system to autoinflammatory diseases," *Cell*, vol. 117, no. 5, pp. 561–574, 2004.

[47] K. Schroder and J. Tschopp, "The Inflammasomes," *Cell*, vol. 140, no. 6, pp. 821–832, 2010.

[48] R. F. Hamilton, S. A. Thakur, and A. Holian, "Silica binding and toxicity in alveolar macrophages," *Free Radical Biology and Medicine*, vol. 44, no. 7, pp. 1246–1258, 2008.

[49] S. A. Thakur, R. F. Hamilton, and A. Holian, "Role of scavenger receptor A family in lung inflammation from exposure to environmental particles," *Journal of Immunotoxicology*, vol. 5, no. 2, pp. 151–157, 2008.

[50] D. Upadhyay and D. W. Kamp, "Asbestos-induced pulmonary toxicity: role of DNA damage and apoptosis," *Experimental Biology and Medicine*, vol. 228, no. 6, pp. 650–659, 2003.

[51] K. Bhattacharya, E. Dopp, P. Kakkar et al., "Biomarkers in risk assessment of asbestos exposure," *Mutation Research*, vol. 579, no. 1-2, pp. 6–21, 2005.

[52] G. Liu, R. Beri, A. Mueller, and D. W. Kamp, "Molecular mechanisms of asbestos-induced lung epithelial cell apoptosis," *Chemico-Biological Interactions*, vol. 188, no. 2, pp. 309–318, 2010.

[53] M. Philip, D. A. Rowley, and H. Schreiber, "Inflammation as a tumor promoter in cancer induction," *Seminars in Cancer Biology*, vol. 14, no. 6, pp. 433–439, 2004.

[54] S. C. Robinson and L. M. Coussens, "Soluble mediators of inflammation during tumor development," *Advances in Cancer Research*, vol. 93, pp. 159–187, 2005.

[55] A. Federico, F. Morgillo, C. Tuccillo, F. Ciardiello, and C. Loguercio, "Chronic inflammation and oxidative stress in human carcinogenesis," *International Journal of Cancer*, vol. 121, no. 11, pp. 2381–2386, 2007.

[56] WHO website, "Hepatitis C," http://www.who.int/csr/disease/hepatitis/whocdscsrlyo2003/en/index4.html#incidence.

[57] WHO website, "New and under-utilized vaccines implementation (NUVI), human papillomavirus (HPV)," http://www.who.int/nuvi/hpv/en/.

[58] Y. Hiraku, S. Kawanishi, T. Ichinose, and M. Murata, "The role of iNOS-mediated DNA damage in infection-and asbestos-induced carcinogenesis," *Annals of the New York Academy of Sciences*, vol. 1203, pp. 15–22, 2010.

[59] S. Toyokuni, "Mechanisms of asbestos-induced carcinogenesis," *Nagoya Journal of Medical Science*, vol. 71, no. 1-2, pp. 1–10, 2009.

[60] S. Toyokuni, "Role of iron in carcinogenesis: cancer as a ferrotoxic disease," *Cancer Science*, vol. 100, no. 1, pp. 9–16, 2009.

[61] H. Nagai and S. Toyokuni, "Biopersistent fiber-induced inflammation and carcinogenesis: lessons learned from asbestos toward safety of fibrous nanomaterials," *Archives of Biochemistry and Biophysics*, vol. 502, no. 1, pp. 1–7, 2010.

[62] Y. Miura, Y. Nishimura, H. Katsuyama et al., "Involvement of IL-10 and Bcl-2 in resistance against an asbestos-induced apoptosis of T cells," *Apoptosis*, vol. 11, no. 10, pp. 1825–1835, 2006.

[63] T. Otsuki, M. Maeda, S. Murakami et al., "Immunological effects of silica and asbestos," *Cellular & Molecular Immunology*, vol. 4, no. 4, pp. 261–268, 2007.

[64] M. Maeda, Y. Miura, Y. Nishimura et al., "Immunological changes in mesothelioma patients and their experimental detection," *Clinical Medicine. Circulatory, Respiratory and Pulmonary*, vol. 2, pp. 11–17, 2008.

[65] P. Romagnani, L. Lasagni, F. Annunziato, M. Serio, and S. Romagnani, "CXC chemokines: the regulatory link between inflammation and angiogenesis," *Trends in Immunology*, vol. 25, no. 4, pp. 201–209, 2004.

[66] U. P. Singh, C. Venkataraman, R. Singh, and J. W. Lillard, "CXCR3 axis: role in inflammatory bowel disease and its therapeutic implication," *Endocrine, Metabolic and Immune Disorders*, vol. 7, no. 2, pp. 111–123, 2007.

[67] A. M. Fulton, "The chemokine receptors CXCR4 and CXCR3 in cancer," *Current Oncology Reports*, vol. 11, no. 2, pp. 125–131, 2009.

[68] S. Lacotte, S. Brun, S. Muller, and H. Dumortier, "CXCR3, inflammation, and autoimmune diseases," *Annals of the New York Academy of Sciences*, vol. 1173, pp. 310–317, 2009.

[69] M. Maeda, Y. Nishimura, H. Hayashi et al., "Decreased CXCR3 expression in CD4+ T cells exposed to asbestos or derived from asbestos-exposed patients," *American Journal of Respiratory Cell and Molecular Biology*, vol. 45, no. 4, pp. 795–803, 2011.

[70] M. Maeda, Y. Nishimura, H. Hayashi et al., "Reduction of CXCR3 in an in vitro model of continuous asbestos exposure on a human T-cell line, MT-2," *American Journal of Respiratory Cell and Molecular Biology*, vol. 45, no. 3, pp. 470–479, 2011.

[71] S. Murakami, Y. Nishimura, M. Maeda et al., "Cytokine alteration and speculated immunological pathophysiology in silicosis and asbestos-related diseases," *Environmental Health and Preventive Medicine*, vol. 14, no. 4, pp. 216–222, 2009.

[72] T. Motoyama, T. Honma, H. Watanabe, S. Honma, T. Kumanishi, and S. Abe, "Interleukin 6-producing malignant mesothelioma," *Virchows Archiv B*, vol. 64, no. 6, pp. 367–372, 1993.

[73] N. Kimura, T. Ogasawara, S. Asonuma, H. Hama, T. Sawai, and T. Toyota, "Granulocyte-colony stimulating factor-and interleukin 6-producing diffuse deciduoid peritoneal mesothelioma," *Modern Pathology*, vol. 18, no. 3, pp. 446–450, 2005.

[74] S. Eikawa, Y. Ohue, K. Kitaoka et al., "Enrichment of Foxp3+ CD4 regulatory T cells in migrated T cells to IL-6-and IL-8-expressing tumors through predominant induction of CXCR1

Asbestos-Induced Cellular and Molecular Alteration of Immunocompetent Cells and Their Relationship with Chronic Inflammation and Carcinogenesis

147

by IL-6," *Journal of Immunology*, vol. 185, no. 11, pp. 6734–6740, 2010.

[75] D. R. Hodge, E. M. Hurt, and W. L. Farrar, "The role of IL-6 and STAT3 in inflammation and cancer," *European Journal of Cancer*, vol. 41, no. 16, pp. 2502–2512, 2005.

[76] A. M. W. Petersen and B. K. Pedersen, "The role of IL-6 in mediating the anti-inflammatory effects of exercise," *Journal of Physiology and Pharmacology*, vol. 57, no. 10, pp. 43–51, 2006.

[77] K. Heikkilä, S. Ebrahim, and D. A. Lawlor, "Systematic review of the association between circulating interleukin-6 (IL-6) and cancer," *European Journal of Cancer*, vol. 44, no. 7, pp. 937–945, 2008.

[78] T. Kishimoto, "IL-6: from its discovery to clinical applications," *International Immunology*, vol. 22, no. 5, pp. 347–352, 2010.

[79] T. Yokoyama, H. Osada, H. Murakami et al., "YAP1 is involved in mesothelioma development and negatively regulated by Merlin through phosphorylation," *Carcinogenesis*, vol. 29, no. 11, pp. 2139–2146, 2008.

[80] H. Murakami, T. Mizuno, T. Taniguchi et al., "LATS2 is a tumor suppressor gene of malignant mesothelioma," *Cancer Research*, vol. 71, no. 3, pp. 873–883, 2011.

[81] Y. Sekido, "Inactivation of Merlin in malignant mesothelioma cells and the Hippo signaling cascade dysregulation," *Pathology International*, vol. 61, no. 6, pp. 331–344, 2011.

16

How Do Cytokines Trigger Genomic Instability?

Ioannis L. Aivaliotis,[1] Ioannis S. Pateras,[1] Marilena Papaioannou,[1] Christina Glytsou,[1] Konstantinos Kontzoglou,[2] Elizabeth O. Johnson,[3] and Vassilis Zoumpourlis[4]

[1] *Molecular Carcinogenesis Group, Department of Histology and Embryology, School of Medicine,*
National and Kapodistrian University of Athens, 11527 Athens, Greece
[2] *2nd Department of Propedeutic Surgery, Laikon General Hospital, School of Medicine,*
National and Kapodistrian University of Athens, 11527 Athens, Greece
[3] *Department of Anatomy, School of Medicine, National and Kapodistrian University of Athens, 11527 Athens, Greece*
[4] *Institute of Biology, Medicinal Chemistry and Biotechnology, National Hellenic Research Foundation, 11635 Athens, Greece*

Correspondence should be addressed to Ioannis L. Aivaliotis, ioaival@gmail.com and Vassilis Zoumpourlis, vzub@eie.gr

Academic Editor: Vassilis Gorgoulis

Inflammation is a double-edged sword presenting a dual effect on cancer development, from one hand promoting tumor initiation and progression and from the other hand protecting against cancer through immunosurveillance mechanisms. Cytokines are crucial components of inflammation, participating in the interaction between the cells of tumor microenvironment. A comprehensive study of the role of cytokines in the context of the inflammation-tumorigenesis interplay helps us to shed light in the pathogenesis of cancer. In this paper we focus on the role of cytokines in the development of genomic instability, an evolving hallmark of cancer.

1. Introduction

Contemporary approaches in cancer research have been influenced by the accumulating data unveiling the importance of inflammatory components in the tumor microenvironment. It is becoming more clearly evident that inflammation demonstrates a dualism effect on cancer development in close resemblance to a *ying-yang* pattern. Inflammation may exhibit either a pro- or an antitumorigenic effect. Cytokines possess a central role in the inflammatory component implicated in the interplay between the host's stromal cells and the tumor cells during tumorigenesis. In this paper we are shedding light on the molecular pathways linking cytokines with the induction of genomic instability, an evolving hallmark of cancer.

2. Interrelation of Inflammation and Carcinogenesis

Rudolf Virchow was the first to observe, back in the nineteenth, the presence of leukocytes inside tumors and this observation was the first indication of a possible linkage between inflammation and cancer. The last decade intensive research has focused on the molecular pathways involved in the above linkage and it is now well understood that chronic inflammation plays a significant role in the carcinogenesis process [1].

In 1909, Paul Ehrlich proposed the immunosurveillance theory, later established by Thomas and Burnet, which supports the tumor suppressive role of the immune system [2–4]. Dunn and his colleagues suggested in 2004 that a new theory should be adopted to describe the relationship between the immune response and tumorigenesis, called immunoediting [5]. According to this theory, three distinct stages exist describing the interrelation between immunity and carcinogenesis. The first stage, termed elimination, represents the period in which the immune system, through successful immunosurveillance, destroys precancerous and cancerous cells. In equilibrium, the second stage, cancer cells have begun to develop abilities to avoid immunosurveillance mechanisms but the balance between "immune patrol" and tumorigenesis is still preserved. In the third stage, named escape, the cancer cells manage to evade the surveillance system of the organism, resulting in aberrant cell proliferation and tumor development. Interestingly, it seems that the immune response to the tumor causes an "immunosculpting" effect on cancer cells that enables them to resist immunological recognition or to exert enhanced defense mechanisms against immunosurveillance [5].

Recent advances in cancer biology research have demonstrated that a chronic indolent inflammation environment harbors potential tumor promoting mechanisms [1]. According to Hanahan and Weinberg, one of the emerging hallmarks of cancer is the ability to escape immunosurveillance and an enabling characteristic for the acquisition of these capabilities is the inflammation propagated by the tumor [6]. Compelling evidence of the last decade supports the notion that the inflammatory microenvironment is important for the survival of tumors [1]. It seems that inflammatory cells of the innate immunity usually display a tumor promoting role whereas cells of adaptive immunity appear to have a tumor suppressive effect [1, 7].

Inadequate pathogen eradication or continuous exposures to chemical carcinogens preserve a chronic inflammation environment that may enhance tumorigenesis [8]. There is evidence supporting that several unresolved inflammatory reactions following persistent pathogen infection promote human malignancies [9]. Pathogens contain specific patterns, known as pathogen-associated molecular patterns (PAMPs), which are recognized by host receptors, named pattern recognition receptors (PRRs), including Toll-like receptors (TLRs), nucleotide-binding oligomerization domain-like receptors (NOD-like) receptors, C-type lectin receptors (CLRs), and triggering receptors expressed on myeloid cells (TREMs) [10, 11]. The binding between PAMPs and PRRs leads to inflammation-related cell activation and triggers host immune defense mechanisms against foreign pathogens [10]. In relation to the previous part, it is well established that chronic viral hepatitis B and C is strongly associated with the development of hepatocellular carcinoma. In this case, excessive host reaction towards the viral infection is believed to play a significant role for the inflammation-mediated liver carcinoma. On the other hand and not mutually exclusive there are several viral infections in which the virus itself through its oncogenic potential is mainly responsible for the cancer development [12]. Human Papilloma Virus (HPV) infection is associated with cervical cancers and Epstein-Barr infection bears significant association with Burkitt lymphoma and nasopharyngeal carcinoma. Particular types of HPV produce the E6 and E7 oncoproteins which interfere with the p53 and Retinoblastoma protein (pRb) pathways, respectively. Epstein-Barr virus (EBV) latent membrane protein 1 (LMP1) is critical for EBV-induced cellular transformation through the activation of NF-κB (analyzed hereinafter), a transcription factor promoting cell survival [13].

In the context of chemically induced carcinogenesis, a typical example involves cigarette smoking, in which the tumorigenic activity is partially attributed to its ability to induce chronic inflammation [14]. Also asbestos or silica exposure may cause inflammation of the lung and

subsequent bronchial carcinoma [15]. The mechanism of the induced inflammation by the above inhaled particles occurs by means of prointerleukin-1β (IL-1β) processing by the inflammasome [1, 16]. The inflammasome is a protein complex which includes two caspase-1 molecules, is related to cryopyrin protein, and leads to IL-1β cross-activation and maturation [17].

3. Cellular Context and Cytokine Signaling in the Tumor Microenvironment

The cellular context of the tumor's microenvironment includes cancer cells and surrounding stromal cells (involving fibroblasts, endothelial cells, pericytes, and mesenchymal cells) along with the infiltrating cells of the innate and adaptive immunity [1]. Innate immune cells include macrophages, myeloid-derived suppressor cells (MDSCs), neutrophils, and mast, dendritic, and natural killer (NK) cells, while adaptive immune cells consist of T and B lymphocytes. The only immune cells with no known tumor promoting role to date are NK cells [1]. MDSCs share common characteristics with macrophages, neutrophils, and dendritic cells and they help in tumor angiogenesis as well as suppression of antioncogenic immune responses [18, 19]. Tumor-associated macrophages (TAMs) and T cells are among the most frequently observed cells within the tumor microenvironment. TAMs are a heterogeneous cell population which evidently exerts a very significant role in tumor cell survival, growth, and progression and can be considered obligate partners for tumor cell migration, invasion, and metastasis [20, 21].

The interaction between the epithelial and stromal cells comprising the inflammatory microenvironment is mediated by a class of molecules named as cytokines. Cytokines are cell-signaling protein molecules with effects on intercellular communication; they include interleukins, lymphokines, and chemokines. Interleukins were initially observed to be expressed by leukocytes and present immunoregulatory action. Lymphokines are produced by lymphocytes and include IL-2, IL-3, IL-4, IL-5, IL-6, GMC-SF (Granulocyte-macrophage colony-stimulating factor), and IFN-γ (Interferon-gamma). Chemokines are chemoattractant cytokines and are named this way due to their ability to control leukocyte migration to inflammation sites through chemotaxis [22, 23]. The nomenclature of chemokines is based on the number and location of the N-terminal cysteine residues, divided in four different groups: CXC, CC, CX3C, and C [23].

In accordance with increased expression of pro- or anti-inflammatory cytokines and to tumor progression, TAMs may be classified into M_1 and M_2 types. M_1 macrophages are mostly found in early stage tumor development and may release proinflammatory cytokines (TNF-a, IL-1, IL-6, IL-12, IL-23) and chemokines CXCL19 and CXCL10. Their physiological role involves Th1/Th17 cellular responses and NK cell development and differentiation. M_2 macrophages may be mostly found in more established tumors, show increased expression of IL-10 and transforming growth factor β (TGF-β) as well as chemokines CCL17, CCL22, and CCL24, and are considered to promote tumor angiogenesis. This subtype of TAMs encourages Th2 and regulatory T cells (Tregs) recruitment [1, 7]. According to their effector functions, T cells can be subdivided in CD8+ cytotoxic T cells (CTLs) and CD4+ helper T (Th) cells, including Th1, Th2, and T regulatory (Treg) cells [1]. T cells may exhibit either antitumor immune responses or promote tumor growth [1, 24].

The mechanisms of inflammation-mediated tumor promotion involve secretion of specific cytokines by both inflammatory and tumor cells as well as activation of transcription factors, mainly (Nuclear Factor-κB) NF-κB, (Signal Transducers and Activators of Transcription) STAT3, and (Activator Protein-1) AP 1. NF-κB and STAT3 expression can be detected in most cancers and these transcription factors activate genes responsible for cell survival, proliferation, angiogenesis, invasiveness, and production of cytokines [1, 25, 26]. NF-κB belongs to a family of transcription factors that regulate the secretion of many inflammatory cytokines, specific adhesion molecules, and the prostaglandin biosynthetic pathway. It also regulates the expression of antiapoptotic proteins and angiogenic factors in a tissue-specific manner [15, 27]. There are three distinct activation pathways of NF-κB, the classical, the alternative, and the atypical pathways, and all seem to support tumorigenesis [18, 28, 29]. The classical pathway is triggered by pathogen infection, T-cell receptor (TCR) engagement, and proinflammatory cytokines, such as IL-1 and TNF-α [28, 29]. The NF-κB alternative pathway is triggered by cytokines belonging to the TNF family and involves the IKKα homodimer and the p52/RelB transcription factors [18, 28]. Finally, the atypical NF-κB pathway is IKK independent and is triggered by several stimuli such as hypoxia and hydrogen peroxide attack of the cells [29]. STAT3 activation induces the expression of *Bcl2* and *Bcl-xL* antiapoptotic genes, *Cyclin D1* or *c-Myc* proliferation genes, and *VEGF* (vascular endothelial growth factor) angiogenesis promoting gene [18, 27]. AP-1 is a dimeric transcription factor composed of members of the Jun, Fos, activating transcription factor (ATF) and musculoaponeurotic fibrosarcoma (Maf) protein families. Several growth factors and cytokines induce MAPK signalling pathway which in turn activates AP-1 [30].

Cytokine IFN-γ most frequently produced by cytotoxic CD8+ and CD4+ Th1 T cells has been recognized as a dominant tumor-inhibitory force (Table 1). On the other side, the cytokines interleukin-6 (IL-6), TNF-α, IL-1β, and IL-23 are mostly considered as tumor promoting [7]. Secretion of the latter cytokines is mainly induced by tumor associated macrophages and myeloid-derived suppressive cells [7]. One of the previously mentioned cytokines, macrophage-derived IL-1, was indicated to promote both inflammation and angiogenesis under a hypoxic environment which in turn ascertains the important role of IL-1 in inflammation-mediated tumorigenesis [31]. Additionally, IL-1β has recently been shown to induce a subset of MDSC in the tumor microenvironment capable of suppressing the development and function of NK cells [32]. Proinflammatory cytokines include IL-2, IL-6, IL-11, IL-15, IL-17, IL-23, TNF-α, and chemokine IL-8 and anti-inflammatory cytokines include IL-4, IL-10, IL-13, transforming growth factor β (TGF-β) and interferon (IFN)-α [22, 33]. Proinflammatory cytokines, such as IL-1 and TNF-α which in turn stimulate IL-8, may also stimulate chemokines [22, 34].

An interesting example of cytokine-mediated carcinogenesis is human malignant mesothelioma which may be provoked by chronic exposure to asbestos. TNF-α displays a tumor promoting role in this type of cancer by helping the survival of the mesothelial cells which, in turn, have been damaged by asbestos exposure and may develop mutagenic phenotypes rendering them susceptible to carcinogenesis [10, 35]. TNF-α favors the survival of tumor cells by inducing the expression of antiapoptotic genes encoded in an NF-κB-dependent manner [10]. As a result, TNF-α—which may be released by host as well as cancer cells—significantly influences the initiation and progression stages of all cancer types [10]. Similarly, IL-6 exerts a tumor promoting role by enhancing cell cycle progression and suppressing apoptosis. Its signaling transduction pathway is induced by STAT3 transcription factor [10, 21]. IL-17 triggers the secretion of TNF-α, IL-6, and IL-1β proinflammatory cytokines and is produced by IL-23-dependent STAT3 activation [10, 36]. IL-23 has similar functions to IL-17, induces IFN-γ, and IL-12 production by activated T cells, and leads to enhanced proliferation of memory T cells [10, 37]. IL-10 activates STAT3 transcription factor but has an opposite effect in carcinogenesis to IL-6 [10]. Additionally it inhibits NF-κB activation and in this manner also inhibits production of TNF-α, IL-6, and IL-12 [10, 38]. IL-10 probably exhibits its tumor suppressive role by inhibiting the production of the above mentioned proinflammatory cytokines [10]. IL-10 may also exert protumorigenic activity through STAT3 activation by upregulating *Bcl2* and *Bcl-xL* antiapoptotic genes, therefore confirming its dual role in the process of carcinogenesis [10, 39]. Other cytokines have a known dual function in carcinogenesis (Table 1). Interestingly, TGF-β may exert a tumor suppressive role in the beginning of the process and a tumor promoting role at the late stages of carcinogenesis [40].

Lessons for the mechanisms implicated in inflammation-associated carcinomas have paradoxically been taken by the study of cellular senescence [41]. It has been shown that senescent cells are metabolically active and secrete several factors that may alter their own as well as the tumor microenvironment [42]. The phenomenon of such secretion by the senescent cells has been termed senescence-associated secretory phenotype (SASP) [43]. SASP acts in a cell-autonomous paracrine manner and has both a bright side, favoring senescence in normal or low-grade preneoplastic cells, as well as a dark side, facilitating evasion of senescence in high grade preneoplastic or cancerous cells [44]. The SASP involves a number of inflammatory cytokines, such as IL-6 and chemokine IL-8, which constitute two of its prominent components [43]. It has been demonstrated that the proinflammatory cytokines IL-6 and IL-8 are upregulated by persistent DNA Damage Response (DDR) activation which in turn boost the DDR signaling pathway forming a positive feedback loop [43]. Of note, IL-6 and IL-8 are known to play

TABLE 1: Potential pro- and antitumorigenic roles of cytokines.

Cytokine	Protumorigenic role	Antitumorigenic role	Unspecified yet role in tumorigenesis	References
IL-1 (α and β)	Tumor growth, invasion and metastasis, mainly through the action of IL-1β in promoting local inflammatory responses as well as angiogenesis.	Restraint of tumor growth through activation of innate and specific immune effector mechanisms mainly through the action of IL-1α.		[17]
IL-2		(i) Stimulates growth, differentiation, and survival of cytotoxic T cells (ii) Induces differentiation and proliferation of NK cells		[53]
IL-3			Stimulates the differentiation and growth of multipotent hematopoietic stem cells	[54]
IL-4	(i) Decreases the production of Th1 cells, macrophages, and IFN-gamma (ii) Has been shown to drive dedifferentiation, mitogenesis and metastasis in rhabdomyosarcoma	Stimulation of activated B-cell and T-cell proliferation, and differentiation of CD4+ T-cells into Th2 cells	IgE and class II MHC expression on B cells	[55]
IL-5			(i) Stimulates B cell growth (ii) Stimulates eosinophil growth and function	[56]
IL-6	(i) Promotion of tumor cell proliferation and inhibition of their apoptosis through activation of STAT-3. (ii) Facilitation of senescence evasion in high-grade preoplastic or cancerous cells through mechanisms of SASP. (iii) Favours metastasis	(i) Mediator of the acute phase response (ii) Induction of senescence in normal or low grade preoplastic cells.		[25, 44]
IL-7		Stimulates proliferation of B cells, T cells, and NK cells	Stimulates the differentiation of multipotent hematopoietic stem cells	[57]
IL-8	Significant role in tumor growth, angiogenesis, epithelial to mesenchymal transition (EMT) and invasiveness	(i) Induction of chemotaxis in its target cells (neutrophils, granulocytes) (ii) Induction of senescence in normal or low-grade preoplastic cells.		[43, 44]
IL-9	(i) Potential role in tumorigenesis due to antiapoptotic and growth factor activities (ii) Deregulated IL-9 response may lead to malignant transformation through Jak/STAT activation		Regulation of hematopoietic cells	[58]

TABLE 1: Continued.

Cytokine	Protumorigenic role	Antitumorigenic role	Unspecified yet role in tumorigenesis	References
IL-10	Potential tumor promoting activity through activation of STAT3 and consequent upregulation of BCL-2 or BCL-XL antiapoptotic genes.	(i) Enhances B-cell survival, proliferation, and antibody production. (ii) Inhibition of tumor development and progression through inhibition of NF-κB activation, TNF-α, IL-6, and IL-12. (iii) Suppression of angiogenesis through inhibition of the tumor stroma.		[10, 59]
IL-11			(i) Regulator of haematopoiesis (ii) Stimulation of megakaryocyte maturation	[60]
IL-12		(i) Stimulates the growth and function of T cells (ii) Stimulates the production of IFN-γ, TNF-α (iii) Induces cell-mediated immune responses (iv) Exhibits antiangiogenic activity		[61]
IL-13			Induces IgE secretion	[62]
IL-14			Regulates the growth and proliferation of B cells	[63]
IL-15		(i) Stimulates growth, differentiation and survival of cytotoxic T cells (ii) Induces differentiation and proliferation of NK cells		[53]
IL-16		Chemoattractant for certain immune cells expressing the cell surface molecule CD4.		[64]
IFN-γ		(i) Produced by cytotoxic CD8+ and CD4+ Th1 T cells (ii) Exhibits an overall significant tumor inhibitory action.		[1, 7]
TGF-β	Tumor promoting role at the late stages of carcinogenesis	Tumor suppressive role in the beginning of carcinogenesis		[40]
OPN	Implicated in enhanced metastasis and invasion of tumor cells			[68]
CCL2	(i) Induces the recruitment of macrophages (ii) Induces angiogenesis and matrix remodeling (iii) Promotes prostate cancer cell proliferation, migration, invasion, and survival			[69]
CCL21		(i) Immune-mediated antitumor response (chemoattraction of B cells and NK cells to the lymph nodes) (ii) Angiostatic effect		[70, 71]

TABLE 1: Continued.

Cytokine	Protumorigenic role	Antitumorigenic role	Unspecified yet role in tumorigenesis	References
CCL16		Augments the cytotoxic activities of effector T cells		[72]
CXCL12	(i) Suppress antitumor immunity in the tumor microenvironment (ii) Regulates trafficking of immature and maturing immune cells (iii) Promotes angiogenesis (iv) Facilitates metastasis			[73]

NK cells: Natural Killer cells; IgE: Immunoglobulin E; MHC: Major Histocompatibility Complex; OPN: Osteopontin.

a significant role in tumor growth, angiogenesis, epithelial to mesenchymal transition (EMT), and invasiveness [43, 45]. The aforementioned part indicates that the IL-6/IL-8 duet seems to exert both anti- as well as protumorigenic functions. IL-1 proinflammatory cytokine (both α and β forms) is also a SASP component secreted at lower levels compared to IL-6/IL-8 and interestingly IL-1α regulates the signaling network that leads to the expression of the latter duet [46].

4. Cytokine-Mediated Growth Signaling, Replication Stress, and Genomic Instability

Recently it has been proposed that genomic instability is an evolving hallmark of cancer and Hanahan and Weinberg (2011) established that it constitutes an enabling characteristic for the acquisition of the essential functions (hallmarks) of a cancerous cell [6, 47]. Genomic instability is present in most human cancer types and has various forms. The most frequently occurring one is chromosomal instability (CIN) which involves structural and numerical chromosomal changes that occur in cancer cells over time [47]. Another common form is microsatellite instability (MIN or MSI) which is caused by alterations in DNA mismatch repair genes and leads to changes in the number of oligonucleotide repeats in microsatellite sequences [47]. Genomic instability is observed from early stages of cancer development, even before the acquisition of the cell's cancerous phenotype [48]. CIN is more frequently observed in human cancers compared to MIN and this might be explained by the formation of double-strand breaks (DSBs) in precancerous lesions and cancers according to the oncogene-induced DNA damage model [49].

Recent studies have unveiled the potential effects of enhanced growth signalling in age and/or age-related diseases, such as cancer. The proposed mechanism involves the induction of DNA replication stress which leads to the formation of DNA double-strand breaks (DSBs), favouring genomic instability and tumorigenesis [50, 51]. Indeed, upregulation of growth-signalling pathways in all eukaryotes has been shown to impact cellular processes leading to increased oxidative DNA damage and replication stress in a correlative manner. DNA replication stress may occur by any mechanism causing slow progression or stalling of DNA replication forks and, as a result, compromise proper DNA replication [50]. The structural characteristics of the replicating DNA strands are greatly responsible for the induction of replication stress-induced DNA damage, as any lesions of the single-strand templates within the unfolded DNA at the sites of the replication forks subsequently cause DSBs. Consequently, the genetic sequence harbouring DSBs is rendered highly susceptible to serious gene rearrangements and genomic instability. Contemporary studies have clearly shown the activation of the replication stress-induced DNA damage response (DDR) pathway at the earliest stages of cancer development underlying its significance in carcinogenesis [48, 52].

Several cytokines exhibit growth factor activity. The most significant cytokines to date designated as growth factors are the following: Epidermal Growth Factor (EGF), Platelet-Derived Growth Factor (PDGF), Fibroblast Growth Factor (FGF), TGF-α, TGF-β, Erythropoietin, Insulin-Like Growth Factor 1 (IGF-1), Insulin-Like Growth Factor 2 (IGF-2), IL-1, IL-2, IL-6, IL-8, TNF-α, TNF-β, INF-γ, and Colony Stimulating Factors (CSFs) (Table 1). Within this context a potential mechanism linking persistent chronic inflammation with carcinogenesis is through the activity of growth-promoting cytokines in the inflamed tissue. It has been previously shown that the injection of adenoviral vectors expressing a cocktail of growth factors (including fibroblast growth factor, stem cell factor, and endothelin-3) in human skin xenografts causes allelic imbalance in common fragile sites (CFSs), suggesting the formation of DSBs through replication stress [48]. In accordance with this, the presence of several growth-promoting cytokines within the inflammatory milieu for a prolonged period of time may induce replication stress and subsequent DSBs (Figure 1). In a rat silica model of inflammation-induced lung cancer the presence of the DNA damage response marker γH2AX was observed from early hyperplastic tissues, in bronchiolar hyperplasia, which supports the previous statement [74]. In another study the addition of IGF-1 or IGF-2 in human peripheral blood lymphocytes already incubated with bleomycin further increased the expression of p53 and the rate of chromosome aberrations, suggesting potential implication of DDR activation [75].Overall, these findings indicate a potential role

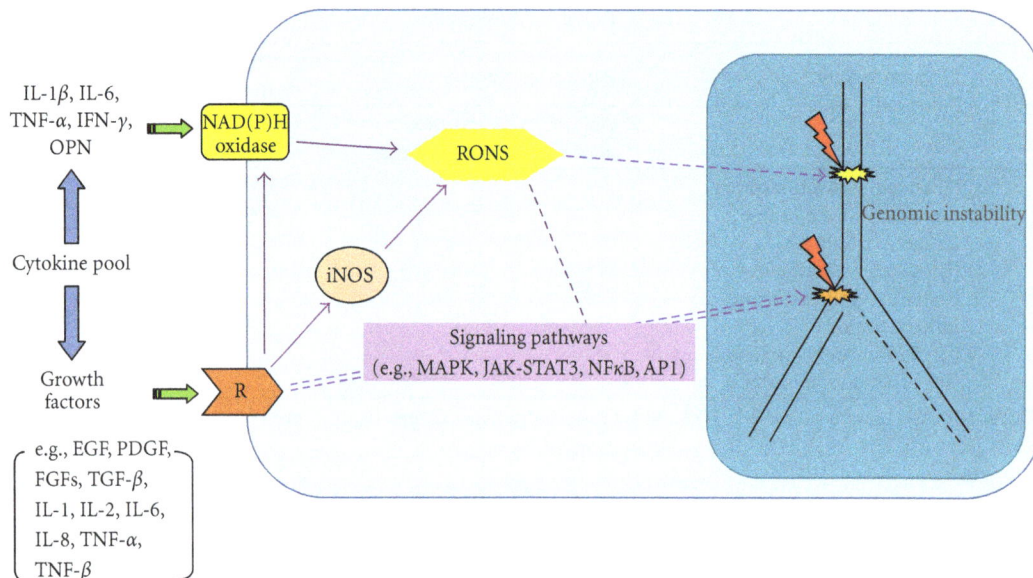

FIGURE 1: Schematic presentation of the proposed mechanism by which cytokines may promote genomic instability in chronic inflammatory conditions. Cytokines with growth factor activity may promote replication stress favouring the formation of DSBs. In addition, several cytokines induce the formation of RONS which in turn may cause DNA damage, including DNA cross-links, single- or double-strand breaks, and oxidative DNA damage with formation of 8-hydroxydeoxyguanosine (8-OHdG). Both mechanisms may provide a mechanism linking cytokine expression with genomic instability. MAPK: mitogen-activated protein kinase; JAK-STAT3: Janus kinase - Signal Transducer and Activator of Transcription 3; NF-κB: Nuclear Factor-κB; AP-1: Activator Protein 1; OPN: Osteopontin; R: Receptor; RONS: Reactive Oxygen/Nitrogen Species; NAD(P)H oxidase: Nicotinamide Adenine Dinucleotide Phosphate-oxidase; iNOS: inducible Nitric Oxide Synthase.

of cytokines with growth factor activity in the promotion of genomic instability through replication stress-induced DNA damage.

5. Cytokines and Oxidative/Nitrosative Stress

Reactive Oxygen and Nitrogen Species (RONS) are the free radical forms of oxygen and nitrogen, respectively. Free radicals contain one or more unpaired electrons rendering them highly reactive molecular metabolites [76, 77]. RONS are produced by endogenous as well as exogenous sources. Endogenous sources include metabolic reactions, such as electron transport reactions in the mitochondrial respiratory chain, metal reactions, and cells of the innate immune system, such as neutrophils and macrophages, during inflammatory responses [33, 77]. Exogenous sources include atmosphere pollutants, ionizing and nonionizing radiation, several carcinogenic compounds, and metal ions [77].

The most commonly generated Reactive Oxygen Species (ROS) in cells are the superoxide anion ($O_2^{-\bullet}$), hydrogen peroxide (H_2O_2), and the hydroxyl radical ($OH^{\bullet}$). The NADPH oxidase is an enzyme that plays the role of an electron donor and generates $O_2^{-\bullet}$ from oxygen in the body [78]. This enzyme can also lead to the production of H_2O_2 on neutrophil membranes [77]. NADPH oxidase exists in both phagocytes and nonphagocytes and has five known isoforms (NOX 1–5) as well as two reported related enzymes (DUOX 1-2) to date [78]. Nitric oxide ($NO^{\bullet}$) is a reactive free radical that generates additional Reactive Nitrogen Species

(RNS) and is more stable in a hypoxic environment than in normal oxygen tension conditions [33, 77]. $NO^{\bullet}$ is generated by the enzyme nitric oxide synthase (NOS) during the metabolization of L-arginine to citrulline [77, 79]. This enzyme exists in three isoforms in mammalian cells which consist of the neuronal (n)NOS, the endothelial (e)NOS, and the inducible (i)NOS [22, 80]. The nNOS and eNOS isoforms are constitutively expressed and produce low levels of $NO^{\bullet}$, while iNOS produces high levels of nitric oxide only upon inflammatory stimuli [22, 33]. The nNOS and eNOS isoforms generate $NO^{\bullet}$ with a neurotransmitting and vasodilating role, respectively, while the iNOS isoform produces $NO^{\bullet}$ as a mediator of the inflammatory response [33, 81].

RONS play a dual role in the cell, at low concentrations they are beneficial exhibiting several physiological functions in cellular responses (such as cellular signaling pathways and mitogenic responses), whereas at high concentrations they are detrimental for its integrity [77, 82, 83]. Excessive ROS and RNS production leads to oxidative and nitrosative stress, respectively. In order to counteract the harmful effects of free radicals the cell has evolved several defense mechanisms. These mechanisms involve the action of enzymatic, such as superoxide dismutase (SOD), catalase, glutathione peroxidase (GPx), and nonenzymatic antioxidants, such as Vitamin C and E, glutathione, carotenoids, and flavonoids [33, 77].

As mentioned earlier, RONS can be released by immune cells as a response to an inflammatory stimulus. Interestingly, during an inflammatory response phagocytic cells can directly produce RONS whereas in nonphagocytic cells their production is triggered by proinflammatory cytokines, such as IL-1 and IL-6 [22, 84]. In 1997, it was the first time reported that specific proinflammatory cytokines, IL-1, TNF-α, and IFN-γ, generate ROS in nonphagocytic cells [77, 85]. Later on, it was shown that IFN-γ, TNF-α, and IL-1β promote ROS through the induction of NOX1 in colon epithelium cells [78]. IFN-γ also increases the expression level of the NOX2 isoform of NADPH oxidase in human macrophages and neutrophils as well as the expression of its DUOXA2-related enzyme in airway epithelial cells [78]. The anti-inflammatory cytokine TGF-β induces NOX4 expression in many cell types, such as cardiac fibroblasts, smooth muscle cells, and hepatocytes. IL-4 and IL-13 anti-inflammatory cytokines augment the expression level of DUOXA1 enzyme in airway epithelial cells [78, 86–89]. Recently, it was shown that the inflammatory mediator leukotriene B$_4$ (LTB$_4$) induces NOX1 in human mast cells [90]. Finally IL-1, TNF-α, and IFN-γ proinflammatory cytokines stimulate the production of iNOS, which is inducibly expressed in macrophages, and thus contribute to the formation of nitric oxide as well as RNS [33, 91]. Overall, both pro- and anti-inflammatory cytokines have been shown to contribute to RNOS production through the induction of NADPH oxidase and iNOS activity. Interestingly, it has been shown that the proinflammatory cytokines seem to play a tumor promoting role while anti-inflammatory cytokines exert an antioncogenic function [22]. The latter leads to the suggestion of a potential link between inflammation, free radical-induced stress, and carcinogenesis.

6. Free Radicals and Inflammatory-Induced Carcinogenesis

Reactive Oxygen and Nitrogen Species (RONS) can be generated during inflammatory responses, and interestingly some inflammation-associated cancers are linked to oxidative and nitrosative stress. Examples include colorectal cancer provoked after active chronic colitis as well as lung, pancreatic, and esophageal cancers provoked after persistent inflammation of the bronchi, pancreas, and oesophagus respectively [33, 92–95]. In the previously mentioned cancer types either excessive free radical production or defective antioxidant mechanisms or both of these were observed.

Inflammation-mediated ROS production can trigger carcinogenesis either directly or indirectly [33]. Direct effects involve the formation of DNA cross-links, single- or double-strand breaks leading to mutations in oncogenes and tumor suppressor genes and ultimately to genomic instability (Figure 1) [33, 77, 84, 96]. An example is the formation of an extremely reactive free radical species during inflammatory response, named peroxynitrite anion (ONOO−), produced by the reaction between the superoxide anion and nitric oxide, which can create DNA fragmentation [77]. Additionally, the hydroxyl radical is known to cause damage to the purine and pyrimidine bases as well as the deoxyribose backbone [77, 97]. The most known DNA damage indicative of oxidative stress is the formation of 8-hydroxydeoxyguanosine

(8-OHdG) that is generated by the oxidative attack of OH$^\bullet$ in the cell DNA [33]. Free radical production during inflammatory process can indirectly lead to carcinogenesis via ROS-mediated activation of signalling pathways (Figure 1) [33]. An important example is the involvement of oxidative stress in the activation of the transcription factor NF-κB which in turn may display a protumorigenic effect, as mentioned earlier. Low to mild levels may lead to NF-κB expression, since H$_2$O$_2$ has been shown to degrade the IκBα subunit, whereas high levels of oxidative stress in the cells may cause inhibition of NF-κB expression [33, 98, 99]. Interestingly, growing evidence supports the fact that ROS act as second messengers in the NF-κB activation through the proinflammatory cytokines TNF and IL-1 [33, 77]. It is also worthwhile mentioning that iNOS can be induced by several stimuli, including cytokines as well as NF-κB transcription factor [22, 100]. Free radicals and cytokines can both either induce or become induced by NF-κB and their generation by tumor cells in inflammation-induced cancers can actually create a positive feedback loop for their own production. The inflammatory microenvironment that develops around the tumors leads to additional production of cytokines and free radicals, which in turn favours the carcinogenesis process [22]. In addition, oxidative stress can cause MIN by reducing the enzymatic activity and expression of the DNA mismatch repair genes mutS homologs 2 and 6 [22, 33]. Free radicals lead to gene silencing of the DNA mismatch repair gene hMLH1 via hypermethylation induced by overexpression of DNA methyltransferases [22, 33].

Cancer development involves three stages, initiation, promotion, and progression, and oxidative stress is involved in all the previous. In the initiation stage, ROS promote DNA damage, such as 8-OHdG formation, that may subsequently lead to gene mutations [33, 77]. During the promotion stage, low levels of oxidative stress can cause modifications in second messenger systems and promote cell division and proliferation [33, 77]. Finally in the progression stage, ROS can create additional genetic alterations fuelling cancer cells with further evolutionary advantages [33, 77]. Overall, the action of RONS is dosedependent within a particular cell context often with tumor-promoting activity in low levels, whereas in high levels they may raise the antitumor barriers, namely, apoptosis and senescence [33].

7. Future Perspectives

Cytokines display pleiotropic actions ranging from tumor-protective to tumor-promoting activity in a spatial and temporal manner. The fact that certain cytokines display growth factor activity as well as the ability to produce RONS suggests that they may promote genomic instability in chronic inflammatory conditions. This is particularly significant taking into account the central role of cytokines as mediators of inflammation, a knowledge that can be exploited in future cancer therapeutic strategies. IL-2 and IL-15 are currently tested for their therapeutic potential in cancer [53]. It has been shown that IL-15 plays an important role in the antitumor efficacy of combination therapy with Imatinib

Mesylate (IM), a tyrosine kinase inhibitor, and IL-2 in a mouse lung metastasis model, inducing a CCL2-dependent chemoattraction of IFN-producing killer dendritic cells (IKDCs) in the tumor microenvironment [101]. The latter finding has been exploited therapeutically by launching phase I clinical trial targeting IM-resistant gastrointestinal sarcomas and TRAIL-sensitive cancers. CCL2 serves as another paradigm of translation of molecular biology to clinical practice [69, 102]. Although there are conflicting data regarding the role of CCL2 in carcinogenesis, it seems that it promotes prostate cancer development. Therefore, a neutralizing antibody against CCL2 is under clinical trials in prostate cancer. Conclusively, elucidation of the molecular mechanisms implicated in tumor-host interactions may provide new insight in understanding tumor development as well as provide additional future prospects for more effective and targeted cancer therapy and prevention.

Acknowledgments

The authors would like to thank Antonia Daleziou for her help and comments in preparing the paper. This work was supported by SARG-NKUA 70/3/8916 Grant.

References

[1] S. I. Grivennikov, F. R. Greten, and M. Karin, "Immunity, inflammation, and cancer," *Cell*, vol. 140, no. 6, pp. 883–899, 2010.

[2] P. Ehrlich, "Ueber den jetzigen stand der karzinomforschung," *Nederlands Tijdschrift voor Geneeskunde*, vol. 5, part 1, pp. 273–290, 1909.

[3] L. Thomas, "Discussion," in *Cellular and Humoral Aspects of the Hypersensitive States*, H. S. Lawrence, Ed., pp. 529–532, Hoeber-Harper, New York, NY, USA, 1959.

[4] F. M. Burnet, "The concept of immunological surveillance," *Progress in Experimental Tumor Research*, vol. 13, pp. 1–27, 1970.

[5] G. P. Dunn, L. J. Old, and R. D. Schreiber, "The three Es of cancer immunoediting," *Annual Review of Immunology*, vol. 22, pp. 329–360, 2004.

[6] D. Hanahan and R. A. Weinberg, "Hallmarks of cancer: the next generation," *Cell*, vol. 144, no. 5, pp. 646–674, 2011.

[7] B. F. Zamarron and W. Chen, "Dual roles of immune cells and their factors in cancer development and progression," *International Journal of Biological Sciences*, vol. 7, no. 5, pp. 651–658, 2011.

[8] J. Han and R. J. Ulevitch, "Limiting inflammatory responses during activation of innate immunity," *Nature Immunology*, vol. 6, no. 12, pp. 1198–1205, 2005.

[9] E. Shacter and S. A. Weitzman, "Chronic inflammation and cancer," *Oncology*, vol. 16, no. 2, pp. 217–230, 2002.

[10] W. W. Lin and M. Karin, "A cytokine-mediated link between innate immunity, inflammation, and cancer," *Journal of Clinical Investigation*, vol. 117, no. 5, pp. 1175–1183, 2007.

[11] S. Akira, S. Uematsu, and O. Takeuchi, "Pathogen recognition and innate immunity," *Cell*, vol. 124, no. 4, pp. 783–801, 2006.

[12] M. M. Markiewski and J. D. Lambris, "Is complement good or bad for cancer patients? A new perspective on an old dilemma," *Trends in Immunology*, vol. 30, no. 6, pp. 286–292, 2009.

[13] V. Bergonzini, C. Salata, A. Calistri, C. Parolin, and G. Palù, "View and review on viral oncology research," *Infectious Agents and Cancer*, vol. 5, no. 1, p. 11, 2010.

[14] H. Takahashi, H. Ogata, R. Nishigaki, D. H. Broide, and M. Karin, "Tobacco smoke promotes lung tumorigenesis by triggering IKKβ- and JNK1-dependent inflammation," *Cancer Cell*, vol. 17, no. 1, pp. 89–97, 2010.

[15] A. Mantovani, P. Allavena, A. Sica, and F. Balkwill, "Cancer-related inflammation," *Nature*, vol. 454, no. 7203, pp. 436–444, 2008.

[16] C. Dostert, V. Pétrilli, R. Van Bruggen, C. Steele, B. T. Mossman, and J. Tschopp, "Innate immune activation through Nalp3 inflammasome sensing of asbestos and silica," *Science*, vol. 320, no. 5876, pp. 674–677, 2008.

[17] R. N. Apte and E. Voronov, "Is interleukin-1 a good or bad 'guy' in tumor immunobiology and immunotherapy?" *Immunological Reviews*, vol. 222, no. 1, pp. 222–241, 2008.

[18] J. Terzić, S. Grivennikov, E. Karin, and M. Karin, "Inflammation and colon cancer," *Gastroenterology*, vol. 138, no. 6, pp. 2101–e5, 2010.

[19] D. I. Gabrilovich and S. Nagaraj, "Myeloid-derived suppressor cells as regulators of the immune system," *Nature Reviews Immunology*, vol. 9, no. 3, pp. 162–174, 2009.

[20] S. K. Biswas and A. Mantovani, "Macrophage plasticity and interaction with lymphocyte subsets: cancer as a paradigm," *Nature Immunology*, vol. 11, no. 10, pp. 889–896, 2010.

[21] J. Condeelis and J. W. Pollard, "Macrophages: obligate partners for tumor cell migration, invasion, and metastasis," *Cell*, vol. 124, no. 2, pp. 263–266, 2006.

[22] A. J. Schetter, N. H. H. Heegaard, and C. C. Harris, "Inflammation and cancer: interweaving microRNA, free radical, cytokine and p53 pathways," *Carcinogenesis*, vol. 31, no. 1, Article ID bgp272, pp. 37–49, 2009.

[23] F. R. Balkwill, "The chemokine system and cancer," *The Journal of Pathology*, vol. 226, no. 2, pp. 148–157, 2012.

[24] J. B. Swann and M. J. Smyth, "Immune surveillance of tumors," *Journal of Clinical Investigation*, vol. 117, no. 5, pp. 1137–1146, 2007.

[25] H. Yu, D. Pardoll, and R. Jove, "STATs in cancer inflammation and immunity: a leading role for STAT3," *Nature Reviews Cancer*, vol. 9, no. 11, pp. 798–809, 2009.

[26] S. I. Grivennikov and M. Karin, "Dangerous liaisons: STAT3 and NF-κB collaboration and crosstalk in cancer," *Cytokine and Growth Factor Reviews*, vol. 21, no. 1, pp. 11–19, 2010.

[27] C. Ferrone and G. Dranoff, "Dual roles for immunity in gastrointestinal cancers," *Journal of Clinical Oncology*, vol. 28, no. 26, pp. 4045–4051, 2010.

[28] M. Karin, "Nuclear factor-κB in cancer development and progression," *Nature*, vol. 441, no. 7092, pp. 431–436, 2006.

[29] N. D. Perkins, "Integrating cell-signalling pathways with NF-κB and IKK function," *Nature Reviews Molecular Cell Biology*, vol. 8, no. 1, pp. 49–62, 2007.

[30] R. Zenz and E. F. Wagner, "Jun signalling in the epidermis: from developmental defects to psoriasis and skin tumors," *International Journal of Biochemistry and Cell Biology*, vol. 38, no. 7, pp. 1043–1049, 2006.

[31] Y. Carmi, E. Voronov, S. Dotan et al., "The role of macrophage-derived IL-1 in induction and maintenance of angiogenesis," *Journal of Immunology*, vol. 183, no. 7, pp. 4705–4714, 2009.

[32] M. Elkabets, V. S. G. Ribeiro, C. A. Dinarello et al., "IL-1β regulates a novel myeloid-derived suppressor cell subset

that impairs NK cell development and function," *European Journal of Immunology*, vol. 40, no. 12, pp. 3347–3357, 2010.

[33] S. Reuter, S. C. Gupta, M. M. Chaturvedi, and B. B. Aggarwal, "Oxidative stress, inflammation, and cancer: how are they linked?" *Free Radical Biology and Medicine*, vol. 49, no. 11, pp. 1603–1616, 2010.

[34] D. J. J. Waugh and C. Wilson, "The interleukin-8 pathway in cancer," *Clinical Cancer Research*, vol. 14, no. 21, pp. 6735–6741, 2008.

[35] H. Yang, M. Bocchetta, B. Kroczynska et al., "TNF-α inhibits asbestos-induced cytotoxicity via a NF-κB-dependent pathway, a possible mechanism for asbestos-induced oncogenesis," *Proceedings of the National Academy of Sciences of the United States of America*, vol. 103, no. 27, pp. 10397–10402, 2006.

[36] M. L. Cho, J. W. Kang, Y. M. Moon et al., "STAT3 and NF-κB signal pathway is required for IL-23-mediated IL-17 production in spontaneous arthritis animal model IL-1 receptor antagonist-deficient mice," *Journal of Immunology*, vol. 176, no. 9, pp. 5652–5661, 2006.

[37] J. S. Hao and B. E. Shan, "Immune enhancement and antitumour activity of IL-23," *Cancer Immunology, Immunotherapy*, vol. 55, no. 11, pp. 1426–1431, 2006.

[38] K. W. Moore, R. W. Malefyt, R. L. Coffman, and A. O'Garra, "Interleukin-10 and the interleukin-10 receptor," *Annual Review of Immunology*, vol. 19, pp. 683–765, 2001.

[39] S. Alas, C. Emmanouilides, and B. Bonavida, "Inhibition of interleukin 10 by Rituximab results in down-regulation of Bcl-2 and sensitization of B-cell non-Hodgkin's lymphoma to apoptosis," *Clinical Cancer Research*, vol. 7, no. 3, pp. 709–723, 2001.

[40] S. B. Jakowlew, "Transforming growth factor-β in cancer and metastasis," *Cancer and Metastasis Reviews*, vol. 25, no. 3, pp. 435–457, 2006.

[41] F. Rodier, J. P. Coppé, C. K. Patil et al., "Persistent DNA damage signalling triggers senescence-associated inflammatory cytokine secretion," *Nature Cell Biology*, vol. 11, no. 8, pp. 973–979, 2009.

[42] G. P. Dimri, "What has senescence got to do with cancer?" *Cancer Cell*, vol. 7, no. 6, pp. 505–512, 2005.

[43] M. Fumagalli and F. d'Adda di Fagagna, "SASPense and DDRama in cancer and ageing," *Nature Cell Biology*, vol. 11, no. 8, pp. 921–923, 2009.

[44] V. G. Gorgoulis and T. D. Halazonetis, "Oncogene-induced senescence: the bright and dark side of the response," *Current Opinion in Cell Biology*, vol. 22, no. 6, pp. 816–827, 2010.

[45] J. C. Acosta and J. Gil, "A role for CXCR2 in senescence, but what about in cancer?" *Cancer Research*, vol. 69, no. 6, pp. 2167–2170, 2009.

[46] A. V. Orjalo, D. Bhaumik, B. K. Gengler, G. K. Scott, and J. Campisi, "Cell surface-bound IL-1α is an upstream regulator of the senescence-associated IL-6/IL-8 cytokine network," *Proceedings of the National Academy of Sciences of the United States of America*, vol. 106, no. 40, pp. 17031–17036, 2009.

[47] S. Negrini, V. G. Gorgoulis, and T. D. Halazonetis, "Genomic instability an evolving hallmark of cancer," *Nature Reviews Molecular Cell Biology*, vol. 11, no. 3, pp. 220–228, 2010.

[48] V. G. Gorgoulis, L. V. F. Vassiliou, P. Karakaidos et al., "Activation of the DNA damage checkpoint and genomic instability in human precancerous lesions," *Nature*, vol. 434, no. 7035, pp. 907–913, 2005.

[49] T. D. Halazonetis, V. G. Gorgoulis, and J. Bartek, "An oncogene-induced DNA damage model for cancer development," *Science*, vol. 319, no. 5868, pp. 1352–1355, 2008.

[50] W. C. Burhans and M. Weinberger, "DNA replication stress, genome instability and aging," *Nucleic Acids Research*, vol. 35, no. 22, pp. 7545–7556, 2007.

[51] K. Aziz, S. Nowsheen, G. Pantelias, G. Iliakis, V. G. Gorgoulis, and A. G. Georgakilas, "Targeting DNA damage and repair: embracing the pharmacological era for successful cancer therapy," *Pharmacology and Therapeutics*, vol. 133, no. 3, pp. 334–350, 2012.

[52] J. Bartkova, Z. Horejsi, K. Koed et al., "DNA damage response as a candidate anti-cancer barrier in early human tumorigenesis," *Nature*, vol. 434, no. 7035, pp. 864–870, 2005.

[53] T. A. Waldmann, "The biology of interleukin-2 and interleukin-15: implications for cancer therapy and vaccine design," *Nature Reviews Immunology*, vol. 6, no. 8, pp. 595–601, 2006.

[54] M. Martinez-Moczygemba and D. P. Huston, "Biology of common β receptor-signaling cytokines: IL-3, IL-5, and GM-CSF," *Journal of Allergy and Clinical Immunology*, vol. 112, no. 4, pp. 653–666, 2003.

[55] T. Hosoyama, M. I. Aslam, J. Abraham et al., "IL-4R drives dedifferentiation, mitogenesis, and metastasis in rhabdomyosarcoma," *Clinical Cancer Research*, vol. 17, no. 9, pp. 2757–2766, 2011.

[56] N. Geijsen, L. Koenderman, and P. J. Coffer, "Specificity in cytokine signal transduction: lessons learned from the IL-3/IL-5/GM-CSF receptor family," *Cytokine and Growth Factor Reviews*, vol. 12, no. 1, pp. 19–25, 2001.

[57] Y. Sawa, Y. Arima, H. Ogura et al., "Hepatic interleukin-7 expression regulates T cell responses," *Immunity*, vol. 30, no. 3, pp. 447–457, 2009.

[58] L. Knoops and J. C. Renauld, "IL-9 and its receptor: from signal transduction to tumorigenesis," *Growth Factors*, vol. 22, no. 4, pp. 207–215, 2004.

[59] T. Blankenstein, "The role of tumor stroma in the interaction between tumor and immune system," *Current Opinion in Immunology*, vol. 17, no. 2, pp. 180–186, 2005.

[60] P. C. Heinrich, I. Behrmann, S. Haan, H. M. Hermanns, G. Müller-Newen, and F. Schaper, "Principles of interleukin (IL)-6-type cytokine signalling and its regulation," *Biochemical Journal*, vol. 374, no. 1, pp. 1–20, 2003.

[61] W. T. Watford, B. D. Hissong, J. H. Bream, Y. Kanno, L. Muul, and J. J. O'Shea, "Signaling by IL-12 and IL-23 and the immunoregulatory roles of STAT4," *Immunological Reviews*, vol. 202, pp. 139–156, 2004.

[62] M. Wills-Karp, "Interleukin-13 in asthma pathogenesis," *Immunological Reviews*, vol. 202, pp. 175–190, 2004.

[63] S. Nogami, S. Satoh, M. Nakano et al., "Taxilin; a novel syntaxin-binding protein that is involved in Ca^{2+}-dependent exocytosis in neuroendocrine cells," *Genes to Cells*, vol. 8, no. 1, pp. 17–28, 2003.

[64] W. Cruikshank and F. Little, "Interleukin-16: the ins and outs of regulating T-cell activation," *Critical Reviews in Immunology*, vol. 28, no. 6, pp. 467–483, 2008.

[65] M. Numasaki, M. Watanabe, T. Suzuki et al., "IL-17 enhances the net angiogenic activity and in vivo growth of human non-small cell lung cancer in SCID mice through promoting CXCR-2-dependent angiogenesis," *Journal of Immunology*, vol. 175, no. 9, pp. 6177–6189, 2005.

[66] F. Benchetrit, A. Ciree, V. Vives et al., "Interleukin-17 inhibits tumor cell growth by means of a T-cell-dependent mechanism," *Blood*, vol. 99, no. 6, pp. 2114–2121, 2002.

[67] S. Lebel-Binay, A. Berger, F. Zinzindohoué et al., "Interleukin-18: biological properties and clinical implications," *European Cytokine Network*, vol. 11, no. 1, pp. 15–25, 2000.

[68] J. K. Kundu and Y. J. Surh, "Emerging avenues linking inflammation and cancer," *Free Radical Biology & Medicine*, vol. 52, no. 9, pp. 2013–2037, 2012.

[69] J. Zhang, L. Patel, and K. J. Pienta, "Targeting chemokine (C-C motif) Ligand 2 (CCL₂) as an example of translation of cancer molecular biology to the clinic," *Progress in Molecular Biology and Translational Science C*, vol. 95, pp. 31–53, 2010.

[70] B. Homey, A. Müller, and A. Zlotnik, "Chemokines: agents for the immunotherapy of cancer?" *Nature Reviews Immunology*, vol. 2, no. 3, pp. 175–184, 2002.

[71] A. P. Vicari, S. Ait-Yahia, K. Chemin, A. Mueller, A. Zlotnik, and C. Caux, "Antitumor effects of the mouse chemokine 6Ckine/SLC through angiostatic and immunological mechanisms," *Journal of Immunology*, vol. 165, no. 4, pp. 1992–2000, 2000.

[72] A. Müller, B. Homey, H. Soto et al., "Involvement of chemokine receptors in breast cancer metastasis," *Nature*, vol. 410, no. 6824, pp. 50–56, 2001.

[73] N. Karin, "The multiple faces of CXCL12 (SDF-1α) in the regulation of immunity during health and disease," *Journal of Leukocyte Biology*, vol. 88, no. 3, pp. 463–473, 2010.

[74] D. Blanco, S. Vicent, M. F. Fragaz et al., "Molecular analysis of a multistep lung cancer model induced by chronic inflammation reveals epigenetic regulation of p16 and activation of the DNA damage response pathway," *Neoplasia*, vol. 9, no. 10, pp. 840–852, 2007.

[75] S. Cianfarani, B. Tedeschi, D. Germani et al., "In vitro effects of growth hormone (GH) and insulin-like growth factor I and II (IGF-I and -II) on chromosome fragility and p53 protein expression in human lymphocytes," *European Journal of Clinical Investigation*, vol. 28, no. 1, pp. 41–47, 1998.

[76] V. J. Thannickal and B. L. Fanburg, "Reactive oxygen species in cell signaling," *American Journal of Physiology-Lung Cellular and Molecular Physiology*, vol. 279, no. 6, pp. L1005–L1028, 2000.

[77] M. Valko, C. J. Rhodes, J. Moncol, M. Izakovic, and M. Mazur, "Free radicals, metals and antioxidants in oxidative stress-induced cancer," *Chemico-Biological Interactions*, vol. 160, no. 1, pp. 1–40, 2006.

[78] M. Katsuyama, "NOX/NADPH oxidase, the superoxide-generating enzyme: its transcriptional regulation and physiological roles," *Journal of Pharmacological Sciences*, vol. 114, no. 2, pp. 134–146, 2010.

[79] P. Ghafourifar and E. Cadenas, "Mitochondrial nitric oxide synthase," *Trends in Pharmacological Sciences*, vol. 26, no. 4, pp. 190–195, 2005.

[80] A. Pautz, J. Art, S. Hahn, S. Nowag, C. Voss, and H. Kleinert, "Regulation of the expression of inducible nitric oxide synthase," *Nitric Oxide*, vol. 23, no. 2, pp. 75–93, 2010.

[81] C. J. Lowenstein and E. Padalko, "iNOS (NOS2) at a glance," *Journal of Cell Science*, vol. 117, no. 14, pp. 2865–2867, 2004.

[82] G. Poli, G. Leonarduzzi, F. Biasi, and E. Chiarpotto, "Oxidative stress and cell signalling," *Current Medicinal Chemistry*, vol. 11, no. 9, pp. 1163–1182, 2004.

[83] M. Liontos, I. S. Pateras, K. Evangelou, and V. G. Gorgoulis, "The tumor suppressor gene ARF as a sensor of oxidative stress," *Current Molecular Medicine*. In press.

[84] S. P. Hussain, L. J. Hofseth, and C. C. Harris, "Radical causes of cancer," *Nature Reviews Cancer*, vol. 3, no. 4, pp. 276–285, 2003.

[85] I. L. C. Chapple, "Reactive oxygen species and antioxidants in inflammatory diseases," *Journal of Clinical Periodontology*, vol. 24, no. 5, pp. 287–296, 1997.

[86] A. Sturrock, B. Cahill, K. Norman et al., "Transforming growth factor-β1 induces Nox4 NAD(P)H oxidase and reactive oxygen species-dependent proliferation in human pulmonary artery smooth muscle cells," *American Journal of Physiology-Lung Cellular and Molecular Physiology*, vol. 290, no. 4, pp. L661–L673, 2006.

[87] A. Sturrock, T. P. Huecksteadt, K. Norman et al., "Nox4 mediates TGF-β1-induced retinoblastoma protein phosphorylation, proliferation, and hypertrophy in human airway smooth muscle cells," *American Journal of Physiology-Lung Cellular and Molecular Physiology*, vol. 292, no. 6, pp. L1543–L1555, 2007.

[88] I. Cucoranu, R. Clempus, A. Dikalova et al., "NAD(P)H oxidase 4 mediates transforming growth factor-β1-induced differentiation of cardiac fibroblasts into myofibroblasts," *Circulation Research*, vol. 97, no. 9, pp. 900–907, 2005.

[89] I. Carmona-Cuenca, C. Roncero, P. Sancho et al., "Upregulation of the NADPH oxidase NOX4 by TGF-beta in hepatocytes is required for its pro-apoptotic activity," *Journal of Hepatology*, vol. 49, no. 6, pp. 965–976, 2008.

[90] G. Y. Kim, J. W. Lee, H. C. Ryu, J. D. Wei, C. M. Seong, and J. H. Kim, "Proinflammatory cytokine IL-1β stimulates IL-8 synthesis in mast cells via a leukotriene B4 receptor 2-linked pathway, contributing to angiogenesis," *Journal of Immunology*, vol. 184, no. 7, pp. 3946–3954, 2010.

[91] M. Valko, D. Leibfritz, J. Moncol, M. T. D. Cronin, M. Mazur, and J. Telser, "Free radicals and antioxidants in normal physiological functions and human disease," *International Journal of Biochemistry and Cell Biology*, vol. 39, no. 1, pp. 44–84, 2007.

[92] A. Roessner, D. Kuester, P. Malfertheiner, and R. Schneider-Stock, "Oxidative stress in ulcerative colitis-associated carcinogenesis," *Pathology Research and Practice*, vol. 204, no. 7, pp. 511–524, 2008.

[93] C. A. Martey, S. J. Pollock, C. K. Turner et al., "Cigarette smoke induces cyclooxygenase-2 and microsomal prostaglandin E2 synthase in human lung fibroblasts: implications for lung inflammation and cancer," *American Journal of Physiology*, vol. 287, no. 5, pp. L981–L991, 2004.

[94] G. Garcea, A. R. Dennison, W. P. Steward, and D. P. Berry, "Role of inflammation in pancreatic carcinogenesis and the implications for future therapy," *Pancreatology*, vol. 5, no. 6, pp. 514–529, 2005.

[95] S. J. Murphy, L. A. Anderson, B. T. Johnston et al., "Have patients with esophagitis got an increased risk of adenocarcinoma? Results from a population-based study," *World Journal of Gastroenterology*, vol. 11, no. 46, pp. 7290–7295, 2005.

[96] R. Visconti and D. Grieco, "New insights on oxidative stress in cancer," *Current Opinion in Drug Discovery and Development*, vol. 12, no. 2, pp. 240–245, 2009.

[97] M. Dizdaroglu, P. Jaruga, M. Birincioglu, and H. Rodriguez, "Free radical-induced damage to DNA: mechanisms and measurement," *Free Radical Biology and Medicine*, vol. 32, no. 11, pp. 1102–1115, 2002.

[98] G. Gloire, S. Legrand-Poels, and J. Piette, "NF-κB activation by reactive oxygen species:fifteen years later," *Biochemical Pharmacology*, vol. 72, no. 11, pp. 1493–1505, 2006.

[99] W. Dröge, "Free radicals in the physiological control of cell function," *Physiological Reviews*, vol. 82, no. 1, pp. 47–95, 2002.

[100] S. P. Hussain and C. C. Harris, "Inflammation and cancer: an ancient link with novel potentials," *International Journal of Cancer*, vol. 121, no. 11, pp. 2373–2380, 2007.

[101] G. Mignot, E. Ullrich, M. Bonmort et al., "The critical role of IL-15 in the antitumor effects mediated by the combination therapy imatinib and IL-2," *Journal of Immunology*, vol. 180, no. 10, pp. 6477–6483, 2008.

[102] M. G. Lechner, S. M. Russell, R. S. Bass, and A. L. Epstein, "Chemokines, costimulatory molecules and fusion proteins for the immunotherapy of solid tumors," *Immunotherapy*, vol. 3, no. 11, pp. 1317–1340, 2011.

17

Novel One-Step Multiplex PCR-Based Method for HLA Typing and Preimplantational Genetic Diagnosis of β-Thalassemia

Raquel M. Fernández,[1,2] **Ana Peciña,**[1,2] **Maria Dolores Lozano-Arana,**[1]
Juan Carlos García-Lozano,[1] **Salud Borrego,**[1,2] **and Guillermo Antiñolo**[1,2]

[1] *Department of Genetics, Reproduction and Fetal Medicine, Institute of Biomedicine of Seville (IBIS),*
 University Hospital Virgen del Rocío/CSIC/University of Seville, Avenida Manuel Siurot, s/n, 41013 Seville, Spain
[2] *Centre for Biomedical Network Research on Rare Diseases (CIBERER), 41013 Seville, Spain*

Correspondence should be addressed to Guillermo Antiñolo; guillermo.antinolo.sspa@juntadeandalucia.es

Academic Editor: Thomas Liehr

Preimplantation genetic diagnosis (PGD) of single gene disorders, combined with HLA matching (PGD-HLA), has emerged as a tool for couples at risk of transmitting a genetic disease to select unaffected embryos of an HLA tissue type compatible with that of an existing affected child. Here, we present a novel one-step multiplex PCR to genotype a spectrum of STRs to simultaneously perform HLA typing and PGD for β-thalassemia. This method is being routinely used for PGD-HLA cycles in our department, with a genotyping success rate of 100%. As an example, we present the first successful PGD-HLA typing in Spain, which resulted in the birth of a boy and subsequent successful HSC transplantation to his affected brother, who is doing well 4 years following transplantation. The advantage of our method is that it involves only a round of single PCR for multiple markers amplification (up to 10 markers within the *HLA* and 6 markers at the *β-globin* loci). This strategy has allowed us to considerably reduce the optimization of the PCR method for each specific PGD-HLA family as well as the time to obtain molecular results in each cycle.

1. Introduction

The thalassemias are a group of autosomal recessive inherited anaemias caused by mutations in the α- and β-globin genes leading to decreased production of one of the globin chains of haemoglobin (Hb). Mutations in such globin genes are by far the most common group of monogenic disorders worldwide. Recent surveys suggest that between 300,000 and 400,000 babies are born with a serious hemoglobin disorder each year, and although information about the precise world distribution and frequency of these disorders is still limited, there is no doubt that they are going to pose an increasing burden on global health resources in the future [1]. Specifically, over 200 different mutations of the β-globin gene cluster have been described, and they result in either absent or reduced synthesis of the β-globin chains leading to β-thalassemia [2] (http://globin.bx.psu.edu/hbvar/). β-thalassemia (OMIM no. 613985) major patients suffer from severe anaemia with marked ineffective erythropoiesis,

crythroid marrow expansion, osteopenia, and bone deformities. The high transfusion requirements result in iron overload, and the patients need chelating therapy. The majority of cases will ultimately develop organ damage, in particular, of the heart and liver, with reduced life quality and expectancy [3].

The human leukocyte antigen (HLA) system is the name of the major histocompatibility complex (MHC) in humans. The super locus resides on chromosome 6 and contains a large number of genes that encode cell-surface antigen-presenting proteins that, among several functions, play a major role in the immune system function in humans. This complex consists of three regions that contain genes encoding class I, class II, and class III antigens and represents one of the most polymorphic regions of the human genome. HLAs corresponding to MHC class I present peptides from inside the cell. These peptides are produced from digested proteins that are broken down in the proteasomes. In general, these particular peptides are small polymers, about 9 amino acids in length. Foreign antigens attract killer T-cells (also called CD8 positive- or cytotoxic T-cells) that destroy cells. HLAs corresponding to MHC class II present antigens from outside of the cell to T-lymphocytes. These particular antigens stimulate the multiplication of T-helper cells, which in turn stimulate antibody-producing B-cells to produce antibodies to that specific antigen. Self-antigens are suppressed by suppressor T-cells. Finally, HLAs corresponding to MHC class III encode components of the complement system. Diversity of HLAs in the human population is one aspect of disease defense, and, as a result, the chance of two unrelated individuals with identical HLA molecules on all loci is very low. HLA genes have historically been identified as a result of the ability to successfully transplant organs between HLA-similar individuals. In other words, HLA complex is responsible for rejection following organ/tissue transplantation. To date, the only available definitive cure for β-thalassemia major is haematopoietic stem cell transplantation (HSCT) from an HLA-identical donor. Matched sibling HSCT for patients under the age of 17 years without major organ damage has been regarded as the best therapy option with a 90% of success, because treatment-related mortality (TRM) is low, and serious morbidity is unusual. Results from matched unrelated and mismatched HSCT are less successful because of a high risk of graft rejection and TRM [4].

Preimplantation genetic diagnosis (PGD) of single gene disorders, combined with HLA matching (PGD-HLA), represents one of the most relevant challenges in reproductive medicine. This strategy has emerged as a tool for couples at risk of transmitting a genetic disease to select unaffected embryos of an HLA tissue type compatible with that of an existing affected child. At delivery, HSC from the newborn umbilical cord blood can be used to treat the affected sibling. This approach is particularly valuable for β-thalassemia and other similar life-threatening disorders that require an HLA-compatible HSC donor, where HLA identity seems to provide the best chance of avoiding graft rejection and other serious complications of bone marrow transplantation. To date, very few cases of successful pregnancies and births of healthy HLA compatible donors for patients have been reported [5–10]. Here, we present a new one-step multiplex PCR to simultaneously perform HLA typing and PGD for β-thalassemia, which provides important advantages with respect to other methods. To date, we have successfully applied it for 4 couples with children affected by β-thalassemia, accounting for a total of 16 PGD-HLA cycles. As an example, we also present the results of the first successful PGD with HLA typing performed in Spain after the application of such method. Subsequent HSCT was performed successfully, and the transplanted sibling is currently doing well, 4 years following transplantation. All the procedures, included HSCs umbilical cord transplantation, were performed at the University Hospital Virgen del Rocio in Sevilla.

2. Materials and Methods

2.1. Selection of β-Globin and HLA Markers.
A panel of six polymorphic short tandem repeats (STRs) located in the neighboring regions to the β-globin gene was selected (Figure 1). The selection was based on the heterozygosity values (>30%) detected for each marker in a group of 30 normal controls and in their specific location with respect to the β-globin gene (3 of them in the 5′ region and the other 3 in the 3′ region surrounding the gene), warranting the possibility to detect any recombination event at the known recombination hotspot in the beta-globin cluster [11].

Regarding HLA typing, a first selection of up to 10 STRs was initially made according to their localization along the HLA locus (Figure 2). The policy was to select, for the subsequent PGD, the maximum number of informative STR markers evenly spaced throughout the HLA complex to obtain an accurate haplotyping, allowing identification of double recombination events, which if not detected may lead to misdiagnosis in HLA typing. Using this panel, we achieved the first successful PGD with HLA typing performed in Spain, which we will also present in this paper. Subsequently and following the ESHRE PGD guidelines [12], the method has been updated with the inclusion of a selection of another 10 markers along the HLA locus (Figure 3).

Primers for the amplification of those markers were designed in order to let all of them be amplified with the same annealing temperature and conditions, using the Primer3 software (http://primer3.sourceforge.net/) (Table 1). As a previous requisite for the final inclusion of the couples in our PGD-HLA program, informativity testing for segregation analyses is always developed on the DNA samples from the corresponding family members (father, mother, and affected child) using standard PCR protocols to identify the "disease haplotypes" and the specific HLA combinations carried by the affected children in the context of their corresponding families.

2.2. Multiplex PCR Protocol.
A one-step multiplex single-cell fluorescent PCR is used for the simultaneous amplification of several combinations of markers at the HLA and β-globin loci, using the QIAGEN multiplex PCR kit (QIAGEN, GmbH; Hilden, Germany) and an adaptation from a protocol previously described [13]. Optimal cell lysis protocol and PCR conditions, further described, were set up on single cells biopsied from supernumerary IVF embryos not suitable for transfer or cryopreservation. Although such cells might give falsely high rates of PCR failure and/or ADO events, achievement of specific optimal PCR conditions under these "not-optimal" circumstances warranted to have a protocol suitable for single cells in real PGD cycles (with a much better quality). The reaction mix contains $0.2\,\mu$M each primer, 5× Sol Q, and 2× QIAGEN multiplex PCR master mix, for a final volume of $25\,\mu$L. The PCR program is as follows: 15 minutes at $95°$C, 10 cycles of 30 seconds at $96°$C, 30 seconds at $55°$C, 30 seconds at $72°$C, followed by 30 cycles of 30 seconds at $94°$C, 30 seconds at $55°$C, 30 seconds at $72°$C, and a final extension of 15 minutes at $60°$C. PCR products are analyzed on an ABI3730 automated sequencer (Applied Biosystems, Foster City, CA).

2.3. Assisted Reproductive Techniques and Embryo Biopsy.
Controlled ovarian stimulation is performed through a long protocol as previously described [14]. Oocytes are carefully

TABLE 1: Primers sequences for the β-thalassemia and HLA markers included in our method.

Marker	Primer	Dye	Sequence	Size of fragment (bp)
HLA typing				
D6S1571	D6S1571F	PET	GGACCTACGCATCTGGTG	160–180
	D6S1571R		TGGCTCTAATGGTTACTTTTTACA	
D6S1260	D6S1260F	NED	ACTGCTCCTGGGCATGGTTG	130–160
	D6S1260R		GTACATGCCTTTGTTAACATC	
D6S1624	D6S1624F	6FAM	GGAAGTCTTCAGTGGAGAGAGT	200–220
	D6S1624R		ACTCCAGGTGTTTGTGGTTT	
MOG-CA	MGCAF	VIC	TCACCTCGAGTGAGTCTCTTT	205–235
	MOGCAR		ACCATGGGTAACTGAAGCAT	
MOG-TAAA	MOG3F	6FAM	GAAATGTGAGAATAAAGGAGA	125–150
	MOG3R		GATAAAGGGGAACTACTACA	
D6S265	D6S265F	6FAM	ACGTTCGTACCCATTAACCT	110–125
	D6S265R		CGAGGTAAACAGCAGAAAGA	
HLAC-CA	HLACCAF	VIC	TCCCTAGTAGCTGGGATTACA	155–175
	HLACCAR		CGGCAAGAGACTCTGATGA	
MIB	MIBF	6FAM	CCACAGTCTCTATCAGTCCAGA	155–185
	MIBR		TCAGCCTGCTAGCTTATCCT	
D6S273	D6S273F	6FAM	GGCCAAAGTTAAAACCAAAC	135–145
	D6S273R		GCAACTTTTCTGTCAATCCA	
DRA-CA	DRACAF	VIC	GATACTTTCCTAATTCTCCTCCTTC	120–140
	DRACAR		ATGGAATCTCATCAAGGTCAG	
D6S2443	D6S2443F	6FAM	CCATACCAAAGTAAAACCCAG	150–200
	D6S2443R		GAGGATGAAGGGAAATTAGAG	
DQCAR	DQCARF	PET	CTTGGCCAATCAGAATCTTT	150–175
	DQCARR		CTGCATTTCTCTTCCTTATCAC	
TAP1-CA	TAP1CAF	PET	GGACAATATTTTGCTCCTGA	195–215
	TAP1CAR		TCATACATCTGCTTTGATCTCC	
RING3-CA	RING3CAF	6FAM	GGGCCGCAGTTTAAGTAAC	125–135
	RING3CAR		TGTTAGGTCAGAACCACAGAA	
D6S1560	D6S1560F	6FAM	CTCCAGTCCCCACTGC	225–250
	D6S1560R		CCCAAGGCCACATAGC	
D6S1583	D6S1583F	VIC	GCCCCTAACCTGCTTCTACTGA	150–200
	D6S1583R		GCAGATGGCCCCACTGAC	
D6S1618	D6S1618F	NED	GGCCTGAGCAGTGCAT	130–170
	D6S1618R		TGATTCCTAATCTGCGGG	
D6S1611	D6S1611F	PET	GAGCAAGACTCCATCTCAAA	225–250
	D6S1611R		ACCTAAGTTCTCTGAAGGGC	
D6S1610	D6S1610F	PET	CCTGGTGAGATAGATGCTTG	100–140
	D6S1610R		ATTTCCAGCAGAGCCTTG	
D6S1552	D6S1552F	VIC	AGCCTGAACGACAGAACAAG	160–200
	D6S1552R		CTGCTTAACTTAGATCTTTGGTAT	
β-thalassemia locus				
D11S4181	D11S4181F	6FAM	AAGCTTCCTTCACATTCTTACAG	200–225
	D11S4181R		GAACTGAGACCAAGAACATTATTCC	
D11S2351	D11S2351F	6FAM	GGGCACCTGTAATCCCA	150–180
	D11S2351R		AGGAGTCACTGGATCTACTC	
D11S1871	D11S1871F	PET	AAGAAGTTGCCCTGATGTCT	160–200
	D11S1871R		TAAAAGGAGCTGAATGCACA	
D11S4891	D11S4891F	6FAM	GGAAATGGACCTCTGTCTC	75–100
	D11S4891R		CTTTTATTCCAGCCCCAC	
D11S1760	D11S1760F	PET	GATCTCAAGTGTTTCCCCAC	75–100
	D11S1760R		AAACGATGTCTGTCCACTCA	
D11S1338	D11S1338F	6FAM	GACGGTTTAACTGTATATCTAAGAC	250–280
	D11S1338R		TAATGCTACTTATTTGGAGTGTG	

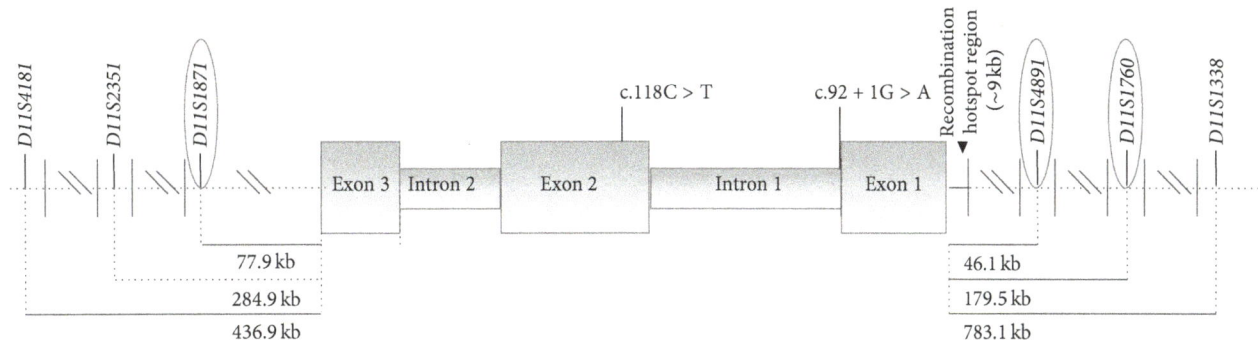

FIGURE 1: Map of human β-globin gene, showing the location of the polymorphic markers that can be tested with our method. The informative STRs for the family here reported are represented as included within an ellipse. The mutations carried by the child and responsible for his clinical picture of β-thalassemia are also shown.

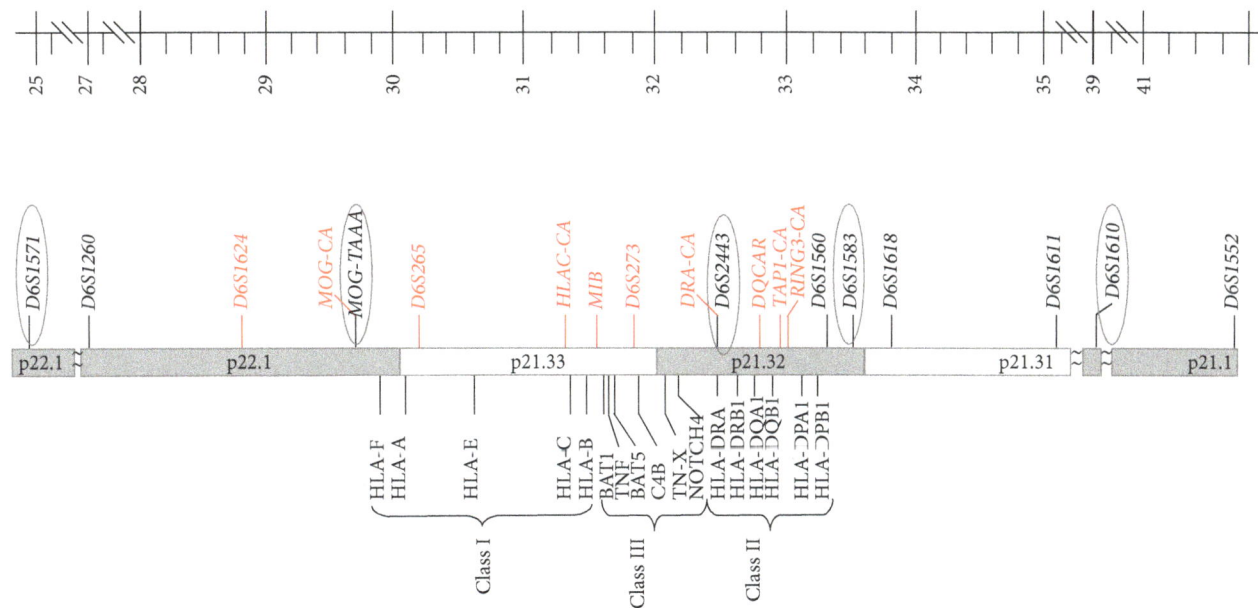

FIGURE 2: STR markers at the HLA locus that can be simultaneously tested with our method. In red, the new STRs included after the update of the method are indicated following ESHRE recommendations. The informative STRs for the family here reported are represented as included within an ellipse.

denuded from cumulus cells, and intracytoplasmic sperm injection (ICSI) is used to prevent contamination with residual sperm adhered to the zona pellucida [15, 16]. Blastomere biopsy is performed the morning of day three after fertilization. Laser technology (Octax Laser) is used to create an opening in the zona pellucida, and one blastomere is gently aspirated for each embryo. Cells are transferred into thin walled 0.2 mL PCR tubes containing 2.5 μL of proteinase K/SDS lysis buffer and frozen at −80°C before cell lysis [13].

2.4. β-Globin and HLA Haplotyping of the Embryos.
After 30 min at −80°C, cells are lysed by incubation for 90 min at 37°C and 15 min at 65°C.

HLA typing as well as β-globin genetic analysis of the embryos are subsequently performed using the previously

selected combination of markers and the described one-step multiplex fluorescent PCR protocol at the single-cell level.

Prior to the analysis, it is established that embryos showing monosomy, trisomy, or uniparental disomy of chromosomes, 6 or 11 will be considered to be abnormal. The embryos with a recombination pattern at the HLA locus are considered to be HLA nonidentical and, therefore, not suitable for transfer.

2.5. Case Description of the First Successful PGD with HLA Typing Performed in Spain.
A couple of a 29-year-old female and a 35-year-old male, with a 4-year-old son, affected with major β-thalassemia, were referred to our department for PGD in combination with HLA typing. Examination of the patient revealed that clinical status of the disease

(a)

(b)

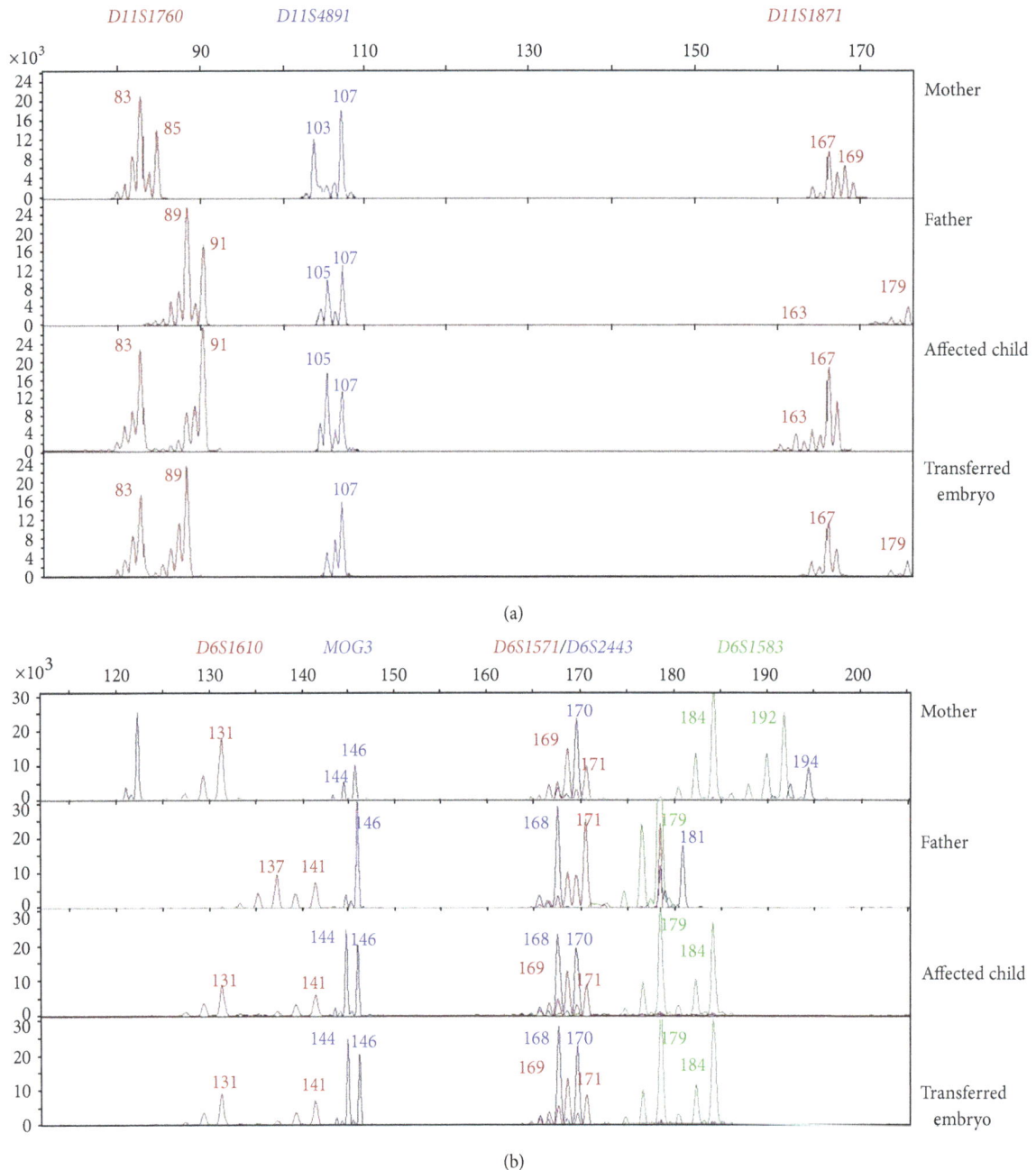

FIGURE 3: Profiles for the selected informative markers employed in the PGD for β-thalassemia (a) with HLA typing (b) in the family here reported.

allowed HSCT. After extensive genetic counselling including information about the PGD procedures, success rate and possibility of misdiagnosis inherent to technique, the couple decided to go on with the treatment and was included in our PGD Program. Informed consent concerning PGD and related procedures as well as the fate of the nontransferred embryos was given.

DNA samples from the affected child and from his parents were obtained with standard protocols and used to perform molecular analyses on both the β-globin and HLA loci.

The disease in this child was found to be due to two different mutations: the IVS-1-nt1[G > A] (HBB:c.92 + 1G > A)

from paternal origin, and the CD39[C > T] (HBB:c.118C > T) inherited from his mother (HbVar ID817 and HbVar ID845, resp., according to the human hemoglobin variants and thalassemias database, http://globin.bx.psu.edu/hbvar/) (Figure 1).

3. Results

3.1. Analysis of the Efficiency of the Multiplex Protocol. After optimization of the corresponding independent PCR procedures for each marker, we proceeded to perform various assays combining different selections of STRs at both

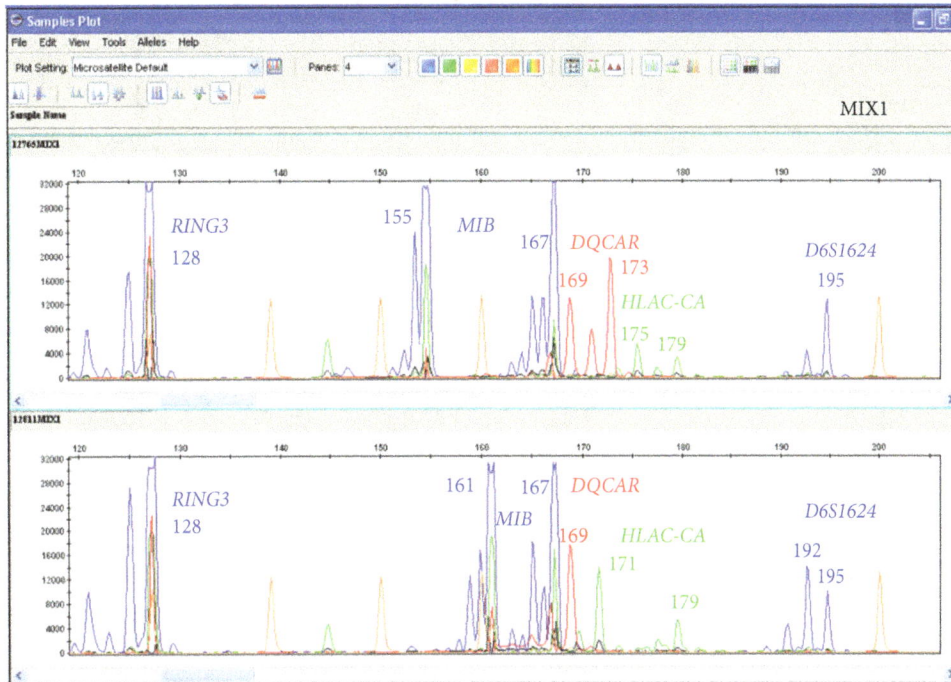

(a)

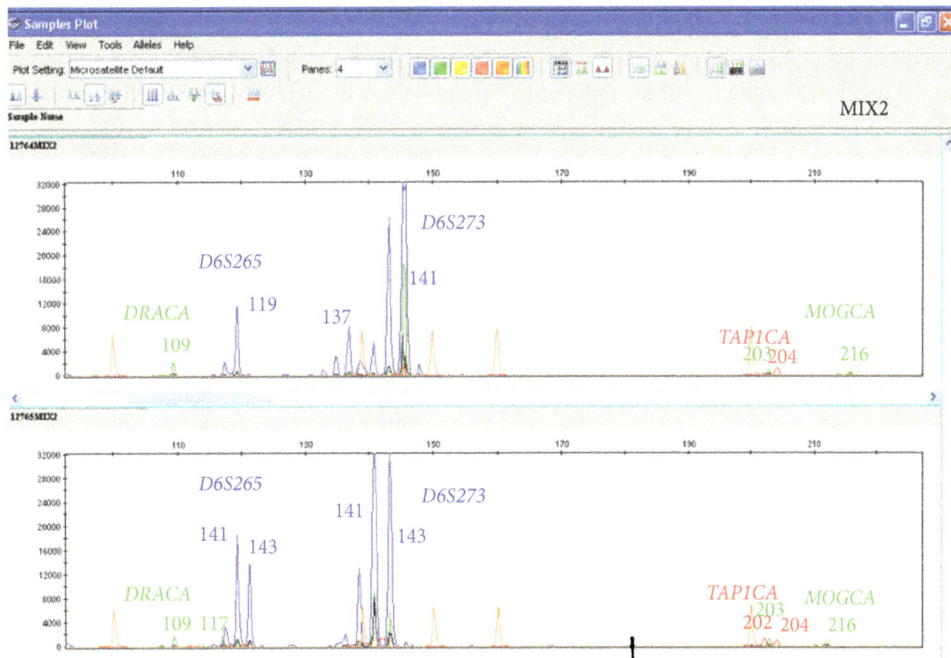

(b)

FIGURE 4: Profiles of 2 different combinations of markers throughout the HLA region after the update of the method on single blastomeres biopsied from nonrelated embryos. Although the efficiency of the PCR varies depending on the marker, the amplification levels obtained for them are adequate to perform the analysis simultaneously. Appropriate zoom of the y-axis of the fluorescence level (proportional to the amplification level), let us to observe and analyze all the peaks, corresponding to the different markers.

the *β-globin* and the *HLA* loci, to analyze the efficiency of the multiplex protocol. Initially, these multiplex protocols were tested using different serial dilutions of genomic DNA extracted from peripheral blood (from 100 ng/mL to 5 pg/mL). Finally, once the final multiplex PCR conditions were adjusted, we tested them on single cells consisting in blastomeres biopsied from supernumerary IVF embryos not suitable for transfer or cryopreservation. Our results showed that different combinations were possible, covering markers at the 5′ and 3′ regions of the *β-globin* locus as well as in

TABLE 2: Results of the clinical PGD-HLA cycles.

PGD	Cycle 1	Cycle 2
Number of oocytes retrieved	20	25
MII oocytes	17	22
2-pronuclei zygotes	14	18
Embryos biopsied	11	17
Affected	2	8
Carriers	7	3
Noncarriers	2	6
HLA nonidentical	9	14
HLA-identical	2	3
HLA nonidentical and affected	1	7
HLA nonidentical and carriers	6	1
HLA nonidentical and noncarriers	2	6
HLA-identical and affected	1	1
HLA-identical and carriers	1	2
HLA-identical and noncarriers	0	0
Embryos transferred	1	1
Ongoing pregnancy	No	Yes

the different regions of the HLA complex (Figure 3). We tested the combination of up to eight different markers (5 at the HLA locus and 3 in the β-globin). Since we had no available DNA from the parents of the embryos donated to this purpose, it was not possible to discriminate between ADO effects or homozygosis for some of the markers tested. Although the efficiency of the PCR varied depending on the marker, no PCR failure was detected for any of them, and the corresponding amplification levels with the final optimized conditions were adequate enough to perform the analysis.

3.2. First Successful PGD with HLA Typing Performed in Spain.
After analysis of the STR markers for HLA and β-globin haplotypes in the context of the family, we selected five markers for HLA typing (*D6S1571, MOG-TAAA, D6S2443, D6S1583,* and *D6S1610*) and another three for the typing of β-globin (*D11S1871, D11S4891,* and *D11S1760*). The selection of such STRs was made according to their amplification efficiency at the single-cell level, the informativity in the family, and their localization along the *HLA* and β-globin loci, respectively (see Supplemental Figure available online at http://dx.doi.org/10.1155/2013/585106). Primers sequences for all the markers susceptible to be analysed with our method are provided in Table 1.

The results of the clinical PGD-HLA cycles for this couple are shown in Table 2. All the embryos generated in the PGD-HLA cycles could be successfully diagnosed and HLA-typed. More specifically, two PGD cycles were performed, resulting in a successful pregnancy, with the birth at term of a healthy boy (Figure 2). Cord blood hematopoietic stem cells were obtained and frozen for later use. The stem cells number in the cord blood was high and HSCT was performed three months later, when the child was 7 years old. The child is currently doing well and off all treatments 4 years following transplantation.

No allele dropout (ADO) and/or contaminations were detected. Recombination events were detected in the paternal β-globin allele in 1 out of the 28 tested embryos, between markers *D11S1871* and *D11S4891*. Regarding HLA typing, a total of 2 recombinant HLA alleles were identified of either maternal or paternal origin, both of them between markers *D6S1583* and *D6S1610*.

The remaining unaffected embryos resulting from both cycles that did not achieve enough quality to be cryopreserved, as well as the affected embryos, were retested for both β-globin and HLA loci, and the initial results were confirmed in all of them. A total of 12 unaffected and non-HLA-identical embryos suitable to be cryopreserved were vitrified using the VitKit Freeze kit (Irvine Scientific) and the protocol provided by manufacturers.

3.3. Update of the Multiplex Genotyping Method.
In 2011, Harton et al. published the ESHRE PGD guidelines in which the general recommendations for HLA typing included a minimum one marker located upstream of HLA-A, minimum one marker between HLA-A and HLA-B, minimum one marker between HLA-B and HLA-DRA, one marker between HLA-DRA and HLA-DQB1, and minimum one marker downstream of HLA-DQB1 [12]. As our initial method did not cover the HLA-HLA-B areas and the HLA-B-HLA-DRA areas, these recommendations prompted us to update our protocol with the inclusion of 10 new markers. We proceeded to perform the same kind of assays previously described for the simultaneous amplification of different combinations of markers, taking into account the policy of covering all the HLA regions recommended by the ESHRE. Figure 4 shows as an example on the analysis of 2 different combinations of the new HLA markers after performance of the multiplex protocol on single blastomeres. Currently, up to 6 different combinations, including 5 HLA and 2 β-globin markers, have been tested and successfully amplified on single blastomeres (Table 3). The goal was to follow the ESHRE PGD guidelines covering all the 5 HLA regions recommended, together with at least 2 markers linked to the β-globin locus (5′ and 3′). Of note, we have applied 3 of those combinations (mixes 1 and 2 from Table 3) for the PGD-HLA typing of 3 other couples (14 cycles, 84 embryos), and all the embryos were successfully typed and diagnosed. Again, neither ADO events nor PCR failures were detected, although the efficiency of amplification varied among the markers.

4. Discussion

Preimplantation genetic diagnosis in combination with human leukocyte antigen typing offer to families not only the possibility of having unaffected children, but also a new therapeutic option for an affected sibling who has HSCT as the unique option of curative treatment. HLA typing on one cell is complex because the HLA locus is highly polymorphic and large (4 Mb) and recombination within the locus has been observed [17, 18]. Worldwide, HLA testing on preimplantation embryos is routinely performed using STRs. Multiple STRs throughout the HLA region allow 100%

TABLE 3: Combinations of markers successfully amplified with our one-step multiplex PCR protocol on a single cell, after the update of the method.

	β-Globin			HLA			
	5′Region	3′Region	Upstream HLAA	HLAA-HLAB	HLAB-HLADRA	HLADRA-HLADQB1	Downstream DQB1
1	D11S1338	D11S2351	D6S1624	HLAC-CA	MIB	DQCAR	RING3-CA
2	D11S1338	D11S2351	MOG-CA	D6S265	D6S273	DRA-CA	TAP1A-CA
3	D11S4891	D11S4181	D6S1624	HLAC-CA	MIB	DRA-CA	TAP1A-CA
4	D11S1760	D11S4181	MOG-CA	D6S265	D6S273	DQCAR	RING3-CA
5	D11S4891	D11S1871	D6S1571	D6S265	MIB	D6S2443	D6S1583
6	D11S1760	D11S1871	MOG-TAAA	HLAC-CA	D6S273	D6S2443	D6S1610

accuracy HLA typing and detect possible recombination events, as well as the copy number of chromosome 6 [19]. Efforts should be devoted to the development of a flexible and reliable methodology for PGD/HLA molecular analysis that would shorten as possible the preclinical development time necessary for future cases. For practical and cost-effective reasons, as well as for clinical and psychological ones, it is important that the time to develop a family-specific PCR protocol should be as short as possible. It is critical to design compatible multiple primer sequences and conditions when several PCR reactions may interfere reciprocally in a tube. In this sense, we have optimized a single-cell multiplex PCR method to simultaneously amplify a wide spectrum of STRs within the HLA and β-globin loci, which minimizes the preclinical development time for the families. Usually, multiplex PCR consists of two rounds of amplification. In the first amplification round, all the primers are added together in the reaction mix; in the second amplification round, the first mix is split into distinct mixtures containing two or three primer combinations. Alternatively, multiple displacement amplification (MDA) has been used for whole genome amplification as a previous step to genetic diagnosis and HLA typing of the embryos [8]. The advantage of our approach is that it involves only a round of single PCR for multiple markers amplification, with a considerable reduction of contamination possibility (more probable in two-rounds methods), and workup time to finally obtain a proper typing and diagnosis. Moreover, no ADO events are detected in comparison with other methods reported elsewhere. In addition, and very importantly, this method has demonstrated to be highly versatile, making the incorporation of new markers at both loci feasible. This has let us to include additional markers for subsequent PGD-HLA cycles, taking into consideration the recommendations provided in the ESHRE PGD consortium best practice guidelines [12], which were published after the initial optimization of our method, and even after its application to several PGD cycles in our unit which led to the first successful PGD with HLA typing performed in Spain.

5. Conclusions

This strategy has allowed us to considerably reduce the optimization of the PCR method for each specific PGD-HLA family as well as the time to obtain molecular results in each cycle (approximately 4 hours after embryo biopsy).

Disclosure

The authors declare no financial relation with the trademarks mentioned in this paper as well as no other conflict of interests.

Authors' Contribution

R. M. Fernández and A. Peciña contributed equally.

References

[1] T. N. Williams and D. J. Weatherall, "World distribution, population genetics, and health burden of the hemoglobinopathies," *Cold Spring Harbor Perspectives in Medicine*, vol. 2, no. 9, Article ID a011692, 2012.

[2] R. C. Hardison, D. H. K. Chui, C. R. Riemer et al., "Access to a syllabus of human hemoglobin variants (1996) via the world wide web," *Hemoglobin*, vol. 22, no. 2, pp. 113–127, 1998.

[3] N. F. Olivieri, "The β-thalassemias," *The New England Journal of Medicine*, vol. 341, no. 2, pp. 9–109, 1999.

[4] G. La Nasa, F. Argiolu, C. Giardini et al., "Unrelated bone marrow transplantation for β-thalassemia patients: the experience of the Italian bone marrow transplant group," *Annals of the New York Academy of Sciences*, vol. 1054, pp. 186–195, 2005.

[5] A. Kuliev, S. Rechitsky, O. Verlinsky et al., "Preimplantation diagnosis and HLA typing for haemoglobin disorders," *Reproductive BioMedicine Online*, vol. 11, no. 3, pp. 362–370, 2005.

[6] J. Reichenbach, H. D. van Velde, M. de Rycke et al., "First successful bone marrow transplantation for X-linked chronic granulomatous disease by using preimplantation female gender typing and HLA matching," *Pediatrics*, vol. 122, no. 3, pp. e778–e782, 2008.

[7] H. van de Velde, M. de Rycke, C. de Man et al., "The experience of two European preimplantation genetic diagnosis centres on human leukocyte antigen typing," *Human Reproduction*, vol. 24, no. 3, pp. 732–740, 2009.

[8] A. M. Hellani, S. M. Akoum, E. S. Fadel, H. M. Yousef, and K. K. Abu-Amero, "Successful pregnancies after combined human leukocyte antigen direct genotyping and preimplantation genetic diagnosis utilizing multiple displacement amplification," *Saudi Medical Journal*, vol. 33, pp. 1059–1064, 2012.

[9] F. Lamazou, J. Steffann, N. Frydman et al., "Preimplantation diagnosis with HLA typing: birth of the first double hope child in France," *Journal de Gynécologie Obstétrique et Biologie de la Reproduction*, vol. 40, no. 7, pp. 682–686, 2011.

[10] T. El-Toukhy, H. Bickerstaff, and S. Meller, "Preimplantation genetic diagnosis for haematologic conditions," *Current Opinion in Pediatrics*, vol. 22, no. 1, pp. 28–34, 2010.

[11] R. A. Smith, P. J. Ho, J. B. Clegg, J. R. Kidd, and S. L. Thein, "Recombination breakpoints in the human β-globin gene cluster," *Blood*, vol. 92, no. 11, pp. 4415–4421, 1998.

[12] G. L. Harton, M. de Rycke, F. Fiorentino et al., "ESHRE PGD consortium best practice guidelines for amplification-based PGD," *Human Reproduction*, vol. 26, no. 1, pp. 33–40, 2011.

[13] A. Peciña, M. D. Lozano-Arana, J. C. García-Lozano, S. Borrego, and G. Antiñolo, "One-step multiplex polymerase chain reaction for preimplantation genetic diagnosis of Huntington disease," *Fertility and Sterility*, vol. 93, no. 7, pp. 2411–2412, 2010.

[14] R. Carrillo-Vadillo, J. C. García-Lozano, M. D. Lozano-Arana, J. L. M. Rivera, P. Sánchez Martín, and G. Antiñolo, "Two sets of monozygotic twins after intracytoplasmic sperm injection and transfer of two embryos on day 2," *Fertility and Sterility*, vol. 88, no. 6, pp. 1676.e3–1676.e5, 2007.

[15] I. Liebaers, K. Sermon, C. Staessen et al., "Clinical experience with preimplantation genetic diagnosis and intracytoplasmic sperm injection," *Human Reproduction*, vol. 13, supplement 1, pp. 186–195, 1998.

[16] A. R. Thornhill, C. E. deDie-Smulders, J. P. Geraedts et al., "ESHRE PGD Consortium 'best practice guidelines for clinical preimplantation genetic diagnosis (PGD) and preimplantation genetic screening (PGS)'," *Human Reproduction*, vol. 20, no. 1, pp. 35–48, 2005.

[17] L. Malfroy, M. P. Roth, M. Carrington et al., "Heterogeneity in rates of recombination in the 6-Mb region telomeric to the human major histocompatibility complex," *Genomics*, vol. 43, no. 2, pp. 226–231, 1997.

[18] M. Martin, D. Mann, and M. Carrington, "Recombination rates across the HLA complex: use of microsatellites as a rapid screen for recombinant chromosomes," *Human Molecular Genetics*, vol. 4, no. 3, pp. 423–428, 1995.

[19] H. van de Velde, I. Georgiou, M. de Rycke et al., "Novel universal approach for preimplantation genetic diagnosis of β-thalassaemia in combination with HLA matching of embryos," *Human Reproduction*, vol. 19, no. 3, pp. 700–708, 2004.

18

Distribution of Genes and Repetitive Elements in the *Diabrotica virgifera virgifera* Genome Estimated Using BAC Sequencing

Brad S. Coates,[1,2] **Analiza P. Alves,**[3] **Haichuan Wang,**[3]
Kimberly K. O. Walden,[4] **B. Wade French,**[5] **Nicholas J. Miller,**[3] **Craig A. Abel,**[1]
Hugh M. Robertson,[4] **Thomas W. Sappington,**[1,2] **and Blair D. Siegfried**[3]

[1] *Corn Insects and Crop Genetics Research Unit, ARS, USDA, Ames, IA 50011, USA*
[2] *Department of Entomology, Iowa State University, Ames, IA 50011, USA*
[3] *Department of Entomology, University of Nebraska, Lincoln, NE 68583, USA*
[4] *University of Illinois, Champaign-Urbana, IL 61801, USA*
[5] *North Central Agricultural Research Laboratory, Brookings, ARS, USDA, SD 57006, USA*

Correspondence should be addressed to Brad S. Coates, brad.coates@ars.usda.gov

Academic Editor: Yong Lim

Feeding damage caused by the western corn rootworm, *Diabrotica virgifera virgifera*, is destructive to corn plants in North America and Europe where control remains challenging due to evolution of resistance to chemical and transgenic toxins. A BAC library, DvvBAC1, containing 109,486 clones with 104 ± 34.5 kb inserts was created, which has an ∼4.56X genome coverage based upon a 2.58 Gb (2.80 pg) flow cytometry-estimated haploid genome size. Paired end sequencing of 1037 BAC inserts produced 1.17 Mb of data (∼0.05% genome coverage) and indicated ∼9.4 and 16.0% of reads encode, respectively, endogenous genes and transposable elements (TEs). Sequencing genes within BAC full inserts demonstrated that TE densities are high within intergenic and intron regions and contribute to the increased gene size. Comparison of homologous genome regions cloned within different BAC clones indicated that TE movement may cause haplotype variation within the inbred strain. The data presented here indicate that the *D. virgifera virgifera* genome is large in size and contains a high proportion of repetitive sequence. These BAC sequencing methods that are applicable for characterization of genomes prior to sequencing may likely be valuable resources for genome annotation as well as scaffolding.

1. Introduction

Bacterial artificial chromosome (BAC) libraries are composed of physical constructs that contain large genomic DNA inserts and provide a tool for the molecular genetic research of organisms of interest. For instance, anonymous genetic markers linked to genes that control insecticide resistance traits have been identified on BAC clones, and, following subsequent sequencing of cloned inserts, allowed the characterization of gene(s) that influence the expression of these traits [1]. Furthermore, sequence data from BAC inserts provide a means to evaluate genome structure, including the estimation of repetitive element densities [2]

and the relative gene content of a species [3]. BAC clones are also useful for the construction of physical maps that represent contiguous sequence from an entire genome or genomic regions, and these assemblies have proven useful for determination of minimum tiling paths prior to BAC-by-BAC sequencing of large or highly repetitive genomes [4]. Scaffolding takes advantage of paired BAC end sequence (BES) data which provide direct physical linkages between sequence tags [5] and may assist in the scaffolding of contigs assembled from mate paired reads from next generation sequencing technologies.

The western corn rootworm, *Diabrotica virgifera virgifera* (Coleoptera: Chrysomelidae), is a beetle native to North

America, which is adapted to feeding on a limited number of grasses including corn [6]. The ancestral geographic range of *D. virgifera virgifera* extended from present day Mexico into the Southwest United States and Great Plains, but an eastward range expansion began in the 1940s that coincided with the widespread cultivation of continuously planted corn in the central United States [7, 8]. *D. virgifera virgifera* was accidentally introduced into Central Europe in the early 1990s [9], and subsequent transatlantic and intra-European introductions have contributed to its contemporary geographic range in Europe [10, 11]. *D. virgifera virgifera* has one generation per year. Individuals overwinter as diapausing eggs which hatch in the spring, and subterranean larvae feed on corn roots [12]. Root damage caused by *D. virgifera virgifera* reduces the plant's ability to absorb soil nutrients and compromises structural stability [13]. Upon pupation and emergence from the soil, adult corn rootworm beetles can persist in fields for up to 4 weeks, can reduce seed pollination rates through feeding damage to corn silks (stamen) and can vector maize chlorotic mottle virus [14] and stalk rot fungus [15].

Larval feeding damage can be suppressed by systemic seed treatments, soil-applied insecticides, or transgenic corn hybrids that express *Bacillus-thuringiensis*-(Bt) derived insecticidal proteins. However, resistance to both chemical and Bt toxins are documented [16–19]. *D. virgifera virgifera* populations have also been managed by an alternating corn-soybean crop rotation (grass-legume rotation) [20], which negates the need for insecticide applications. This control strategy is based on a strong female preference to oviposit in soil at the base of corn plants, and the cospecialization of larvae for feeding on grass roots. However, female *D. virgifera virgifera* phenotypes have evolved that are no longer specifically attracted to cornfields but will lay eggs near other plants [21–23]. In the subsequent crop production year, progeny of these variant *D. virgifera virgifera* females will emerge in and damage first-year corn crops. This adaptive loss of adult fidelity in oviposition behavior has defeated the use of corn-soybean rotation as an effective control practice in many corn growing regions of the United States [22].

The propensity for corn rootworm to adapt to control measures has raised concern among producers, scientists, and regulatory agencies, and the need to investigate the underlying genetic mechanisms for adaptation is critical to developing sustainable pest management approaches [7, 24, 25]. In anticipation of a recently initiated whole genome sequencing (WGS) effort for *D. virgifera virgifera* that aims to build a foundation for future genetic and genomics research [25], we have determined the haploid genome size and have estimated gene and repetitive fraction densities from BAC sequencing data. These data and resources will facilitate annotation and contig scaffolding efforts of the upcoming WGS project.

2. Materials and Methods

2.1. Genome Size Estimation. Three males from the inbred nondiapausing *D. virgifera virgifera* colony at USDA-ARS, North Central Agricultural Research Laboratory in Brookings, SD [26] were starved for 24 hr, and homogenized with a razor blade in 0.5 ml chopping buffer (15 mM HEPES, 1 mM EDTA, 80 mM KCl, 20 mM NaCl, 300 mM sucrose, 0.2% Triton X-100, 0.5 mM spermine tetrahydrochloride, 0.25 mM PVP). Homogenate was filtered through 20 um nylon mesh, centrifugated at $100 \times g$ for 5 minutes, and nuclei suspended in 0.5 ml propidium iodide (PI) staining buffer (10 mM MgSO$_2$, 50 mM KCl, 5 mM HEPES, 0.1% DL-dithiothreitol, 2.5% Triton X-100, 100 ug/mL propidium iodide). Nuclei from the *Zea mays* inbred line B73 (genome size 2.5 Gb) and *Glycine max* line Williams 83 (1.115 Gb) were similarly prepared. Propidium iodide stained nuclei were analyzed on a BD Biosciences (San Jose, CA, USA) FACSCanto flow cytometer equipped with a 488 nm laser and 610/620 emission filter. Estimates for standards (B73 and Williams 83) and *D. virgifera virgifera* were performed in triplicate. The estimated *D. virgifera virgifera* genome size was calculated from PI signals [27] and converted to pg estimates [28].

2.2. BAC Library Construction, End Sequencing, and Annotation. Genomic DNA was extracted from ~100 individuals of the *D. virgifera virgifera* nondiapausing strain, pooled, and fractionated by partial digestion with *Hind*III, fragments between ~100 and 150 kb were excised, and these inserts were ligated into the pCC1 BAC vector (Epicentre, Madison, WI, USA). Constructs were used to transform the *Escherichia coli* strain DH10B T1 by electroporation. Transformants were plated on LB agar (12.5 μg mL^{-1} chloramphenicol, 80 μg mL^{-1} X-Gal, and 100 0.5 mM IPTG) and a total of 110,592 BAC clones were arrayed on 288 individual 384-well plates to comprise the DvvBAC1 library. The mean insert size within DvvBAC1 was estimated by contour-clamped homogeneous electric field (CHEF) electrophoresis of *Not*I digested BAC DNA from 96 clones on a 0.9% agarose in 0.5X TAE buffer gel ramp run with a pulse time 5–15 s at 5 V/cm for 24 hrs and 4°C. Insert size estimates were made by comparison to the MidRange II PFG Marker (New England Biolabs, Ipswich, MA, USA), and the fold genome coverage of DvvBAC1 was estimated according to Clark and Carbon [29]. BAC DNA from 1152 DvvBAC1 clones (plates 217, 218, and 227) were purified and sequenced and annotated as described by Coates et al. [2], and sequence data deposited into the GenBank genome survey sequence (GSS) database (accession numbers JM104642–JM106797).

2.3. BAC Screening and Full Insert Sequencing. DNA from DvvBAC1 clones were pooled into matrix, row, and column pools according to Yim et al. [30] and used in PCR reactions as described by Coates et al. [31]. DNA from DvvBAC1 clones was purified using the Large Construct Purification Kit (Qiagen, Valencia, CA, USA) according to the manufacturers instructions, and DNA preparations run on 0.8% agarose gel electrophoresis. BAC DNA was used to create individual mid-tagged libraries (RL1 to RL10) and each was sequenced on Roche GS-FLX at the William H. Keck Center for Comparative and Functional Genomics at the University of Illinois.

Cross-match (http://www.phrap.org/), Roche-provided sff tools (http://454.com/products/analysis-software/index.asp) and custom Java scripts were used to identify trim sequencing adaptors within sff file data and remove sequences of <50 nucleotides or with homopolymer stretches ≥60% of the raw read length. Processed sequence data was assembled into contigs using the Roche GS De Novo Assembler v 1.1.03 using default parameters (seed step: 12, seed length: 16, min overlap length: 40, min overlap identity: 90%, alignment identity score: 2, and alignment difference score: 23).

Cloning vector sequence was identified using VecScreen (http://www.ncbi.nlm.nih.gov/VecScreen/VecScreen.html) and masked using Maskseq [32]. Contigs assembled from contaminating *Escherichia coli* DNA were identified by querying against the K-12 reference genome (GenBank accession NC_000913) using the blastn algorithm, and contigs that produced E-values $\leq 10^{-15}$ were removed manually. The remaining filtered BAC contigs were annotated using the MAKER 2 genome annotation pipeline [33] using coding sequence evidence from 17,778 *D. virgifera virgifera* ESTs (GenBank dbEST accessions EW761110.1– EW777358.1 [34] and CN497248.1–CN498776.1 [35]), protein homology by blastx searches of the UniProt/Swiss-Prot databases, and *T. castaneum* gene models using the AUGUSTUS web server [36]. Prior to any annotation, RepeatMasker and RepeatRunner were used to identify retroelement-like regions within the BAC full inserts by running against predefined RepBase and RepeatRunner te_proteins provided by MAKER 2 [33]. MAKER 2 output was imported into the Apollo Genome Annotation and Curation Tool [37], where additional annotations were performed via blastx searches of the NCBI nr protein database (E-values $\leq 10^{-15}$).

Contigs from clones 142B02 and 156M20 represent partially overlapping homologous sequence and were assembled into a single reference using CAP3 [38] (default parameters). Processed 454 read data from libraries RL003 and RL007 were mapped to this assembled reference using the program LASTZ [39] (default parameters). LASTZ output was made in Sequence Alignment/Map (SAM) format which was used to create an indexing sorted alignment file (.BAI file) with the command line index tool from SAMtools [40]. Mapped read data was visualized using BAMview in the Artemis Genome Viewer [41].

2.4. Comparative Genomics and Annotation of Repetitive Elements. Genomic and EST sequences were separately aligned for cadherin orthologs from *D. virgifera virgifera* (GenBank accessions; mRNA EF531715.1 with DNA EF541349.1) and *T. castaneum* (gene model XM_966295.2 with DNA scaffold NC_007417 positions 19,140,127 to 19,1330,052) using the program Splign with the discontinuous megablast option [42]. Splign tab-delimited output was used to estimate mean intron and exon size. Additionally, a *de novo* prediction of *D. virgifera virgifera* repetitive sequence was made by assembling BAC end sequence (BES) data using CAP3 [38] (default parameters), and subsequently used to query the *D. virgifera virgifera* cadherin genomic DNA sequence (EF541349.1) for putative repetitive sequence using the

blastn algorithm. Blastn output was filtered for E-values $\geq 10^{-40}$. *De novo* prediction of *D. virgifera virgifera* repetitive sequences were also made within our assembled BAC full inserts (GenBank accessions JQ581035–JQ581043) by querying accession EF541349.1 using identical parameters. BAC insert regions with similarity to the EF541349.1 sequence were excised from BES contigs and BAC full insert sequences (using a custom PERL script), mapped to EF541349.1 using LASTZ [39] (default parameters) and output handled as described previously.

A computational prediction of short repetitive DNA elements known as miniature inverted repeat transposable elements (MITEs) was made for *D. virgifera virgifera* BES contigs and singletons as well as GenBank accession EF541349.1 and *T. castaneum* scaffold NC_007417 positions 19,140,127 to 19,1330,052 using the MITE Uncovering SysTem (MUST; http://csbl1.bmb.uga.edu/ffzhou/MUST/) [43] (default parameters except max DR length = 4 and Min MITE length = 150). The secondary structures of putative MITEs were confirmed by using the Mfold DNA Server (http://mfold.rna.albany.edu/?q=mfold/DNA-Folding-Form) [44] with conditions 25°C and 1.0 mM Mg^{2+}.

3. Results

3.1. Genome Size Estimation. The *D. virgifera virgifera* haploid genome size was estimated at $71,144 \pm 537$ fluorescent units from propidium iodide (PI) stained nuclei, which compared to $69,319 \pm 491$ and $35,631 \pm 687$ units for the internal standards of known genome size, *Zea mays* (2.50 Gbp) and *Glycine max* (1.115 Gbp), respectively. Populations of nuclei from *Z. mays* and *D. virgifera virgifera* produced overlapping PI signals on a flow cytometer, but the size scatter component (SSC-A) indicative of nucleosome densities was used to separate the signals of independent PI readings (Figure 1). Subsequent calculations of PI to genome size ratios indicate an estimated *D. virgifera virgifera* haploid genome size of ~2.58 Gb or 2.80 pg.

3.2. BAC Library Construction, End Sequencing, and Annotation. Blue-white screening indicated the ligation efficiency with the pCC1 vector was ~99.25% and arraying of clones onto 384-well plates with ~99.75% of the wells being successfully filled (Amplicon Express, personal comm.). From these data, ~109,486 genomic clones were estimated within the 288 × 384 well plates of DvvBAC1. Insert DNA was isolated, digested with *Not*I, and separated by CHEF electrophoresis from 93 of 96 DvvBAC1 clones (96.9%), which indicated a mean pCC1 insert size of 104.4 ± 34.5 kb (Figure 2; not all data shown). From these data, we estimate that $11,496 \pm 3,758$ Mbp are within DvvBAC1, translating to ~4.56 ± 1.49-fold genome coverage (1.49- and 0.97-fold genome coverage at 95 and 99% probability thresholds, resp.).

Paired end sequencing of 1152 DvvBAC1 clones generated 2304 raw reads, of which 2156 produced high quality sequence data (PHRED scores ≥20; NCBI dbGSS; accessions JM104642–JM106797). Paired BAC end sequence (BES) data was obtained from 1037 of the 1152 clones (90.0%). Filtering for reads >100 bp resulted in 1999 sequences averaging 579.0 ± 141.1 bp (1.17 Mb total; ~0.05% of the 2.58 Gb *D. virgifera virgifera* genome). Functional annotations were obtained for 599 of 1999 filtered BES reads (30.0%) using blastx results, of which 167 sequences received 620 gene ontology (GO) annotations (3.75 ± 1.83 GO annotations per annotated sequence; see Table S1 in Supplementary Material available online at doi:10.1155/2012/604076 which provides a list of putative genes, biochemical functions, and pathway assignments). At level 2, the distribution of GO terms among biological process (P), cellular component (C), and molecular function (F), respectively, showed cellular process, cellular component, and binding activity as most prominent (Figure S1). A total of 447 unique InterPro annotations were made (Table S1; 12 most frequent are listed in Table 1). Predicted functional gene annotations within catalytic activities at GO level F were corroborated by 154 reverse transcriptase (IPR015706 and IPR000477), 12 endonuclease/exonuclease (IPR005135), 17 ribonuclease (IPR012337 and IPR002156), and 5 integrase (IPR017853) annotations in the InterProScan output. An analogous blastn search indicated that 210 sequences (14.3%) showed $\geq 68.0\%$ similarity to the complete *D. virgifera virgifera* cadherin gene (GenBank Accession EF531715.1; E-values $\leq 1.31 \times 10^{-11}$), and 23 (1.6%) showed $\geq 69.0\%$ similarity to the *D. virgifera virgifera*, *D. barberi*, and *D. virgifera zeae* microsatellite sequences (E-values $\leq 1.14 \times 10^{-11}$; Figure S2). In addition, a total of 45 annotations of DvvBAC1 BES reads (3.1% of total) indicate an origin from the proteobacterial endosymbiont, *Wolbachia* (Figure S2; Table S1).

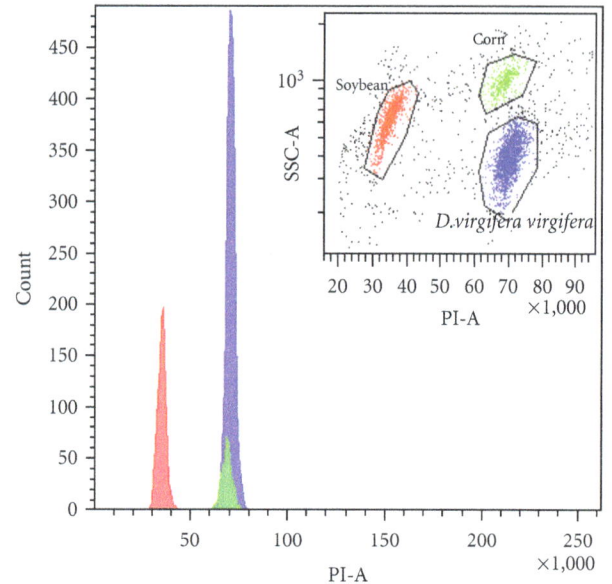

Population	Number of events	Percentage of parent	PI-A mean	Genome (Gb)
Soybean	978	13.8	35,631	1.115
Corn	397	5.6	69,319	2.5
D.virgifera virgifera	2,453	34.6	71,144	2.58

FIGURE 1: Flow cytometry estimate of the *D. virgifera virgifera* genome size compared to internal standards of *Zea mays* (inbred line B73; 2.500 Gb) and *Glycine max* (isoline Williams 83; 1.115 Gb).

3.3. BAC Screening and Full Insert Sequencing. Screening of DvvBAC1 identified clones containing sequence from eight EST markers (5.29 ± 2.98 hits; range of 1 to 9 hits per marker; data not shown). Eight of the 9 BAC inserts (88.9%) were successfully sequenced on the Roche GS-FLX. After raw data filtering, a total of 240,586 reads were assembled into 39 contigs that contained 642.0 kb of sequence (16.5 ± 18.9 kb per contig; Table 2). The annotation of BAC inserts using MAKER 2 predicted 37 putative genes and 48 retrotransposon-like protein coding intervals with 3 and 31 of these sequences supported by EST evidence, respectively.

Contigs from clones 142B02 and 156M20 represent homologous genomic regions within different clones and provide a measure of haplotype variation within the library. Sequences from these two clones shared 11 endogenous and 5 retroelement-like protein coding sequences, which represent homologous genome intervals from unique BAC inserts. Six contigs totaling 31.9 kb were aligned (Figure 3), and haplotype variation between inserts was shown via 3 SNPs within the 22.5 kb of CAP3 aligned sequence (SNP frequency ~1.3 $\times 10^{-4}$), protein coding sequences were 100% conserved, and no indels were present. Compared to the consensus, 2564 and 5467 bp regions were not represented within clones 142B02 and 156M20, respectively, and was verified by mapping reads to the CAP3 scaffolds (Figure 3). *Hind*III restriction site

mapping showed that cut sites used in cloning may not have been the cause of sequence disparity. Additionally, the entire pCC1 cloning vector sequence was sequenced and masked from both clone 142B02 and 156M20 assemblies, indicating that insert boundaries did not give rise to the two gaps. Retroelement-like sequences were annotated within the two haplotype sequence gaps. These results also suggest that structural variation based on the integration/excision or random deletion of repetitive DNA elements may exist among *D. virgifera virgifera* haplotypes.

3.4. Comparative Genomics and Annotation of Repetitive Elements. Comparison of the cadherin gene intron and exon structure from the 94.6 kb *D. virgifera virgifera* and 7.1 kb *T. castaneum* orthologs indicated that the ~13.3-fold increase in the former is accounted for by intron sequence. Specifically, the *T. castaneum* cadherin has a mean intron size of 0.085 ± 0.189 kb, compared to 2.9 ± 1.5 kb in the *D. virgifera virgifera* cadherin, whereas respective total exon sizes of 4.9 kb (mean 180 ± 72.8) and 5.4 kb (mean 173 ± 71.2) were similar between species (Tables S2 and S3). The de novo prediction of repetitive elements by alignment of *D. virgifera virgifera* BES data and BAC full insert sequences resulted in 226 contigs and 1089 singletons (mean length of 761.0 ± 236.3 bp; maximum 2002 bp, mean depth = 3.3 ± 4.3 reads). Mapping *de novo* repetitive genome regions (150 from BES

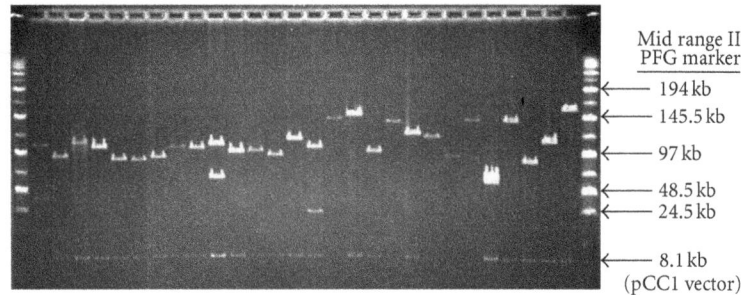

FIGURE 2: Estimated BAC genomic insert sizes using contour-clamped homogeneous electric field (CHEF) electrophoresis. DNA preparations were digested with *Not*I prior to separation on a 0.9% agarose gel in 0.5X TAE buffer for 24 h at 4°C.

TABLE 1: The number of InterPro accessions obtained during annotation of *D. v. virgifera* BAC end sequences.

InterPro entry	Number	InterPro functional description(s)
IPR015706	87	RNA-directed DNA polymerase (reverse transcriptase)
IPR000477	67	Reverse transcriptase
IPR009072	18	Histone-fold
IPR012337	17	Ribonuclease H-like
IPR007125	16	Histone core
IPR000558	14	Histone H2B
IPR005135	12	Endonuclease/exonuclease/phosphatase
IPR011991	8	Winged helix-turn-helix transcription repressor
IPR001878	8	Zinc finger, CCHC-type
IPR005819	5	Histone H5
IPR001584	5	Integrase, catalytic core
IPR002156	5	Ribonuclease H domain
IPR005818	5	Histone H1/H5

data (mean 287.4 ± 131.9 bp) and 146 from BAC full inserts (348.2 ± 201.0 bp)) to the *D. virgifera virgifera* cadherin gene sequence resulted in alignments mostly within introns, where the greatest read depth of 37 and 42 were in introns 2 and 12, respectively (Figure 4). Annotation of *de novo* assembled repetitive sequences indicated that 36 (15.9%) encoded reverse transcriptase, gag-pol, or other retrovirus-associated proteins. Histone-like proteins were encoded by contig 87 (histone H1), contig 110 and 149 (histone H2a), contigs 11 and 173 (histone H2b), and contig 214 (histone H3; remaining data not shown).

Predictions of transposable elements by MUST indicated 88 putative MITE-like sequences with direct repeats (DRs) of 2 nucleotides were located within the *D. virgifera virgifera* cadherin gene (Table S4), where 22, 18, and 11 of the DRs involved AT/TA, AA, and TT dinucleotides. Putative MITEs that occupy a total of 2.4 kb (mean = 278.9 ± 124.5 bp) are composed of 65.7 ± 0.1% A or T nucleotides and have predicted terminal inverted repeat (TIR) lengths of 11.9 ± 5.5 bp. Positions of MITE-like inserts were predicted to be within intron regions (Figure 4). Comparatively, the *T.*

castaneum cadherin gene contained 12 putative MITE-like elements that were all predicted within intron regions (Table S5).

4. Discussion

4.1. Genome Size Estimation. The haploid *D. virgifera virgifera* genome size of 2.58 Gb (2.80 pg) is one of the largest estimated among beetle species ([45]; mean 0.891 ± 0.795 pg), which range from ∼0.15 for *Oryzaephilus surinamensis* (Coleoptera: Silvanidae) [46] to 3.40 Gb for *Chrysolina carnifex* (Coleoptera: Chysromelidae) [45]. Genome size heterogeneity among beetle species does not appear to be correlated with organism "complexity" (*C*-value paradox) [47], specialization [48], or increased gene content [49]. The relation between repetitive DNA content and genome size in Coleoptera is only available for the model species *Tribolium castaneum* (Coleoptera: Tenebrionidae), where the ∼0.200 Gb genome has an estimated 5110 repetitive elements [30] which comprise ∼13 of the 0.160 Gb assembled sequence [49]. In contrast, our data suggest that the proportion of the *D. virgifera virgifera* genome consisting of repetitive DNA is much higher.

4.2. BAC Library Construction, End Sequencing, and Annotation. BAC libraries are genomic tools that are useful for the isolation of genes linked to a trait [50] as well as the generation of end sequences that provide estimates of genome structure and TE densities [2, 51, 52]. Despite their utility in genomics research, only one coleopteran BAC library has previously been reported, for *T. castaneum* [53]. The prediction of gene-coding regions from BAC end sequence from nonmodel species rely on functional annotation by homology-based identification with related genes in model organisms. This can result in vague or inaccurate gene definitions for nonmodel species [54], such as our *D. virgifera virgifera* dbGSS dataset. Despite the relative dearth in gene discovery by *D. virgifera virgifera* BES, 179 novel protein coding regions were annotated which will provide a resource for annotation of future WGS efforts. Studies with similarly low genome sequence coverage have been useful for initial descriptions of functional and repetitive elements [55].

TABLE 2: Summary of contigs per BAC that were assembled from Roche-454 sequencing data.

Mid-tag library	BAC clone	Marker	Raw data (reads/kb)	Assembled data (reads/kb)	GenBank accession	Contig size (kb)
RL001	40F02	1304	23,300/10,298	21,525/97.8	JQ581035	19.6 ± 22.2
RL002	89B10	1224	8,701/7,408	7,277/118.7	JQ581036	29.7 ± 20.6
RL003	142B02	1203	29,444/12,894	19,447/30.5	JQ581037	4.3 ± 3.6
RL005	191G22	1125	64,410/28,909	8,348/104.0	JQ581038	17.3 ± 18.8
RL006	163F14	1304	9,495/3,523	22,530/101.7	JQ581039	25.4 ± 7.2
RL007	156M20	1203	16,427/7,179	9,435/24.6	JQ581040	4.1 ± 2.8
RL008	222P02	1345	43,702/20,431	0/0.0	FAILED	NA
RL009	213A05	1411	25,196/11,345	18,868/74.5	JQ581041	74.5 ± 0.0
RL010	188M01	1300	19,912/9,238	18,478/90.6	JQ581042	15.1 ± 9.7
		Total	240,587/111,225	125,908 (15,738 ± 6286)/642.3 (80.3 ± 34.9)		13.6 ± 20.1

NA: not applicable due to DNA sequencing failure.

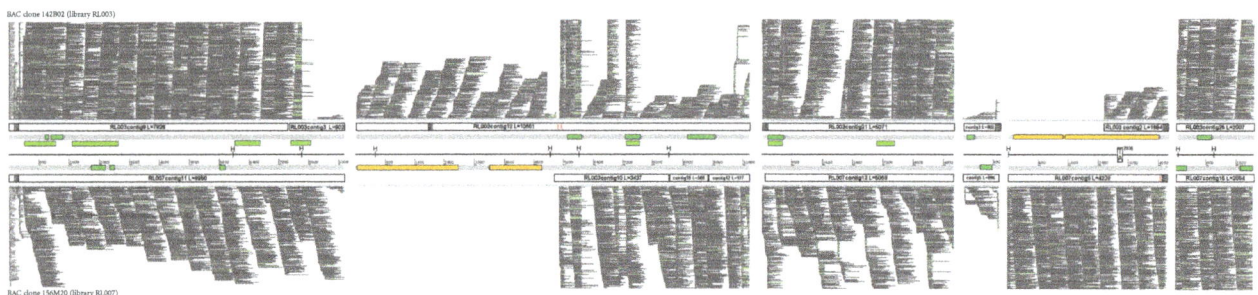

FIGURE 3: Comparison of haplotypes between assembled full BAC insert sequences of clones 142B02 and 156 M20. Homologous regions are aligned and representative read depths are indicated above for 142B02 and below for 156M20. Annotated genes (dark green), expressed sequence tag (EST; light green), and repetitive element sequences (orange) are indicated. Microsatellite repeat motifs are shown as (|||||).

Proteins encoded by DNA-based TEs and retrotransposons totaled ~16.0% of BES reads and outnumbered endogenous genes by ~1.8-fold. Extrapolation suggests that retroelement-like TE genes might occupy ~0.41 Gb of the 2.58 Gb genome. Compared to *T. castaneum* which has ~3.7% of the genome assembly occupied by LTR- and non-LTR-retrotransposons [30], the *D. virgifera virgifera* genome may have an ~4.3-fold higher retroelement content. Our investigations also indicate that small nonautonomous miniature inverted repeat transposable elements (MITEs) are present within the *D. virgifera virgifera* genome.

4.3. BAC Screening and Full Insert Sequencing. The Roche-454 GS-FLX provides a robust method for rapid sequence generation, from which single end read data were sufficient to assemble 8 of the 9 BAC plasmids we sequenced. Assembly of *D. virgifera virgifera* BAC inserts into an average of ~5 contigs per clone and encompassing 80.3 kb of total sequence was greater than that obtained following assembly of BACs from barley [23]. Annotations indicated that the number of TE-derived genes in assembled contigs were 1.3-fold higher than endogenous protein coding genes. This result differs from our estimate from BES data but may be influenced by sample number or by the effect of large TE-derived gene sizes on the probability of sampling from BES data. Regardless, full BAC insert sequences indicate that the *D. virgifera*

virgifera genome is comprised of a high proportion of TE-derived sequence but also suggests that DNA-based and retroelement-like TEs are localized within intergenic space. This preliminary genome sequencing evidence suggests that genic regions of the *D. virgifera virgifera* genome can be assembled from short single-end NGS read data, but the use of longer read lengths and paired-end or mate-pair NGS strategies may result in increased contig size and/or scaffolding by the spanning of repetitive elements.

Comparison of the homologous regions within contigs from clones 142B02 and 156M20 provided a direct measure of haplotype variation within DvvBAC and also within the *D. virgifera virgifera* nondiapause strain. SNP variation between haplotypes was low, which may be the result of a genetic bottleneck and subsequent inbreeding within the colony. These results are consistent with a microsatellite marker-based estimate of 15–39% allele diversity reduction in the nondiapause colony compared to wild populations [56]. Comparison of *D. virgifera virgifera* haplotypes suggested that local genome variation based upon insertion/deletion of large DNA regions may occur. Evidence suggests that these variations are not likely due to differences in read depth or effects of cloning due to variation in *Hind*III restriction sites. Interestingly, retroelement-like sequences were annotated within regions of haplotype variation and may indicate that microsynteny is altered through TE integration. Analogous

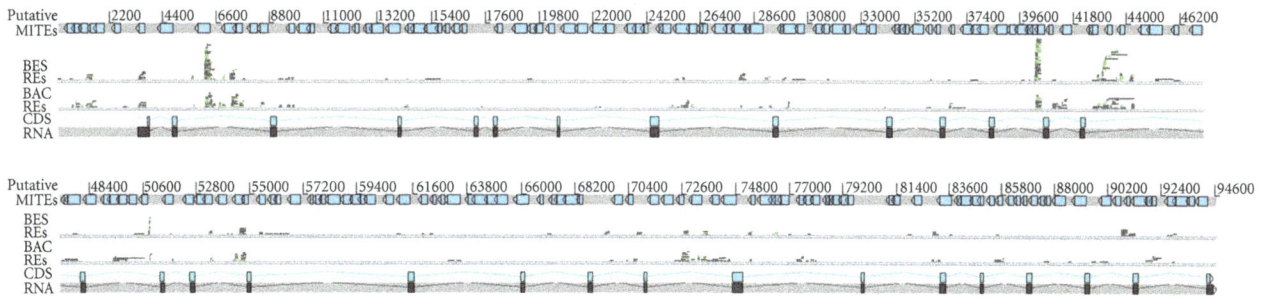

FIGURE 4: Identification of putative miniature inverted repeat transposable elements (MITEs) (blue rectangles indicating direction), and *de novo* mapped repetitive elements identified from BAC end sequences (BES REs) and BAC full insert sequences (BAC REs), within the gene protein coding sequence (CDS) and transcript sequence (RNA).

haplotype variation was caused by movement of *Helitron*-like TEs in maize and SINEs in canine genomes. Our results similarly suggest that retroelement movement may be a source of haplotype variation in the *D. virgifera virgifera* genome but will require further investigation to realize the extent to which movements affect genome structure and function.

4.4. Comparative Genomics and Annotation of Repetitive Elements. Compared to *T. castaneum*, the orthologs of intron-less histone encoding genes show no size increase within the *D. virgifera virgifera* genome, although intron-containing genes tend to show a dramatic increased size in *D. virgifera virgifera*. For example, the 94.6 kb *D. virgifera virgifera* cadherin gene is ~13.3-fold larger than the *T. castaneum* ortholog despite the coding regions being approximately the same length. Mapping of BES reads and computational prediction of MITE-like elements within the *D. virgifera virgifera* cadherin gene indicated that TEs and other repetitive elements have inserted within intron regions and are the cause of the comparative increase in gene size. TE integrations within introns are known to affect splicing efficiencies [57], but this remains to be investigated in *D. virgifera virgifera*. As stated previously, the insertion of large retroelements within gene coding regions was not predicted. The insertion of a repetitive DNA in the *D. virgifera virgifera* cadherin 5′-UTR suggests that the movement of TEs within the genome could alter gene expression and regulation. TE integrations are also known to cause chromosomal changes that alter gene expression [58]. The accumulation of these changes across the genome can lead to differential selection among local environments [59] or even contribute to the evolution of new species [60]. Knowledge of TE composition within a genome is a fundamental step in the study of relationships between structure and function that may form a basis for future comparative studies. We defined 296 small repetitive DNA elements and 48 large retroelement-like coding sequences within the *D. virgifera virgifera* genome. Although these elements were defined from only 1.15 Mb of genomic sequence, these predictions represent an initial resource for understanding the proliferation and phenotypic effects of repetitive DNA. The DvvBAC1 library has proven useful for the description of gene and repetitive element densities in the *D. virgifera virgifera* genome and will be a

tool for the investigation of the genetic basis of problematic insecticide resistance and behavioral traits expressed by this crop pest species.

Acknowledgments

This research was a joint contribution from the USDA Agricultural Research Service (CRIS Project 3625-22000-017-00D), Iowa Agriculture and Home Economics Experiment Station, Iowa State University, Ames, IA (Project 3543), the University of Nebraska, and the University of Illinois at Champaign-Urbana. This paper reports the results of research only. Mention of a proprietary product does not constitute an endorsement or a recommendation by the parties herein for their use.

References

[1] B. Grisart, W. Coppieters, F. Farnir et al., "Positional candidate cloning of a QTL in dairy cattle: identification of a missense mutation in the bovine DGAT1 gene with major effect on milk yield and composition," *Genome Research*, vol. 12, no. 2, pp. 222–231, 2002.

[2] B. S. Coates, D. V. Sumerford, R. L. Hellmich, and L. C. Lewis, "Repetitive genome elements in a European corn borer, *Ostrinia nubilalis*, bacterial artificial chromosome library were indicated by bacterial artificial chromosome end sequencing and development of sequence tag site markers: implications for lepidopteran genomic research," *Genome*, vol. 52, no. 1, pp. 57–67, 2009.

[3] T. Wicker, E. Schlagenhauf, A. Graner, T. J. Close, B. Keller, and N. Stein, "454 sequencing put to the test using the complex genome of barley," *BMC Genomics*, vol. 7, article 275, 2006.

[4] K. Osoegawa, A. G. Mammoser, C. Wu et al., "A bacterial artificial chromosome library for sequencing the complete human genome," *Genome Research*, vol. 11, no. 3, pp. 483–496, 2001.

[5] J. M. Kelley, C. E. Field, M. B. Craven et al., "High throughput direct end sequencing of BAC clones," *Nucleic Acids Research*, vol. 27, no. 6, pp. 1539–1546, 1999.

[6] J. L. Krysan and T. F. Branson, "Biology, ecology and distribution of *Diabrotica*," in *Proceedings of the International Maize Virus Disease Colloquium and Workshop*, O. H. Wooster, D. T. Gordon, J. K. Knoke, L. R. Nault, and R. M. Ritter, Eds., pp. 144–150, August 1982.

[7] M. E. Gray, T. W. Sappington, N. J. Miller, J. Moeser, and M. O. Bohn, "Adaptation and invasiveness of western corn rootworm: intensifying research on a worsening pest," *Annual Review of Entomology*, vol. 54, pp. 303–321, 2009.

[8] L. J. Meinke, T. W. Sappington, D. W. Onstad et al., "Western corn rootworm (*Diabrotica virgifera virgifera* LeConte) population dynamics," *Agricultural and Forest Entomology*, vol. 11, no. 1, pp. 29–46, 2009.

[9] F. Baca, "New member of the harmful entomofauna of Yugoslavia *Diabrotica virgifera virgifera* LeConte (Coleoptera: Chrysomelidae)," *Zaštita Bilja*, vol. 45, no. 2, pp. 125–131, 1994.

[10] N. Miller, A. Estoup, S. Toepfer et al., "Multiple transatlantic introductions of the western corn rootworm," *Science*, vol. 310, no. 5750, p. 992, 2005.

[11] M. Ciosi, N. J. Miller, K. S. Kim, R. Giordano, A. Estoup, and T. Guillemaud, "Invasion of Europe by the western corn rootworm, *Diabrotica virgifera virgifera*: multiple transatlantic introductions with various reductions of genetic diversity," *Molecular Ecology*, vol. 17, no. 16, pp. 3614–3627, 2008.

[12] H. C. Chiang, "Bionomics of the northern and western corn rootworms," *Annual Review of Entomology*, vol. 18, pp. 47–72, 1973.

[13] A. L. Kahler, A. E. Olness, O. R. Sutter, C. D. Dybing, and O. J. Devine, "Root damage by western corn rootworm and nutrient content in maize," *Agronomy Journal*, vol. 77, pp. 769–774, 1985.

[14] S. G. Jensen, "Laboratory transmission of maize chlorotic mottle virus by three species of corn rootworms," *Plant Disease*, vol. 69, no. 10, pp. 864–868, 1985.

[15] R. L. Gilbertson, W. M. Brown, E. G. Ruppel, and J. L. Capinera, "Association of corn stalk rot *Fusarium* spp. and western corn rootworm beetles in Colorado," *Phytopathology*, vol. 76, no. 12, pp. 1309–1314, 1986.

[16] S. A. Lefko, T. M. Nowatzki, S. D. Thompson et al., "Characterizing laboratory colonies of western corn rootworm (Coleoptera: Chrysomelidae) selected for survival on maize containing event DAS-59122-7," *Journal of Applied Entomology*, vol. 132, no. 3, pp. 189–204, 2008.

[17] L. N. Meihls, M. L. Higdon, B. D. Siegfried et al., "Increased survival of western corn rootworm on transgenic corn within three generations of on-plant greenhouse selection," *Proceedings of the National Academy of Sciences of the United States of America*, vol. 105, no. 49, pp. 19177–19182, 2008.

[18] K. J. Oswald, B. W. French, C. Nielson, and M. Bagley, "Selection for Cry3Bb1 resistance in a genetically diverse population of nondiapausing western corn rootworm (Coleoptera: Chrysomelidae)," *Journal of Economic Entomology*, vol. 104, no. 3, pp. 1038–1044, 2011.

[19] A. J. Gassmann, J. L. Petzold-Maxwell, R. S. Keweshan, and M. W. Dunbar, "Field-evolved resistance to Bt maize by Western corn rootworm," *PLoS ONE*, vol. 6, no. 7, Article ID e22629, 2011.

[20] J. L. Krysan, D. E. Foster, T. F. Branson, K. R. Ostlie, and W. S. Cranshaw, "Two years before the hatch: rootworms adapt to crop rotation," *Bulletin of the Entomological Society of America*, vol. 32, no. 4, pp. 250–253, 1986.

[21] J. T. Shaw, J. H. Paullus, and W. H. Luckmann, "Corn rootworm oviposition in soybeans," *Journal of Economic Entomology*, vol. 71, no. 2, pp. 189–191, 1978.

[22] A. E. Sammons, C. R. Edwards, L. W. Bledsoe, P. J. Boeve, and J. J. Stuart, "Behavioral and feeding assays reveal a western corn rootworm (Coleoptera: Chrysomelidae) variant that is attracted to soybean," *Environmental Entomology*, vol. 26, no. 6, pp. 1336–1342, 1997.

[23] M. E. O'Neal, C. D. DiFonzo, and D. A. Landis, "Western corn rootworm (Coleoptera: Chrysomelidae) feeding on corn and soybean leaves affected by corn phenology," *Environmental Entomology*, vol. 31, no. 2, pp. 285–292, 2002.

[24] T. W. Sappington, B. D. Siegfried, and T. Guillemaud, "Coordinated *Diabrotica* genetics research: accelerating progress on an urgent insect pest problem," *American Entomologist*, vol. 52, no. 2, pp. 90–97, 2006.

[25] N. J. Miller, S. Richards, and T. W. Sappington, "The prospects for sequencing the western corn rootworm genome," *Journal of Applied Entomology*, vol. 134, no. 5, pp. 420–428, 2010.

[26] T. F. Branson, "The selection of a non-diapause strain of *Diabrotica virgifera* (Coleoptera: Chrysomelidae)," *Entomologia Experimentalis et Applicata*, vol. 19, no. 2, pp. 148–154, 1976.

[27] J. Doležel and J. Bartoš, "Plant DNA flow cytometry and estimation of nuclear genome size," *Annals of Botany*, vol. 95, no. 1, pp. 99–110, 2005.

[28] J. Doležel, J. Bartoš, H. Voglmayr, J. Greilhuber, and R. A. Thomas, "Nuclear DNA content and genome size of trout and human," *Cytometry Part A*, vol. 51, no. 2, pp. 127–129, 2003.

[29] L. Clarke and J. Carbon, "A colony bank containing synthetic Col El hybrid plasmids representative of the entire *E. coli* genome," *Cell*, vol. 9, no. 1, pp. 91–99, 1976.

[30] Y. S. Yim, P. Moak, H. Sanchez-Villeda et al., "A BAC pooling strategy combined with PCR-based screenings in a large, highly repetitive genome enables integration of the maize genetic and physical maps," *BMC Genomics*, vol. 8, article 47, 2007.

[31] B. S. Coates, D. V. Sumerford, N. J. Miller et al., "Comparative performance of single nucleotide polymorphism and microsatellite markers for population genetic analysis," *Journal of Heredity*, vol. 100, no. 5, pp. 556–564, 2009.

[32] P. Rice, I. Longden, and A. Bleasby, "EMBOSS: the European molecular biology open software suite," *Trends in Genetics*, vol. 16, no. 6, pp. 276–277, 2000.

[33] B. L. Cantarel, I. Korf, S. M. C. Robb et al., "MAKER: an easy-to-use annotation pipeline designed for emerging model organism genomes," *Genome Research*, vol. 18, no. 1, pp. 188–196, 2008.

[34] L. M. Knolhoff, K. K. O. Walden, S. T. Ratcliffe, D. W. Onstad, and H. M. Robertson, "Microarray analysis yields candidate markers for rotation resistance in the western corn rootworm beetle, *Diabrotica virgifera virgifera*," *Evolutionary Applications*, vol. 3, no. 1, pp. 17–27, 2010.

[35] B. D. Siegfried, N. Waterfield, and R. H. Ffrench-Constant, "Expressed sequence tags from *Diabrotica virgifera virgifera* midgut identify a coleopteran cadherin and a diversity of cathepsins," *Insect Molecular Biology*, vol. 14, no. 2, pp. 137–143, 2005.

[36] M. Stanke, R. Steinkamp, S. Waack, and B. Morgenstern, "AUGUSTUS: a web server for gene finding in eukaryotes," *Nucleic Acids Research*, vol. 32, pp. W309–W312, 2004.

[37] S. E. Lewis, S. M. J. Searle, N. Harris et al., "Apollo: a sequence annotation editor," *Genome Biology*, vol. 3, no. 12, Article ID R0082, 2002.

[38] X. Huang and A. Madan, "CAP3: a DNA sequence assembly program," *Genome Research*, vol. 9, no. 9, pp. 868–877, 1999.

[39] R. S. Harris, *Improved pairwise alignment of genomic DNA [Ph.D. thesis]*, The Pennsylvania State University, 2007.

[40] H. Li, B. Handsaker, A. Wysoker et al., "The sequence alignment/map format and SAMtools," *Bioinformatics*, vol. 25, no. 16, pp. 2078–2079, 2009.

[41] K. Rutherford, J. Parkhill, J. Crook et al., "Artemis: sequence visualization and annotation," *Bioinformatics*, vol. 16, no. 10, pp. 944–945, 2000.

[42] Y. Kapustin, A. Souvorov, T. Tatusova, and D. Lipman, "Splign: algorithms for computing spliced alignments with identification of paralogs," *Biology Direct*, vol. 3, article 20, 2008.

[43] Y. Chen, F. Zhou, G. Li, and Y. Xu, "A recently active miniature inverted-repeat transposable element, Chunjie, inserted into an operon without disturbing the operon structure in *Geobacter uraniireducens* Rf4," *Genetics*, vol. 179, no. 4, pp. 2291–2297, 2008.

[44] M. Zuker, "Mfold web server for nucleic acid folding and hybridization prediction," *Nucleic Acids Research*, vol. 31, no. 13, pp. 3406–3415, 2003.

[45] E. Petitpierre, C. Segarra, and C. Juan, "Genome size and chromosomal evolution in leaf beetles (Coleoptera, Chrysomelidae)," *Hereditas*, vol. 119, no. 1, pp. 1–6, 1993.

[46] K. Sharaf, L. Horová, T. Pavlíček, E. Nevo, and P. Bureš, "Genome size and base composition in *Oryzaephilus surinamensis* (Coleoptera: Sylvanidae) and differences between native (feral) and silo pest populations in Israel," *Journal of Stored Products Research*, vol. 46, no. 1, pp. 34–37, 2010.

[47] C. A. Thomas Jr., "The genetic organization of chromosomes," *Annual Review of Genetics*, vol. 5, pp. 237–256, 1971.

[48] R. Hinegardner, "Evolution of genome size," in *Molecular Evolution*, F. Ayala, Ed., pp. 179–199, Sinauer, Sunderland, Mass, USA, 1976.

[49] Tribolium Genome Sequencing Consortium, "The genome of the model beetle and pest *Tribolium castaneum*," *Nature*, vol. 452, no. 6782, pp. 949–955, 2008.

[50] S. Wang, M. D. Lorenzen, R. W. Beeman, and S. J. Brown, "Analysis of repetitive DNA distribution patterns in the *Tribolium castaneum* genome," *Genome Biology*, vol. 9, no. 3, article R61, 2008.

[51] L. Mao, T. C. Wood, Y. Yu et al., "Rice Transposable elements: a survey of 73,000 sequence-tagged-connectors," *Genome Research*, vol. 10, no. 7, pp. 982–990, 2000.

[52] S. R. Cornman, M. C. Schatz, S. J. Johnston et al., "Genomic survey of the ectoparasitic mite Varroa destructor, a major pest of the honey bee Apis mellifera," *BMC Genomics*, vol. 11, no. 1, article 602, 2010.

[53] M. D. Lorenzen, Z. Doyungan, J. Savard et al., "Genetic linkage maps of the red flour beetle, *Tribolium castaneum*, based on bacterial artificial chromosomes and expressed sequence tags," *Genetics*, vol. 170, no. 2, pp. 741–747, 2005.

[54] Y. Pauchet, P. Wilkinson, H. Vogel et al., "Pyrosequencing the *Manduca sexta* larval midgut transcriptome: messages for digestion, detoxification and defence," *Insect Molecular Biology*, vol. 19, no. 1, pp. 61–75, 2010.

[55] D. A. Rasmussen and M. A. F. Noor, "What can you do with 0.1× genome coverage? A case study based on a genome survey of the scuttle fly *Megaselia scalaris* (Phoridae)," *BMC Genomics*, vol. 10, article 382, 2009.

[56] K. S. Kim, B. W. French, D. V. Sumerford, and T. W. Sappington, "Genetic diversity in laboratory colonies of western corn rootworm (Coleoptera: Chrysomelidae), including a nondiapause colony," *Environmental Entomology*, vol. 36, no. 3, pp. 637–645, 2007.

[57] M. B. Davis, J. Dietz, D. M. Standiford, and C. P. Emerson, "Transposable element insertions respecify alternative exon splicing in three *Drosophila myosin* heavy chain mutants," *Genetics*, vol. 150, no. 3, pp. 1105–1114, 1998.

[58] M. G. Kidwell, "Transposable elements and the evolution of genome size in eukaryotes," *Genetica*, vol. 115, no. 1, pp. 49–63, 2002.

[59] J. González, T. L. Karasov, P. W. Messer, and D. A. Petrov, "Genome-wide patterns of adaptation to temperate environments associated with transposable elements in *Drosophila*," *PLoS Genetics*, vol. 6, no. 4, 2010.

[60] M. A. F. Noor and A. S. Chang, "Evolutionary Genetics: jumping into a New Species," *Current Biology*, vol. 16, no. 20, pp. R890–R892, 2006.

19

Role of Nitrative and Oxidative DNA Damage in Inflammation-Related Carcinogenesis

Mariko Murata,[1] Raynoo Thanan,[1, 2] Ning Ma,[3] and Shosuke Kawanishi[2]

[1] *Department of Environmental and Molecular Medicine, Mie University Graduate School of Medicine, Tsu, 514-8507, Japan*
[2] *Faculty of Pharmaceutical Sciences, Suzuka University of Medical Science, Suzuka, 513-8670, Japan*
[3] *Faculty of Health Science, Suzuka University of Medical Science, Suzuka, 510-0293, Japan*

Correspondence should be addressed to Shosuke Kawanishi, kawanisi@suzuka-u.ac.jp

Academic Editor: Vassilis Gorgoulis

Chronic inflammation induced by biological, chemical, and physical factors has been found to be associated with the increased risk of cancer in various organs. We revealed that infectious agents including liver fluke, *Helicobacter pylori*, and human papilloma virus and noninfectious agents such as asbestos fiber induced iNOS-dependent formation of 8-nitroguanine and 8-oxo-7, 8-dihydro-2′-deoxyguanosine (8-oxodG) in cancer tissues and precancerous regions. Our results with the colocalization of phosphorylated ATM and γ-H2AX with 8-oxodG and 8-nitroguanine in inflammation-related cancer tissues suggest that DNA base damage leads to double-stranded breaks. It is interesting from the aspect of genetic instability. We also demonstrated IL-6-modulated iNOS expression via STAT3 and EGFR in Epstein-Barr-virus-associated nasopharyngeal carcinoma and found promoter hypermethylation in several tumor suppressor genes. Such epigenetic alteration may occur by controlling the DNA methylation through IL-6-mediated JAK/STAT3 pathways. Collectively, 8-nitroguanine would be a useful biomarker for predicting the risk of inflammation-related cancers.

1. Introduction

Chronic inflammation induced by biological, chemical, and physical factors has been found to be associated with the increased risk of cancer in various organs [1–3] (Table 1). Inflammation activates a variety of inflammatory cells, which trigger oxidant-generating enzymes such as inducible nitric oxide synthase (iNOS), NADPH oxidase, and myeloperoxidase to produce high concentrations of free radicals including reactive nitrogen species (RNS) and reactive oxygen species (ROS) [1]. Overproduction of RNS and ROS can change the balance of oxidants and antioxidants and cause nitrative and oxidative stress which contributes to the damage of biomolecules such as DNA, RNA, lipid and proteins, leading to an increase in mutations, genomic instability, epigenetic changes, and protein dysfunction and play roles in the multistage carcinogenic process.

ROS generate 8-oxo-7,8-dihydro-2′-deoxyguanosine (8-oxodG, also known as 8-hydroxydG (8-OHdG)), a marker of oxidative DNA damage [4, 5]. 8-OxodG, a potentially muta-genic DNA lesion, leading to the transversion of G : C to T : A (G $\rightarrow$ T transversion) [6], has been implicated in cancers triggered by infections [7]. The generation of ROS is not confined to inflammatory processes. Carcinogenic chemicals and their metabolites as well as electron transport chains in mitochondria are able to generate ROS. On the other hand, nitric oxide (NO), a primary initiator of RNS, is generated specifically during inflammation via iNOS in inflammatory and epithelial cells [5, 8]. Overproduction of NO participates in the generation of peroxynitrite (ONOO$^-$), which can lead to the formation of 8-nitroguanine, an indicator of nitrative DNA damage [9, 10]. 8-Nitroguanine undergoes spontaneous depurination in DNA, resulting in the formation of an apurinic site [11]. Incorporated adenine can form a pair with apurinic sites during DNA replication, leading to the G $\rightarrow$ T transversion [12] (Figure 1). Moreover, apurinic sites might represent major damage that requires error-prone DNA polymerase ζ for efficient *trans*-lesion DNA synthesis. It was reported that DNA polymerase ζ can efficiently bypass abasic sites by extending from nucleotides

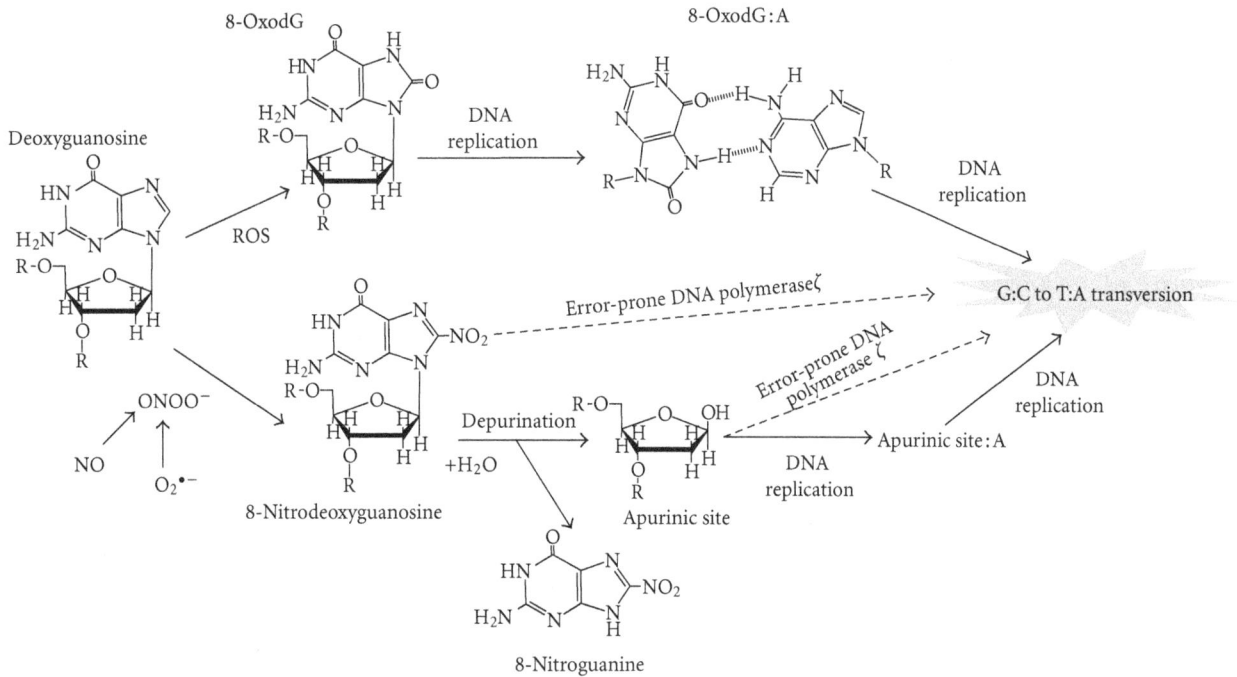

FIGURE 1: Proposed mechanism of point mutation induced by 8-nitroguanine and 8-oxodG through induction of the $G:C \rightarrow T:A$ transversion.

TABLE 1: Nitrative and oxidative DNA damage in inflammation-induced carcinogenesis.

Etiologic agent/pathologic condition	IARC classification[a]	Cancer site	Associated neoplasm	Detection of DNA lesions[c] [reference no.]
(I) *Infection agent*				
Viruses				
HPV[b]				
High-risk types	1	Cervix and other site	Cervical carcinoma	IHC [38]
Low-risk types	2A			
HCV, HBV[b]	1	Liver	Hepatocellular carcinoma	IHC [56–59]
EBV[b]	1	Nasopharynx	Nasopharyngeal carcinoma	IHC [38, 49, 50], ELISA [49]
Bacterium				
Helicobacter pylori	1	Stomach	Gastric cancer	IHC [36]
Parasites				
Opisthorchis viverrini	1	Intra- and extrahepatic bile duct	Cholangiocarcinoma	IHC [17, 22–26], HPLC-ECD [23, 27]
Schistosoma haematobium	1	Bladder	Bladder cancer	IHC [60]
(II) *Inflammatory disease*				
Asbestos fiber	1	Lung	Mesothelioma, lung carcinoma	IHC [61]
Reflux oesophagitis Barrett's oesophagitis		Oesophagus	Oesophageal carcinoma	IHC (In prep.)
Lichen planus		Oral	Oral squamous cell carcinoma	IHC [62]
Inflammatory bowel disease		Colon	Colorectal carcinoma	IHC [63]
Crohn's disease				
Chronic ulcerative colitis				IHC (this paper)
Unknown		Soft tissue	Malignant fibrous histiocytoma	IHC [64, 65]

This table was adapted and modified from the IARC [2] and Coussens and Werb [1].
IARC: International Agency for Research on Cancer. [a]IARC classification: Group 1: carcinogenic to humans; Group 2A: probably carcinogenic to humans.
[b]HPV: human papilloma virus; HBV: hepatitis B virus; HCV: hepatitis C virus; EBV: Epstein-Barr virus.
[c]DNA lesions: IHC, 8-nitroguanine and 8-oxodG detected by immunohistochemistry; HPLC-ECD: 8-oxodG detected by HPLC-ECD; ELISA: serum 8-oxodG detected by ELISA.

inserted opposite the lesion by other DNA polymerases [13]. Wu et al. suggested that cells deficient in subunits of DNA polymerase ζ were hypersensitive to nitrative stress, and *trans*-lesion DNA synthesis mediated by this polymerase contributes to extensive point mutations [14]. Additionally, DNA polymerases η and κ were also found to be involved in the incorporation of adenine opposite 8-nitroguanine during DNA synthesis in a cell-free system associated with *trans*-lesion DNA synthesis leading to the G → T transversion [15]. Therefore, 8-nitroguanine is a potential mutagenic DNA lesion involved in inflammation-mediated carcinogenesis. Relevantly, systematic and comprehensive genome-scale approaches by using the immunoprecipitation-based technique combined with high-density microarrays may be useful to investigate roles of DNA lesions in carcinogenesis [16].

We focus on the roles of nitrative and oxidative DNA damage in infection- and inflammation-related carcinogenesis. We produced a specific anti-8-nitroguanine antibody [17] and examined the localization of DNA lesions by immunohistochemical analysis in animal models and clinical samples (Table 1). Here, we review the effects of RNS-/ROS-mediated DNA damage on genomic instability and epigenetic change in relation to carcinogenesis.

2. DNA Damage in Infection-Related Carcinogenesis

2.1. Liver Fluke Infection and Cholangiocarcinoma. Liver fluke infections of *Opisthorchis viverrini* (*O. viverrini*) are a risk factor for cholangiocarcinoma in Southeast Asia [18]. *O. viverrini* infestations are endemic in Khon Kaen province, northeastern Thailand, and Khon Kaen has the highest incidence of cholangiocarcinoma in the world [19]. *O. viverrini* infections induce inflammation in both animal models [20] and humans [21]. Our previous studies showed that 8-oxodG and 8-nitroguanine levels were increased in *O. viverrini*-infected hamsters compared with uninfected control groups [17, 22–24]. In addition, DNA damage was significantly increased in reinfected hamsters compared with animals infected just once [23]. Notably, repeated infection increased iNOS expression and 8-nitroguanine production in the epithelium of bile ducts even after a decrease in inflammatory cells. To elucidate the mechanism involved, we examined the expression of iNOS, NF-κB, and Toll-like receptor (TLR) 2 in mouse macrophage cell lines treated with *O. viverrini* crude antigens [25], suggesting that *O. viverrini* infection induced TLR2 activation with NF-κB-dependent transcription and iNOS expression. Treatment with an antiparasitic drug (praziquantel) significantly improved the DNA lesions [22]. These findings in hamsters were confirmed by the observation that 8-oxodG and 8-nitroguanine accumulated more in cancerous areas than in intrahepatic areas adjacent to tumors in surgical specimens [26]. Furthermore, an epidemiological study of *O. viverrini*-infected subjects and cholangiocarcinoma patients demonstrated that urinary 8-oxodG levels were significantly higher in cholangiocarcinoma patients than in *O. viverrini*-infected patients and healthy subjects and higher in *O. viverrini*-infected

subjects than in healthy subjects [27]. The urinary 8-oxodG levels in *O. viverrini*-infected patients significantly decreased two months after praziquantel treatment and were comparable to levels in healthy subjects one year after treatment [27]. These results indicate that *O. viverrini* causes chronic and recurrent inflammation followed by the accumulation of oxidative and nitrative DNA lesions, which may participate in the development of cholangiocarcinomas.

2.2. H. pylori and Gastric Cancer. Helicobacter pylori is the main cause of chronic gastritis and a potential risk factor for gastric carcinoma [28]. The molecular mechanisms behind *H. pylori*-induced production of ROS/RNS were wide ranging from activated neutrophils to *H. pylori* itself, as nicely reviewed by Handa et al. [29]. *H. pylori* infections promote the secretion of various inflammatory cytokines, contributing directly to the pronounced inflammatory response. Lipopolysaccharide, a component of Gram-negative bacteria such as *H. pylori*, is a TLR4 ligand that induces inflammatory responses via NF-κB expression [30]. NF-κB, which is involved in the regulation of iNOS, had been reported to function as a tumor promoter in inflammation-associated cancer [31, 32]. In patients with *H. pylori*-induced gastritis or gastric ulcers, iNOS is expressed in the infiltrating inflammatory cells [33]. The expression of iNOS mRNA and protein was significantly increased in the epithelial cells of *H. pylori*-positive gastritis patients compared to *H. pylori*-negative patients [34]. Recently, it was also found that *H. pylori* in a Korean isolate induced the expression of iNOS via AP-1 activation [35]. Our previous study [36] demonstrated that levels of 8-nitroguanine and 8-oxodG in gastric gland epithelium were significantly higher in gastritis patients with *H. pylori* infections than in those without infections. A significant accumulation of proliferating cell nuclear antigen (PCNA) was observed in gastric gland epithelial cells in patients infected with *H. pylori* in comparison to those not infected. Interestingly, the accumulation of PCNA was closely correlated with the formation of 8-nitroguanine and 8-oxodG. Collectively, the host response to *H. pylori* mediated NF-κB expression, resulting in iNOS expression accompanied by 8-nitroguanine and 8-oxodG production in the gastric epithelium. 8-Nitroguanine could be not only a promising biomarker for inflammation but also a useful indicator of the risk of developing gastric cancer in response to chronic *H. pylori* infection.

2.3. HPV and Cervical Carcinoma. Cervical cancer is the second most common cancer among women worldwide and the most common cancer among women in many developing countries [37]. Inflammation is proposed to play an integral role in the development of human papilloma virus (HPV)-induced cervical cancer [1]. Our previous study [38] examined the formation of 8-nitroguanine and 8-oxodG in cells of cervical intraepithelial neoplasia (CIN, grades 1–3) and condyloma acuminatum samples and compared it with the expression of the cyclin-dependent kinase inhibitor p16, considered a biomarker for cervical neoplasia [39–42]. Double immunofluorescence labeling revealed that 8-nitroguanine and 8-oxodG immunoreactivities correlated significantly with CIN grade. There were no statistically significant differences in p16 expression between CIN and condyloma acuminatum samples. These results suggest that high-risk HPV types promote iNOS-dependent DNA damage, which leads to dysplastic changes and carcinogenesis. Therefore, 8-nitroguanine is a more suitable and promising biomarker for evaluating the risk of inflammation-mediated cervical carcinogenesis than p16.

2.4. EBV and Nasopharyngeal Carcinoma. Nasopharyngeal carcinoma (NPC) is strongly associated with Epstein-Barr virus (EBV) infections [43]. Various transcription factors are known to participate in iNOS expression including signal transducers and activators of transcription (STATs), such as STAT1α and STAT3 [44, 45]. Epidermal growth factor receptor (EGFR) physically interacts with STAT3 in the nucleus, leading to transcriptional activation of iNOS [44]. STAT3 is repeatedly activated through phosphorylation via the expression of latent membrane protein 1 (LMP1) as well as EGFR [46, 47], and interleukin-6 (IL-6) is required for LMP1-mediated STAT3 activation [46]. In addition, LMP1-mediated iNOS expression was reported in EBV-infected epithelium cell lines, which play a role in colonization independent of anchorage and tumorigenicity in nude mice [48]. Using biopsy and surgical specimens of nasopharyngeal tissues from NPC patients in southern China, we performed double immunofluorescent staining to examine the formation of 8-nitroguanine and 8-oxodG [49, 50]. Intensive immunoreactivity to iNOS was detected in the cytoplasm of 8-nitroguanine-positive cancer cells. DNA lesions and iNOS expression were also observed in epithelial cells of EBV-positive patients with chronic nasopharyngitis but weaker than those in NPC patients. No or few DNA lesions were observed in EBV-negative subjects. EGFR and phosphorylated STAT3 were strongly expressed in cancer cells of NPC patients, suggesting that the STAT3-dependent mechanism is important to the carcinogenesis [50]. IL-6 was expressed mainly in inflammatory cells of nasopharyngeal tissues of EBV-infected patients. We also found that serum levels of 8-oxodG were significantly higher in NPC patients than control subjects [49]. Collectively, these findings indicate that the nuclear accumulation of EGFR and activation of STAT3 by IL-6 play a key role in iNOS expression and resultant DNA damage, leading to EBV-related NPC.

2.5. HCV and Hepatocellular Carcinoma. Hepatitis C virus (HCV) is a major cause of chronic hepatitis, liver cirrhosis, and hepatocellular carcinoma throughout the world [51]. Hepatocellular carcinoma arises through genetic alterations in hepatocytes during a chronic HCV infection [52–55]. We investigated the extent of nucleic acid damage in HCV-infected individuals and its change after interferon treatment [56]. Immunoreactivities of 8-nitroguanine and 8-oxodG were strongly detected in the liver of patients with chronic hepatitis C, but not control subjects. 8-Nitroguanine was found to be accumulated in hepatocytes particularly in the periportal area. In the sustained virological responder group after interferon therapy, the accumulation of 8-nitroguanine and 8-oxodG was markedly decreased in the liver. We observed a strong correlation between hepatic 8-oxodG staining and serum ferritin levels, suggesting the iron content to be a strong mediator of oxidative stress and iron reduction to reduce the incidence of hepatocellular carcinoma in patients with chronic hepatitis C [57, 58]. We also demonstrated that oxidative DNA damage widely occurred in the livers of patients with chronic viral hepatitis especially chronic hepatitis C, and the iron load and 8-oxodG-positive hepatocytic count was significantly higher in HCV-infected than in HBV-infected livers [59]. It is plausible that ROS production during chronic HCV infection is the result of high iron levels in hepatic tissues, which lead to progressive liver inflammation and an increased risk of developing liver cancer. These findings indicate that 8-nitroguanine and 8-oxodG are useful as biomarkers for evaluating the severity of HCV-induced chronic inflammation leading to hepatocellular carcinoma and the efficacy of chronic hepatitis C treatment.

3. DNA Damage in Inflammation-Related Carcinogenesis

3.1. Asbestos and Lung Carcinoma. Excessive and persistent production of ROS/RNS by inflammatory cells is considered as a hallmark of the secondary genotoxicity of nonfibrous and fibrous particles including asbestos [66]. Asbestos is a carcinogen (IARC Group1) causing lung cancer and malignant mesothelioma of the pleura and peritoneum [67]. Among the different types of asbestos, crocidolite (blue asbestos) and amosite (brown asbestos) are more potent carcinogens than chrysotile (white asbestos) [67]. Inflammation is a hallmark of the response to exposure to asbestos in both animal and human models [68, 69]. NO and nitrative stress were reported to be involved in the asbestos-derived inflammatory response via myeloperoxidase, a major constituent of neutrophils which generates hypochlorous acid and RNS [70–73]. Myeloperoxidase plays a significant role in asbestos-induced carcinogenesis [74]. However, the precise mechanisms of nitrative DNA damage remain to be clarified. We performed an immunohistochemical analysis to examine the formation of 8-nitroguanine and the expression of iNOS and its transcription factor (NF-κB) in the lungs of mice intratracheally administered asbestos fibers, including crocidolite and chrysolite [61]. 8-Nitroguanine was significantly detected in bronchial epithelial cells of asbestos-exposed groups compared with the untreated group. Interestingly, the immunoreactivities of 8-nitroguanine, iNOS, and NF-κB were significantly higher in the crocidolite-exposed group than in the chrysotile-exposed group. Therefore, the formation of nitrative DNA damage could be one of the mechanisms responsible for the difference in carcinogenic potential between crocidolite and chrysotile.

3.2. Inflammatory Bowel Disease and Colon Cancer. Ulcerative colitis and Crohn's disease, which are referred to as inflammatory bowel diseases (IBDs), are well known as chronic inflammatory diseases in the lower bowel. Epidemiological studies have shown that the incidence of colorectal cancer in IBD patients is greater than the expected incidence

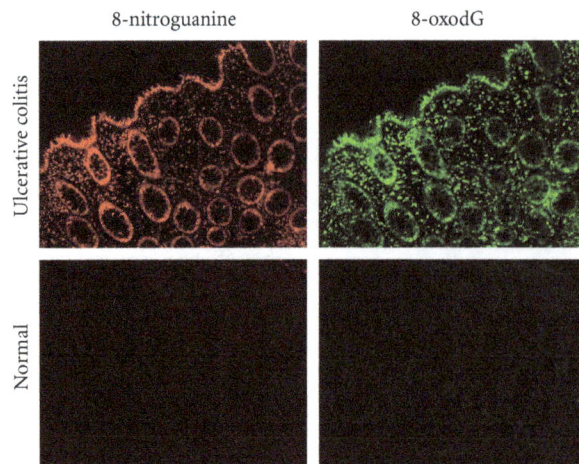

FIGURE 2: 8-Nitroguanine in colon epithelium of a patient with active ulcerative colitis.

in the general population [75]. We hypothesized that an imbalance of helper and regulatory T-cell functions plays a key role in the pathogenesis of IBD. Therefore, we prepared a mouse model of IBD with an imbalance of Th1 and Th2 and, using double immunofluorescence labeling, revealed that both 8-nitroguanine and 8-oxodG were mainly formed in epithelial cells [63]. iNOS, PCNA, and p53 proteins were also expressed in colon epithelium. We observed by using clinical samples that 8-nitroguanine and 8-oxodG were formed in colon epithelium of patients with ulcerative colitis in the active stage (Figure 2). Of relevance, several studies have shown that iNOS is expressed in epithelial cells in colitis patients [76–78]. In noncancerous colon tissues from patients with ulcerative colitis, iNOS protein levels were positively correlated with p53 serine 15 phosphorylation levels [76]. These results suggest that nitrative DNA damage, as well as oxidative DNA damage, participates in colon carcinogenesis in patients with IBD.

3.3. Oral Lichen Planus and Oral Cancer.

Oral lichen planus (OLP) is a chronic inflammatory mucosal disease [79] and a risk factor for oral squamous cell carcinoma (OSCC) [80]. Oral leukoplakia is a precancerous lesion characterized by white plaques and hyperkeratosis [81, 82]. We demonstrated that 8-nitroguanine and 8-oxodG accumulated in oral epithelium of biopsy specimens from patients with OLP, leukoplakia, and OSCC, whereas no immunoreactivity was observed in normal oral mucosa [62, 83]. Colocalization of 8-nitroguanine and iNOS was found in oral epithelium of patients with OLP, leukoplakia, and OSCC. Accumulation of p53 was observed in oral epithelium in OLP and leukoplakia patients, and more prominent expression of this protein was observed in OSCC patients. In addition, the immunoreactivity to PCNA was significantly higher in leukoplakia patients than that in normal mucosa, suggesting an increase in cell proliferation [83]. Lee et al. also reported that PCNA and p53 were highly expressed in oral tissues in OLP patients [84]. We conclude that inflammation-mediated DNA damage and additional epithelial cell proliferation promote oral carcinogenesis.

3.4. DNA Damage in Malignant Fibrous Histiocytoma.

Malignant fibrous histiocytoma (MFH) is one of the most common soft tissue sarcomas [85, 86] and has a poor prognosis [87, 88]. MFH has been proposed to be accompanied by inflammatory responses [89, 90]. However, the mechanism of its inflammation-induced carcinogenesis is still unclear. We investigated DNA lesions and inflammatory-related molecules including iNOS, NF-κB, and COX-2 [64]. Immunohistochemical staining revealed that the formation of 8-nitroguanine and 8-oxodG occurred to a much greater extent in MFH tissue specimens from deceased patients than in live patients. iNOS, NF-κB, and COX-2 were colocalized with 8-nitroguanine in MFH tissues. It is worth noting that a statistical analysis using the Kaplan-Meier method demonstrated strong 8-nitroguanine staining to be associated with a poor prognosis. Furthermore, our study demonstrated significantly higher levels of both 8-nitroguanine and HIF-1α in the tissue specimens of deceased patients than in those of living subjects. Survival curves analyzed by the Kaplan-Meier method differed significantly between the groups with high and low staining of 8-nitroguanine as well as HIF-1α [65]. These results suggest a significant role for the iNOS-dependent formation of 8-nitroguanine via HIF-1α and NF-κB in the progression of inflammation-related cancer. These results indicate that 8-nitroguanine is involved in not only the initiation of carcinogenesis but also its progression and prognosis in cases of MFH.

4. DNA Damage in relation to Genomic Instability

Genomic instability is a defining characteristic of most carcinogenesis through the accumulation of mutations in several tumor suppressor genes, oncogenes, and genes that are involved in maintaining genomic stability [91]. Events resulting in chromosomal instability, such as amplification and deletions of large segments of DNA, reciprocal and non-reciprocal translocations, aneuploidy, and polyploidy, constitute the large-scale genomic aberrations that characterize the majority of human cancer cells and are thought to accelerate

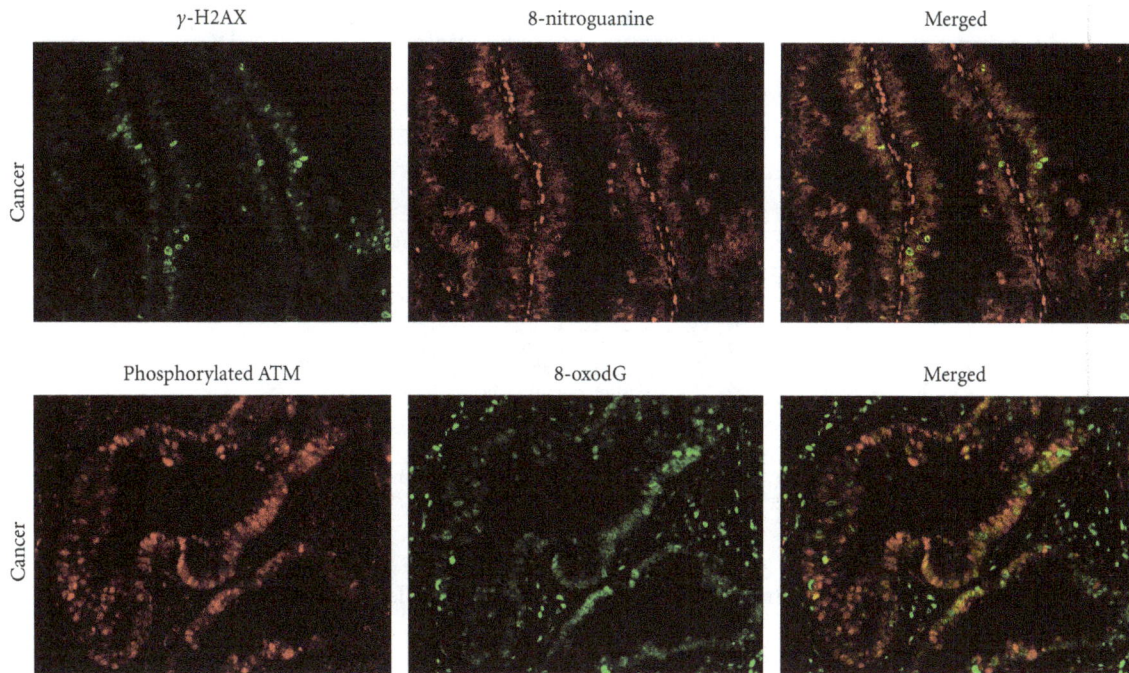

FIGURE 3: Colocalization of DDR proteins and DNA lesions. (a) Colocalization of γ-H2AX (green) and 8-nitroguanine (red). (b) Colocalization of phosphorylated ATM (green) and 8-oxodG (red).

carcinogenesis [91, 92]. Degtyareva et al. demonstrated that chronic oxidative DNA damage due to DNA repair defects induced chromosome instability in a *Saccharomyces cerevisiae* model [92]. Trouiller et al. showed that titanium dioxide, a risk factor for lung cancer, induced oxidative DNA damage, γ-H2AX foci, micronuclei, and DNA deletions, suggesting a link between inflammation-associated DNA damage and genomic instability [93]. The DNA damage response (DDR) is essential for maintaining the integrity of the genome, and a failure of this response results in genomic instability and predisposition to malignancy [94]. Phosphorylated ATM (ataxia telangiectasia mutated) plays a role in DDR to DNA double-stranded breaks. Impaired function of ATM was reported to be involved in DNA damage-induced genomic instability [94, 95]. TNF-α is a proinflammatory cytokine and also acts as an iNOS regulator protein [96, 97]. Natarajan et al. reported that TNF-α induced the formation of 8-oxodG and genomic instability in primary vascular endothelial cells [98]. Yan et al. showed that antioxidants significantly reduced TNF-α-induced genetic damage [99]. Therefore, TNF-α and a dysfunction of ATM could play key roles in the integration between iNOS-mediated DNA damage and genomic instability. Recently, we observed that phosphorylated ATM and γ-H2AX were colocalized with 8-oxodG and 8-nitroguanine in clinical samples of cholangiocarcinoma patients as shown in Figure 3, suggesting that DNA base damage caused double-stranded breaks. DNA lesions were found in infiltrating inflammatory cells, hepatocyte cells of nontumor areas, and cancer cells, whereas γ-H2AX and phosphorylated ATM were expressed in only cancer cells. Moreover, DDR proteins and DNA lesions were detected only very weakly in normal liver tissues, suggesting that the DNA double-stranded breaks were specific to cancer cells. Our observations also support the idea that highly iNOS-dependent DNA damage causes DNA double-stranded breaks and genomic instability, which play important roles in inflammation-induced carcinogenesis via TNF-α signaling and DDR protein dysfunction.

5. DNA Damage in relation to Epigenetic Change

Diverse cellular functions including the regulation of inflammatory gene expression, DNA repair, and cell proliferation are regulated by epigenetic changes [100]. DNA methylation and histone modifications are the major events involved in epigenetic changes. An important proinflammatory cytokine IL-6 has been reported to control DNA methylation through IL-6-mediated Janus kinase (JAK)/STAT3 pathways [101–105]. We demonstrated that IL-6 modulated iNOS expression via STAT3 and EGFR in EBV-associated nasopharyngeal carcinoma [50]. Accumulating evidence makes it increasingly clear that epigenetic silencing plays an important role in EBV-associated neoplasia [106]. We and our colleagues have found promoter hypermethylation in several candidate genes for tumor suppressor genes [107–110]. Histone modifications play a role in the response to DNA double-stranded breaks through ATM signaling to activate γ-H2AX, resulting in histone ubiquitination and acetylation, and destabilization and conformational changes to nucleosomes lead to DNA repair [111]. RNS cause base lesions, abasic sites, and single-stranded breaks, which may be converted into double-strand

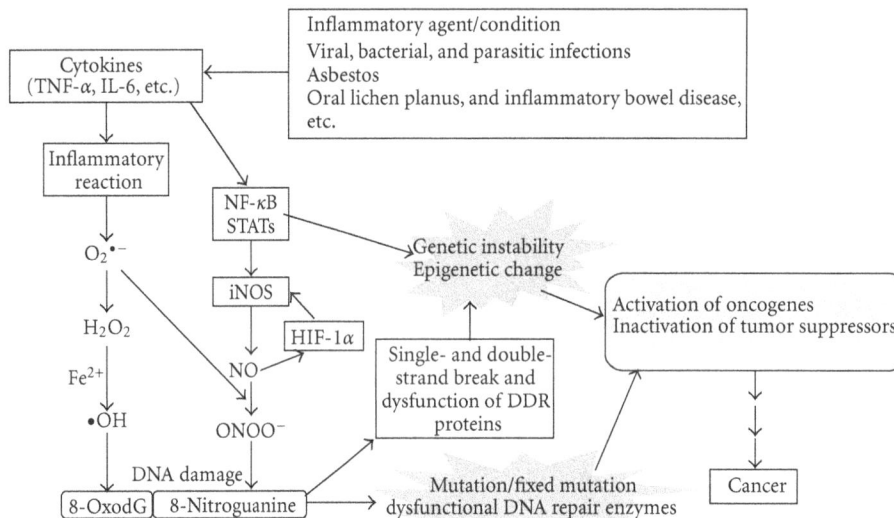

FIGURE 4: Proposed roles of nitrative and oxidative DNA damage in inflammation-related carcinogenesis.

breaks in cells by enzymatic processing, when the damage is in close proximity to or encountered by the replication fork [112]. Collectively, nitrative and oxidative DNA damage may activate epigenetic change via IL-6 signaling and the expression of DDR proteins.

6. Conclusion

We investigated the formation of 8-nitroguanine and 8-oxodG at sites of carcinogenesis in various clinical specimens and animal models in relation to inflammation-related carcinogenesis. We also observed that DNA lesions were formed and significantly increased in *S. haematobium*-induced urinary bladder cancer compared with cancer without such an infection [60]. In addition, Barrett's esophagus, an inflammation-related disease caused by the reflux of gastric acid, also showed greater DNA damage than normal esophageal tissues (unpublished data). Proposed roles of inflammation-related DNA damage in carcinogenesis on the basis of our findings and studies in the literature [94, 113] are summarized in Figure 4. 8-Nitroguanine and 8-oxodG are formed in various inflammation-related cancers and precancerous regions in an iNOS-dependent manner. TNF-α and IL-6 are proinflammatory cytokines which play roles in the control of iNOS expression via the regulation of NF-κB and STAT3 signaling pathways. 8-Nitroguanine and 8-oxodG are mutagenic lesions resulting in the G $\rightarrow$ T transversion. This type of mutation has been found to occur in vivo in the *ras* gene and the *p53* tumor suppressor gene in various cancers [114]. Nitrative and oxidative DNA damage induce not only mutations but also genomic instability and epigenetic change via TNF-α and IL-6 activities and DNA double-stranded breaks resulting in the activation of oncogenes and inactivation of tumor suppressor genes, which may lead to inflammation-related carcinogenesis.

Acknowledgment

The authors thank all coworkers at the Mie University Graduate School of Medicine (Japan), Khon Kaen University (Thailand), and Guangxi Medical University (China).

References

[1] L. M. Coussens and Z. Werb, "Inflammation and cancer," *Nature*, vol. 420, no. 6917, pp. 860–867, 2002.

[2] IARC, "Chronic infections," in *World Cancer Report*, B. W. Stewart and P. Kleihues, Eds., pp. 56–61, IARC Press, Lyon, France, 2003.

[3] S. Kawanishi and Y. Hiraku, "Oxidative and nitrative DNA damage as biomarker for carcinogenesis with special reference to inflammation," *Antioxidants and Redox Signaling*, vol. 8, no. 5-6, pp. 1047–1058, 2006.

[4] S. Kawanishi, Y. Hiraku, and S. Oikawa, "Mechanism of guanine-specific DNA damage by oxidative stress and its role in carcinogenesis and aging," *Mutation Research*, vol. 488, no. 1, pp. 65–76, 2001.

[5] H. Ohshima, M. Tatemichi, and T. Sawa, "Chemical basis of inflammation-induced carcinogenesis," *Archives of Biochemistry and Biophysics*, vol. 417, no. 1, pp. 3–11, 2003.

[6] S. D. Bruner, D. P. G. Norman, and G. L. Verdine, "Structural basis for recognition and repair of the endogenous mutagen 8-oxoguanine in DNA," *Nature*, vol. 403, no. 6772, pp. 859–866, 2000.

[7] F. Farinati, R. Cardin, P. Degan et al., "Oxidative DNA damage in circulating leukocytes occurs as an early event in chronic HCV infection," *Free Radical Biology and Medicine*, vol. 27, no. 11-12, pp. 1284–1291, 1999.

[8] T. Akaike, S. Fujii, A. Kato et al., "Viral mutation accelerated by nitric oxide production during infection *in vivo*," *The FASEB Journal*, vol. 14, no. 10, pp. 1447–1454, 2000.

[9] B. Halliwell, "Oxygen and nitrogen are pro-carcinogens. Damage to DNA by reactive oxygen, chlorine and nitrogen species: measurement, mechanism and the effects of nutrition," *Mutation Research*, vol. 443, no. 1-2, pp. 37–52, 1999.

[10] V. Yermilov, J. Rubio, M. Becchi, M. D. Friesen, B. Pignatelli, and H. Ohshima, "Formation of 8-nitroguanine by the reaction of guanine with peroxynitrite in vitro," Carcinogenesis, vol. 16, no. 9, pp. 2045–2050, 1995.

[11] V. Yermilov, J. Rubio, and H. Ohshima, "Formation of 8-nitroguanine in DNA treated with peroxynitrite in vitro and its rapid removal from DNA by depurination," FEBS Letters, vol. 376, no. 3, pp. 207–210, 1995.

[12] L. A. Loeb and B. D. Preston, "Mutagenesis by apurinic/apyrimidinic sites," Annual Review of Genetics, vol. 20, pp. 201–230, 1986.

[13] L. Haracska, I. Unk, R. E. Johnson et al., "Roles of yeast DNA polymerases δ and ζ of Rev 1 in the bypass of abasic sites," Genes and Development, vol. 15, no. 8, pp. 945–954, 2001.

[14] X. Wu, K. Takenaka, E. Sonoda et al., "Critical roles for polymerase ζ in cellular tolerance to nitric oxide-induced DNA damage," Cancer Research, vol. 66, no. 2, pp. 748–754, 2006.

[15] N. Suzuki, M. Yasui, N. E. Geacintov, V. Shafirovich, and S. Shibutani, "Miscoding events during DNA synthesis past the nitration-damaged base 8-nitroguanine," Biochemistry, vol. 44, no. 25, pp. 9238–9245, 2005.

[16] S. Akatsuka and S. Toyokuni, "Genome-scale approaches to investigate oxidative DNA damage," Journal of Clinical Biochemistry and Nutrition, vol. 47, no. 2, pp. 91–97, 2010.

[17] S. Pinlaor, Y. Hiraku, N. Ma et al., "Mechanism of NO-mediated oxidative and nitrative DNA damage in hamsters infected with Opisthorchis viverrini: a model of inflammation-mediated carcinogenesis," Nitric Oxide, vol. 11, no. 2, pp. 175–183, 2004.

[18] T. Patel, "Increasing incidence and mortality of primary intrahepatic cholangiocarcinoma in the United States," Hepatology, vol. 33, no. 6, pp. 1353–1357, 2001.

[19] S. Sriamporn, P. Pisani, V. Pipitgool, K. Suwanrungruang, S. Kamsa-ard, and D. H. Parkin, "Prevalence of Opisthorchis viverrini infection and incidence of cholangiocarcinoma in Khon Kaen, Northeast Thailand," Tropical Medicine and International Health, vol. 9, no. 5, pp. 588–594, 2004.

[20] B. Sripa and S. Kaewkes, "Localisation of parasite antigens and inflammatory responses in experimental opisthorchiasis," International Journal for Parasitology, vol. 30, no. 6, pp. 735–740, 2000.

[21] M. Riganti, S. Pungpak, B. Punpoowong, D. Bunnag, and T. Harinasuta, "Human pathology of Opisthorchis viverrini infection: a comparison of adults and children," The Southeast Asian Journal of Tropical Medicine and Public Health, vol. 20, no. 1, pp. 95–100, 1989.

[22] S. Pinlaor, Y. Hiraku, P. Yongvanit et al., "iNOS-dependent DNA damage via NF-κB expression in hamsters infected with Opisthorchis viverrini and its suppression by the antihelminthic drug praziquantel," International Journal of Cancer, vol. 119, no. 5, pp. 1067–1072, 2006.

[23] S. Pinlaor, N. Ma, Y. Hiraku et al., "Repeated infection with Opisthorchis viverrini induces accumulation of 8-nitroguanine and 8-oxo-7, 8-dihydro-2′-deoxyguanine in the bile duct of hamsters via inducible nitric oxide synthase," Carcinogenesis, vol. 25, no. 8, pp. 1535–1542, 2004.

[24] S. Pinlaor, P. Yongvanit, Y. Hiraku et al., "8-nitroguanine formation in the liver of hamsters infected with Opisthorchis viverrini," Biochemical and Biophysical Research Communications, vol. 309, no. 3, pp. 567–571, 2003.

[25] S. Pinlaor, S. Tada-Oikawa, Y. Hiraku et al., "Opisthorchis viverrini antigen induces the expression of Toll-like receptor

2 in macrophage RAW cell line," International Journal for Parasitology, vol. 35, no. 6, pp. 591–596, 2005.

[26] S. Pinlaor, B. Sripa, N. Ma et al., "Nitrative and oxidative DNA damage in intrahepatic cholangiocarcinoma patients in relation to tumor invasion," World Journal of Gastroenterology, vol. 11, no. 30, pp. 4644–4649, 2005.

[27] R. Thanan, M. Murata, S. Pinlaor et al., "Urinary 8-oxo-7, 8-dihydro-2′-deoxyguanosine in patients with parasite infection and effect of antiparasitic drug in relation to cholangiocarcinogenesis," Cancer Epidemiology Biomarkers and Prevention, vol. 17, no. 3, pp. 518–524, 2008.

[28] R. M. Peek Jr. and M. J. Blaser, "Helicobacter pylori and gastrointestinal tract adenocarcinomas," Nature Reviews Cancer, vol. 2, no. 1, pp. 28–37, 2002.

[29] O. Handa, Y. Naito, and T. Yoshikawa, "Helicobacter pylori: a ROS-inducing bacterial species in the stomach," Inflammation Research, vol. 59, no. 12, pp. 997–1003, 2010.

[30] S. Maeda, M. Akanuma, Y. Mitsuno et al., "Distinct mechanism of Helicobacter pylori-mediated NF-κB activation between gastric cancer cells and monocytic cells," Journal of Biological Chemistry, vol. 276, no. 48, pp. 44856–44864, 2001.

[31] E. Pikarsky, R. M. Porat, I. Stein et al., "NF-κB functions as a tumour promoter in inflammation-associated cancer," Nature, vol. 431, no. 7007, pp. 461–466, 2004.

[32] Y. J. Surh, K. S. Chun, H. H. Cha et al., "Molecular mechanisms underlying chemopreventive activities of anti-inflammatory phytochemicals: down-regulation of COX-2 and iNOS through suppression of NF-κB activation," Mutation Research, vol. 480-481, pp. 243–268, 2001.

[33] E. E. Mannick, L. E. Bravo, G. Zarama et al., "Inducible nitric oxide synthase, nitrotyrosine, and apoptosis in Helicobacter pylori gastritis: effect of antibiotics and antioxidants," Cancer Research, vol. 56, no. 14, pp. 3238–3243, 1996.

[34] S. Fu, K. S. Ramanujam, A. Wong et al., "Increased expression and cellular localization of inducible nitric oxide synthase and cyclooxygenase 2 in Helicobacter pylori gastritis," Gastroenterology, vol. 116, no. 6, pp. 1319–1329, 1999.

[35] S. O. Cho, J. W. Lim, K. H. Kim, and H. Kim, "Involvement of ras and AP-1 in Helicobacter pylori-induced expression of COX-2 and iNOS in gastric epithelial AGS cells," Digestive Diseases and Sciences, vol. 55, no. 4, pp. 988–996, 2010.

[36] N. Ma, Y. Adachi, Y. Hiraku et al., "Accumulation of 8-nitroguanine in human gastric epithelium induced by Helicobacter pylori infection," Biochemical and Biophysical Research Communications, vol. 319, no. 2, pp. 506–510, 2004.

[37] IARC, "cancer of female reproductive tract," in World Cancer Report, B. W. Stewart and P. Kleihues, Eds., pp. 215–222, IARC Press, Lyon, France, 2003.

[38] Y. Hiraku, T. Tabata, N. Ma, M. Murata, X. Ding, and S. Kawanishi, "Nitrative and oxidative DNA damage in cervical intraepithelial neoplasia associated with human papilloma virus infection," Cancer Science, vol. 98, no. 7, pp. 964–972, 2007.

[39] R. Klaes, T. Friedrich, D. Spitkovsky et al., "Overexpression of p16ink4A as a specific marker for dysplastic and neoplastic epithelial cells of the cervix uteri," International Journal of Cancer, vol. 92, no. 2, pp. 276–284, 2001.

[40] T. Sano, T. Oyama, K. Kashiwabara, T. Fukuda, and T. Nakajima, "Expression status of p16 protein is associated with human papillomavirus oncogenic potential in cervical and genital lesions," American Journal of Pathology, vol. 153, no. 6, pp. 1741–1748, 1998.

[41] M. von Knebel Doeberitz, "New markers for cervical dysplasia to visualise the genomic chaos created by aberrant

oncogenic papillomavirus infections," *European Journal of Cancer*, vol. 38, no. 17, pp. 2229–2242, 2002.

[42] J. L. Wang, B. Y. Zheng, X. D. Li, T. Angstrom, M. S. Lindstrom, and K. L. Wallin, "Predictive significance of the alterations of p16INK4A, p14ARF, p53, and proliferating cell nuclear antigen expression in the progression of cervical cancer," *Clinical Cancer Research*, vol. 10, no. 7, pp. 2407–2414, 2004.

[43] A. L. McDermott, S. N. Dutt, and J. C. Watkinson, "The aetiology of nasopharyngeal carcinoma," *Clinical Otolaryngology and Allied Sciences*, vol. 26, no. 2, pp. 82–92, 2001.

[44] H. W. Lo, S. C. Hsu, M. Ali-Seyed et al., "Nuclear interaction of EGFR and STAT3 in the activation of the iNOS/NO pathway," *Cancer Cell*, vol. 7, no. 6, pp. 575–589, 2005.

[45] E. Tedeschi, M. Menegazzi, D. Margotto, H. Suzuki, U. Forstermann, and H. Kleinert, "Anti-inflammatory actions of St. John's wort: inhibition of human inducible nitric-oxide synthase expression by down-regulating signal transducer and activator of transcription-1α (STAT-1α) activation," *Journal of Pharmacology and Experimental Therapeutics*, vol. 307, no. 1, pp. 254–261, 2003.

[46] H. Chen, L. Hutt-Fletcher, L. Cao, and S. D. Hayward, "A positive autoregulatory loop of LMP1 expression and STAT activation in epithelial cells latently infected with Epstein-Barr virus," *Journal of Virology*, vol. 77, no. 7, pp. 4139–4148, 2003.

[47] Y. Tao, X. Song, X. Deng et al., "Nuclear accumulation of epidermal growth factor receptor and acceleration of G1/S stage by Epstein-Barr-encoded oncoprotein latent membrane protein 1," *Experimental Cell Research*, vol. 303, no. 2, pp. 240–251, 2005.

[48] J. S. Yu, H. C. Tsai, C. C. Wu et al., "Induction of inducible nitric oxide synthase by Epstein-Barr virus B95-8-derived LMP1 in Balb/3T3 cells promotes stress-induced cell death and impairs LMP1-mediated transformation," *Oncogene*, vol. 21, no. 52, pp. 8047–8061, 2002.

[49] Y. J. Huang, B. B. Zhang, N. Ma, M. Murata, A. Z. Tang, and G. W. Huang, "Nitrative and oxidative DNA damage as potential survival biomarkers for nasopharyngeal carcinoma," *Medical Oncology*, vol. 28, no. 1, pp. 377–384, 2011.

[50] N. Ma, M. Kawanishi, Y. Hiraku et al., "Reactive nitrogen species-dependent DNA damage in EBV-associated nasopharyngeal carcinoma: the relation to STAT3 activation and EGFR expression," *International Journal of Cancer*, vol. 122, no. 11, pp. 2517–2525, 2008.

[51] T. Poynard, M. F. Yuen, V. Ratziu, and C. L. Lai, "Viral hepatitis C," *The Lancet*, vol. 362, no. 9401, pp. 2095–2100, 2003.

[52] B. Bressac, M. Kew, J. Wands, and M. Ozturk, "Selective G to T mutations of p53 gene in hepatocellular carcinoma from southern Africa," *Nature*, vol. 350, no. 6317, pp. 429–431, 1991.

[53] W. H. Caselmann and M. Alt, "Hepatitis C virus infection as a major risk factor for hepatocellular carcinoma," *Journal of Hepatology*, vol. 24, no. 2, supplement, pp. 61–66, 1996.

[54] I. C. Hsu, R. A. Metcalf, T. Sun, J. A. Welsh, N. J. Wang, and C. C. Harris, "Mutational hotspot in the p53 gene in human hepatocellular carcinomas," *Nature*, vol. 350, no. 6317, pp. 427–428, 1991.

[55] T. Oda, H. Tsuda, A. Scarpa, M. Sakamoto, and S. Hirohashi, "p53 gene mutation spectrum in hepatocellular carcinoma," *Cancer Research*, vol. 52, no. 22, pp. 6358–6364, 1992.

[56] S. Horiike, S. Kawanishi, M. Kaito et al., "Accumulation of 8-nitroguanine in the liver of patients with chronic hepatitis C," *Journal of Hepatology*, vol. 43, no. 3, pp. 403–410, 2005.

[57] N. Fujita, S. Horiike, R. Sugimoto et al., "Hepatic oxidative DNA damage correlates with iron overload in chronic hepatitis C patients," *Free Radical Biology and Medicine*, vol. 42, no. 3, pp. 353–362, 2007.

[58] H. Tanaka, N. Fujita, R. Sugimoto et al., "Hepatic oxidative DNA damage is associated with increased risk for hepatocellular carcinoma in chronic hepatitis C," *British Journal of Cancer*, vol. 98, no. 3, pp. 580–586, 2008.

[59] N. Fujita, R. Sugimoto, N. Ma et al., "Comparison of hepatic oxidative DNA damage in patients with chronic hepatitis B and C," *Journal of Viral Hepatitis*, vol. 15, no. 7, pp. 498–507, 2008.

[60] N. Ma, R. Thanan, H. Kobayashi et al., "Nitrative DNA damage and Oct3/4 expression in urinary bladder cancer with *Schistosoma haematobium* infection," *Biochemical and Biophysical Research Communications*, vol. 414, no. 2, pp. 344–349, 2011.

[61] Y. Hiraku, S. Kawanishi, T. Ichinose, and M. Murata, "The role of iNOS-mediated DNA damage in infection- and asbestos-induced carcinogenesis," *Annals of the New York Academy of Sciences*, vol. 1203, pp. 15–22, 2010.

[62] P. Chaiyarit, N. Ma, Y. Hiraku et al., "Nitrative and oxidative DNA damage in oral lichen planus in relation to human oral carcinogenesis," *Cancer Science*, vol. 96, no. 9, pp. 553–559, 2005.

[63] X. Ding, Y. Hiraku, N. Ma et al., "Inducible nitric oxide synthase-dependent DNA damage in mouse model of inflammatory bowel disease," *Cancer Science*, vol. 96, no. 3, pp. 157–163, 2005.

[64] Y. Hoki, Y. Hiraku, N. Ma et al., "iNOS-dependent DNA damage in patients with malignant fibrous histiocytoma in relation to prognosis," *Cancer Science*, vol. 98, no. 2, pp. 163–168, 2007.

[65] Y. Hoki, M. Murata, Y. Hiraku et al., "8-nitroguanine as a potential biomarker for progression of malignant fibrous histiocytoma, a model of inflammation-related cancer," *Oncology Reports*, vol. 18, no. 5, pp. 1165–1169, 2007.

[66] R. P. Schins, "Mechanisms of genotoxicity of particles and fibers," *Inhalation Toxicology*, vol. 14, no. 1, pp. 57–78, 2002.

[67] IARC, "Asbestos," *IRAC Mongraphs on the Evaluation of Carcinogenic Risk to Humans*, vol. 7, supplement 1, pp. 106–116, 1987.

[68] J. E. Craighead, J. L. Abraham, A. Churg et al., "The pathology of asbestos-associated diseases of the lungs and pleural cavities: diagnostic criteria and proposed grading schema. Report of the pneumoconiosis committee of the college of American pathologists and the national institute for occupational safety and health," *Archives of Pathology and Laboratory Medicine*, vol. 106, no. 11, pp. 544–596, 1982.

[69] C. B. Manning, V. Vallyathan, and B. T. Mossman, "Diseases caused by asbestos: mechanisms of injury and disease development," *International Immunopharmacology*, vol. 2, no. 2-3, pp. 191–200, 2002.

[70] J. P. Eiserich, M. Hristova, C. E. Cross et al., "Formation of nitric oxide-derived inflammatory oxidants by myeloperoxidase in neutrophils," *Nature*, vol. 391, no. 6665, pp. 393–397, 1998.

[71] J. P. Gaut, J. Byun, H. D. Tran et al., "Myeloperoxidase produces nitrating oxidants *in vivo*," *Journal of Clinical Investigation*, vol. 109, no. 10, pp. 1311–1319, 2002.

[72] S. Tanaka, N. Choe, D. R. Hemenway, S. Zhu, S. Matalon, and E. Kagan, "Asbestos inhalation induces reactive nitrogen species and nitrotyrosine formation in the lungs and pleura of the rat," *Journal of Clinical Investigation*, vol. 102, no. 2, pp. 445–454, 1998.

[73] A. van der Vliet, J. P. Eiserich, M. K. Shigenaga, and C. E. Cross, "Reactive nitrogen species and tyrosine nitration in the respiratory tract: epiphenomena or a pathobiologic mechanism of disease?" *American Journal of Respiratory and Critical Care Medicine*, vol. 160, no. 1, pp. 1–9, 1999.

[74] A. Haegens, A. van der Vliet, K. J. Butnor et al., "Asbestos-induced lung inflammation and epithelial cell proliferation are altered in myeloperoxidase-null mice," *Cancer Research*, vol. 65, no. 21, pp. 9670–9677, 2005.

[75] A. Ekbom, C. Helmick, M. Zack, and H. O. Adami, "Increased risk of large-bowel cancer in Crohn's disease with colonic involvement," *The Lancet*, vol. 336, no. 8711, pp. 357–359, 1990.

[76] L. J. Hofseth, S. Saito, S. P. Hussain et al., "Nitric oxide-induced cellular stress and p53 activation in chronic inflammation," *Proceedings of the National Academy of Sciences of the United States of America*, vol. 100, no. 1, pp. 143–148, 2003.

[77] I. I. Singer, D. W. Kawka, S. Scott et al., "Expression of inducible nitric oxide synthase and nitrotyrosine in colonic epithelium in inflammatory bowel disease," *Gastroenterology*, vol. 111, no. 4, pp. 871–885, 1996.

[78] H. Wiseman and B. Halliwell, "Damage to DNA by reactive oxygen and nitrogen species: role in inflammatory disease and progression to cancer," *Biochemical Journal*, vol. 313, part 1, pp. 17–29, 1996.

[79] C. Scully, M. Beyli, M. C. Ferreiro et al., "Update on oral lichen planus: etiopathogenesis and management," *Critical Reviews in Oral Biology and Medicine*, vol. 9, no. 1, pp. 86–122, 1998.

[80] M. D. Mignogna, S. Fedele, L. Lo Russo, L. Lo Muzio, and E. Bucci, "Immune activation and chronic inflammation as the cause of malignancy in oral lichen planus: is there any evidence?" *Oral Oncology*, vol. 40, no. 2, pp. 120–130, 2004.

[81] B. W. Neville and T. A. Day, "Oral cancer and precancerous lesions," *CA: A Cancer Journal for Clinicians*, vol. 52, no. 4, pp. 195–215, 2002.

[82] J. Reibel, "Prognosis of oral pre-malignant lesions: significance of clinical, histopathological, and molecular biological characteristics," *Critical Reviews in Oral Biology & Medicine*, vol. 14, no. 1, pp. 47–62, 2003.

[83] N. Ma, T. Tagawa, Y. Hiraku, M. Murata, X. Ding, and S. Kawanishi, "8-nitroguanine formation in oral leukoplakia, a premalignant lesion," *Nitric Oxide*, vol. 14, no. 2, pp. 137–143, 2006.

[84] J. J. Lee, M. Y. Kuo, S. J. Cheng et al., "Higher expressions of p53 and proliferating cell nuclear antigen (PCNA) in atrophic oral lichen planus and patients with areca quid chewing," *Oral Surgery, Oral Medicine, Oral Pathology, Oral Radiology and Endodontology*, vol. 99, no. 4, pp. 471–478, 2005.

[85] A. Jemal, R. C. Tiwari, T. Murray et al., "Cancer Statistics, 2004," *CA: A Cancer Journal for Clinicians*, vol. 54, no. 1, pp. 8–29, 2004.

[86] S. W. Weiss and F. M. Enzinger, "Malignant fibrous histiocytoma. An analysis of 200 cases," *Cancer*, vol. 41, no. 6, pp. 2250–2266, 1978.

[87] A. Belal, A. Kandil, A. Allam et al., "Malignant fibrous histiocytoma: a retrospective study of 109 cases," *American Journal of Clinical Oncology*, vol. 25, no. 1, pp. 16–22, 2002.

[88] R. L. Randall, K. H. Albritton, B. J. Ferney, and L. Layfield, "Malignant fibrous histiocytoma of soft tissue: an abandoned diagnosis," *The American Journal of Orthopedics*, vol. 33, no. 12, pp. 602–608, 2004.

[89] M. F. Melhem, A. I. Meisler, R. Saito, G. G. Finley, H. R. Hockman, and R. A. Koski, "Cytokines in inflammatory malignant fibrous histiocytoma presenting with leukemoid reaction," *Blood*, vol. 82, no. 7, pp. 2038–2044, 1993.

[90] K. K. Richter, D. M. Parham, J. Scheele, R. Hinze, and F. W. Rath, "Presarcomatous lesions of experimentally induced sarcomas in rats: morphologic, histochemical, and immuno-histochemical features," *In Vivo*, vol. 13, no. 4, pp. 349–355, 1999.

[91] F. Michor, Y. Iwasa, B. Vogelstein, C. Lengauer, and M. A. Nowak, "Can chromosomal instability initiate tumorigenesis?" *Seminars in Cancer Biology*, vol. 15, no. 1, pp. 43–49, 2005.

[92] N. P. Degtyareva, L. Chen, P. Mieczkowski, T. D. Petes, and P. W. Doetsch, "Chronic oxidative DNA damage due to DNA repair defects causes chromosomal instability in Saccharomyces cerevisiae," *Molecular and Cellular Biology*, vol. 28, no. 17, pp. 5432–5445, 2008.

[93] B. Trouiller, R. Reliene, A. Westbrook, P. Solaimani, and R. H. Schiestl, "Titanium dioxide nanoparticles induce DNA damage and genetic instability *in vivo* in mice," *Cancer Research*, vol. 69, no. 22, pp. 8784–8789, 2009.

[94] K. Mazan-Mamczarz, P. R. Hagner, Y. Zhang et al., "ATM regulates a DNA damage response posttranscriptional RNA operon in lymphocytes," *Blood*, vol. 117, no. 8, pp. 2441–2450, 2011.

[95] E. M. Goetz, B. Shankar, Y. Zou et al., "ATM-dependent IGF-1 induction regulates secretory clusterin expression after DNA damage and in genetic instability," *Oncogene*, vol. 30, no. 35, pp. 3745–3754, 2011.

[96] R. Medeiros, R. D. Prediger, G. F. Passos et al., "Connecting TNF-α signaling pathways to iNOS expression in a mouse model of Alzheimer's disease: relevance for the behavioral and synaptic deficits induced by amyloid β protein," *Journal of Neuroscience*, vol. 27, no. 20, pp. 5394–5404, 2007.

[97] J. Nandi, B. Saud, J. M. Zinkievich, Z. J. Yang, and R. A. Levine, "TNF-α modulates iNOS expression in an experimental rat model of indomethacin-induced jejunoileitis," *Molecular and Cellular Biochemistry*, vol. 336, no. 1-2, pp. 17–24, 2009.

[98] M. Natarajan, C. F. Gibbons, S. Mohan, S. Moore, and M. A. Kadhim, "Oxidative stress signalling: a potential mediator of tumour necrosis factor α-induced genomic instability in primary vascular endothelial cells," *British Journal of Radiology*, vol. 80, supplement 1, pp. S13–S22, 2007.

[99] B. Yan, Y. Peng, and C. Y. Li, "Molecular analysis of genetic instability caused by chronic inflammation," *Methods in Molecular Biology*, vol. 512, pp. 15–28, 2009.

[100] I. M. Adcock, L. Tsaprouni, P. Bhavsar, and K. Ito, "Epigenetic regulation of airway inflammation," *Current Opinion in Immunology*, vol. 19, no. 6, pp. 694–700, 2007.

[101] F. Armenante, M. Merola, A. Furia, and M. Palmieri, "Repression of the IL-6 gene is associated with hypermethylation," *Biochemical and Biophysical Research Communications*, vol. 258, no. 3, pp. 644–647, 1999.

[102] S. Garaud, C. Le Dantec, S. Jousse-Joulin et al., "IL-6 Modulates CD5 expression in B cells from patients with lupus by regulating DNA methylation," *Journal of Immunology*, vol. 182, no. 9, pp. 5623–5632, 2009.

[103] H. Isomoto, J. L. Mott, S. Kobayashi et al., "Sustained IL-6/STAT-3 signaling in cholangiocarcinoma cells due to SOCS-3 epigenetic silencing," *Gastroenterology*, vol. 132, no. 1, pp. 384–396, 2007.

[104] R. Thaler, M. Agsten, S. Spitzer et al., "Homocysteine suppresses the expression of the collagen cross-linker lysyl oxidase involving IL-6, Fli1, and epigenetic DNA methylation," *Journal of Biological Chemistry*, vol. 286, no. 7, pp. 5578–5588, 2010.

[105] K. F. To, M. W. Chan, W. K. Leung et al., "Constitutional activation of IL-6-mediated JAK/STAT pathway through hypermethylation of SOCS-1 in human gastric cancer cell line," *British Journal of Cancer*, vol. 91, no. 7, pp. 1335–1341, 2004.

[106] H. H. Niller, H. Wolf, and J. Minarovits, "Epigenetic dysregulation of the host cell genome in Epstein-Barr virus-associated neoplasia," *Seminars in Cancer Biology*, vol. 19, no. 3, pp. 158–164, 2009.

[107] F. Chen, Y. Mo, H. Ding et al., "Frequent epigenetic inactivation of Myocardin in human nasopharyngeal carcinoma," *Head and Neck*, vol. 33, no. 1, pp. 54–59, 2011.

[108] C. Du, T. Huang, D. Sun et al., "CDH4 as a novel putative tumor suppressor gene epigenetically silenced by promoter hypermethylation in nasopharyngeal carcinoma," *Cancer Letters*, vol. 309, no. 1, pp. 54–61, 2011.

[109] S. Wang, X. Xiao, X. Zhou et al., "TFPI-2 is a putative tumor suppressor gene frequently inactivated by promoter hypermethylation in nasopharyngeal carcinoma," *BMC Cancer*, vol. 10, article 617, 2010.

[110] Z. Zhang, D. Sun, N. Van Do, A. Tang, L. Hu, and G. Huang, "Inactivation of RASSF2A by promoter methylation correlates with lymph node metastasis in nasopharyngeal carcinoma," *International Journal of Cancer*, vol. 120, no. 1, pp. 32–38, 2007.

[111] D. Rossetto, A. W. Truman, S. J. Kron, and J. Cote, "Epigenetic modifications in double-strand break DNA damage signaling and repair," *Clinical Cancer Research*, vol. 16, no. 18, pp. 4543–4552, 2010.

[112] T. Sawa and H. Ohshima, "Nitrative DNA damage in inflammation and its possible role in carcinogenesis," *Nitric Oxide*, vol. 14, no. 2, pp. 91–100, 2006.

[113] F. Colotta, P. Allavena, A. Sica, C. Garlanda, and A. Mantovani, "Cancer-related inflammation, the seventh hallmark of cancer: links to genetic instability," *Carcinogenesis*, vol. 30, no. 7, pp. 1073–1081, 2009.

[114] G. P. Pfeifer and A. Besaratinia, "Mutational spectra of human cancer," *Human Genetics*, vol. 125, no. 5-6, pp. 493–506, 2009.

20

Genome-Wide Analysis of mir-548 Gene Family Reveals Evolutionary and Functional Implications

Tingming Liang,[1] Li Guo,[2] and Chang Liu[1]

[1] Jiangsu Key Laboratory for Molecular and Medical Biotechnology, College of Life Science, Nanjing Normal University, Nanjing 210046, China
[2] Department of Epidemiology and Biostatistics and Ministry of Education Key Lab for Modern Toxicology, School of Public Health, Nanjing Medical University, Nanjing 210029, China

Correspondence should be addressed to Chang Liu, changliu@njnu.edu.cn

Academic Editor: Mouldy Sioud

mir-548 is a larger, poorly conserved primate-specific miRNA gene family. 69 human mir-548 genes located in almost all human chromosomes whose widespread distribution pattern implicates the evolutionary origin from transposable elements. Higher level of nucleotide divergence was detected between these human miRNA genes, which mainly derived from divergence of multicopy pre-miRNAs and homologous miRNA genes. Products of mir-548, miR-548-5p, and miR-548-3p showed inconsistent evolutionary patterns, which partly contributed to larger genetic distances between pre-miRNAs. "Seed shifting" events could be detected among miR-548 sequences due to various 5′ ends. The events led to shift of seed sequences and target mRNAs, even generated to new target mRNAs. Additionally, the phenomenon of miRNA:miRNA interaction in the miRNA gene family was found. The potential interaction between miRNAs may be contributed to dynamic miRNA expression profiles by complementarily binding events to form miRNA:miRNA duplex with 5′-/3′-overhangs. The miRNA gene family had important roles in multiple biological processes, including signaling pathways and some cancers. The potential abundant roles and functional implication further led to the larger and poorly conserved gene family with genetic variation based on transposable elements. The evolutionary pattern of the primate-specific gene family might contribute to dynamic expression profiles and regulatory network.

1. Introduction

MicroRNAs (miRNAs), a distinct class of ~22 nt single-stranded noncoding endogenous RNAs, play pivotal roles in negatively regulating gene expression by targeting mRNAs with an influence on multiple biological processes in plants and animals, including cell growth, differentiation, and apoptosis [1–3]. The primary transcript (pri-miRNA) encoding multiple miRNAs is generated by polymerase II in the nucleus. For example, miRNA members in a gene cluster are cotranscribed as a single polycistronic transcript to coregulate several categories of genes simultaneously [4–6]. Subsequently, the pri-miRNA is converted into pre-miRNA with a hairpin structure by Drosha [7–9]. Pre-miRNA is then translocated to the cytoplasm via exportin-5 [7] where the miRNA:miRNA* duplex is released from the hairpin structure by Dicer [10, 11]. Mature miRNA is loaded

into RISC (RNA induced silencing complex) to mediate mRNA targeting [12, 13]. Although miRNA* is typically degraded as an inactive sequence, accumulating reports indicate that miRNA* also contributes to the gene network regulation as a potential active miRNA [14–17]. Posttranscriptional silencing of target genes by miRNAs occurs either by targeting specific cleavage of homologous mRNAs, or by targeting specific inhibition of protein synthesis [18]. Some miRNAs have higher level of sequence similarity and form the miRNA gene family, and even coregulate complex biological processes.

The small noncoding RNAs are well evolutionary conserved across large phylogenetic distances [19], and they have been subjected to the evolutionary patterns, genetic and phylogenetic analysis [20–22]. Compared with other components in the complex gene regulatory networks, the small RNA sequences show a different evolutionary pattern

[20, 23]. Despite well-conserved sequences, miRNAs show evolutionary stable "seed shifting" events across different animal species, especially toward the 3′ ends [22]. The diverse biological roles of the small noncoding regulatory RNAs have been investigated recently, but the origin and evolution of miRNAs still remain largely obscure. Some literatures show that miRNAs may originate from repetitive elements, especially for transposable elements (TEs) [24–32]. For example, a larger human gene family, hsa-mir-548, are derived from Made1 transposable elements and have important roles in multiple biological processes [28]. However, with more members in the gene family have been identified, it is quite necessary to further study its evolutionary and functional relationship by systematic analysis, especially based on their potential biological roles in complex regulatory network. Here, according to known members in mir-548 gene family, we sought to indicate the evolutionary pattern and functional implication of the larger gene family.

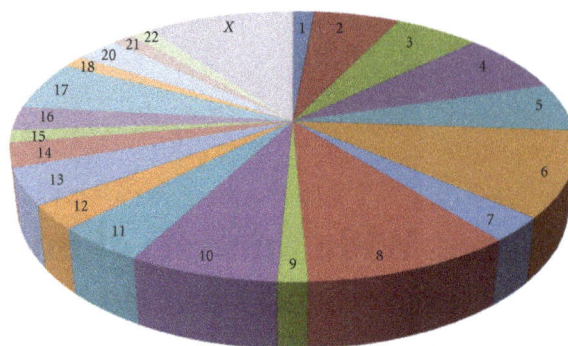

Figure 1: Pie distribution of mir-548 gene family in human chromosomes. The location of mir-548 gene family members is detected in human chromosomes except chromosomes 19 and Y. Note that 30.43% of the members are located in chromosomes 6, 8, and X.

```
hsa-miR-570-5p   - - - A A A G G T A A T T G C A G T T T T T C C C
hsa-miR-548x-5p  T G C A A A A G T A A T T G C A G T T T T T G - -
hsa-miR-548n     - - C A A A A G T A A T T G T G G A T T T T G T -
hsa-miR-548m     - - C A A A G G T A T T T G T G G T T T T T G - -
hsa-miR-548an    - - A A A A G G C A T T - G T G G T T T T T G - -
hsa-miR-548t-5p  - - C A A A A G T G A T C G T G G T T T T T G - -
```

Figure 2: An example of nucleotide divergence patterns of mature miRNAs. Substitution (transition/transversion) and insertion/deletion can be detected among different mature miRNAs. Simultaneously, their seed sequences can be found shifting events because different 5′ ends.

2. Results

According to the annotation in the miRBase database (Release 18.0), mir-548 was a larger primate-specific gene family. There were 100 miRNA members that were distributed in primates, including *Homo sapiens* (hsa, 69), *Pongo pygmaeus* (ppy, 2), *Pan troglodytes* (ptr, 20), and *Macaca mulatta* (mml, 9). Except for hsa-mir-548, the miRNA genes in other primates were mainly predicted by softwares based on sequence similarity. Therefore, we herein mainly analyzed and discussed human mir-548.

As a larger human gene family, hsa-mir-548 was located on almost all of the human chromosomes, especially for chromosomes 6, 8, and X (Figure 1). There were 69 members in the gene family, and higher level of nucleotide divergence could be detected between these homologous miRNA genes. The main reasons were mainly derived from a larger number of miRNA gene members and multicopy pre-miRNAs. Some miRNAs, such as hsa-miR-548f and hsa-miR-548 h, were found 5 multicopy pre-miRNAs that were located on different chromosomes. Although these multicopy miRNA precursors could yield the same miRNA sequence, other regions, including adjacent nucleotides of miRNA sequence, showed inconsistent nucleotide substitution patterns (see Figure S1 in Supplementary Materials available online at doi:10.1155/2012/679563). Similarly, mature miRNAs also showed higher level of nucleotide divergence, including nucleotide substitution and insertion/deletion (Figures 2 and 3). Compared to other larger miRNA gene families, such as hsa-let-7 gene family [16], hsa-mir-548 gene family showed higher level of nucleotide divergence and was a poorly conserved miRNA gene family (Figures 2 and 3). The phenomenon of "seed shifting" events could be detected among miR-548 due to involved in various 5′ ends (Figure 2). The events led to variety of "seed sequences," which also influenced the prediction of target mRNAs.

Based on all hsa-mir-548 sequences, NJ tree showed that some of miRNA genes had larger genetic distances with other members, such as hsa-mir-548e, 548ao, 548i-4 and 548 m (Figure S2). Multicopy pre-miRNAs for a specific miRNA might be reconstructed in different clusters based on phylogenetic tree (Figure S2) and phylogenetic network by using neighbor-net method (data not shown here). Although both arms of hsa-mir-548 could yield mature miRNAs (miR-548-5p and miR-548-3p), hsa-miR-548-5p was more conservation than hsa-miR-548-3p. Nucleotide diversity of miR-548-5p was 0.19 ± 0.01, while nucleotide diversity of miR-548-3p was 0.20 ± 0.02. Average number of nucleotide differences of miR-548-5p and miR-548-3p were 3.81 and 4.30, respectively. Higher frequency of nucleotide substitution and insertion/deletion could be detected between miR-548-3p sequences (Figure 3). Their phylogenetic networks showed different evolutionary patterns. Hsa-miR-548-3p indicated a complex network with more median vectors (Figure 4). According to the predicted target mRNAs of hsa-miR-548, functional enrichment analysis showed that the miRNA gene family played important roles in multiple biological processes, including various human diseases (Table 1). For example, they were involved in regulation of actin cytoskeleton, MAPK signaling pathway, ubiquitin mediated proteolysis, colorectal cancer, glioma, and nonsmall cell lung cancer (Table 1).

Interestingly, we found that some miRNA members were natural or endogenous sense/antisense miRNA genes (Table 2). Hsa-mir-548aa and hsa-mir-548d were located on the sense and antisense strands of the same genomic region, and their products could complementarily bind

TABLE 1: Target genes of hsa-miR-548 gene family enrichment pathway analysis.

Pathway	Count (gene no.)	Enrichment P value	Target genes
Regulation of actin cytoskeleton	14	$2.10E-07$	VAV2;IQGAP2;RRAS2;ITGB1;PAK7;SOS1;DOCK1;PPP1CB;SSH2;BAIAP2;ABI2; APC;ITGB3;SSH1
MAPK signaling pathway	14	$3.33E-06$	DUSP4;PRKCB;RPS6KA3;RRAS2;BDNF;MAP3K2;SOS1;ELK4;TRAF6;STK4; CACNB4;MAP3K13;FOS;PRKCA
Ubiquitin mediated proteolysis	13	$6.77E-09$	UBE2E3;CDC27;RCHY1;UBA2;FBXW7;HERC3;UBR5;UBE2D1;TRAF6;HERC4; UBE2B;VHL;MID1
Focal adhesion	12	$4.04E-06$	VAV2;PRKCB;COL1A1;VASP;ITGB1;PAK7;SOS1;DOCK1;PPP1CB;PTEN;ITGB3; PRKCA
Melanogenesis	11	$2.73E-08$	CREBBP;PRKCB;FZD3;FZD9;CALM2;KITLG;WNT5A;CREB1;CALM1;FZD7; PRKCA
Phosphatidylinositol signaling system	10	$1.40E-08$	PRKCB;CDS2;CALM2;ITPKB;DGKB;PTEN;CDS1;INPP5E;CALM1;PRKCA
Tight junction	10	$3.43E-06$	LLGL2;PRKCB;RRAS2;PPP2R2B;VAPA;PARD6B;PTEN;TJP1;PPP2R2A;PRKCA
Wnt signaling pathway	10	$9.24E-06$	NFAT5;CREBBP;PRKCB;TBL1X;FZD3;FZD9;WNT5A;FZD7;APC;PRKCA
Calcium signaling pathway	10	$4.79E-05$	ERBB3;PRKCB;TRPC1;CALM2;ITPKB;ATP2B4;ERBB4;CALM1;SLC8A1;PRKCA
Cell cycle	9	$8.41E-06$	CCNA2;CREBBP;YWHAZ;CDC27;YWHAE;CDK6;CCNB1;CHEK2;MCM6
Axon guidance	9	$1.72E-05$	EPHA3;NFAT5;EPHB1;EFNB2;ITGB1;PAK7;DCC;SEMA6D;ABLIM3
Insulin signaling pathway	9	$2.76E-05$	PCK1;IRS2;CALM2;SOS1;PPARGC1A;PPP1CB;RHOQ;PRKAR2B;CALM1
p53 signaling pathway	8	$1.08E-06$	CCNG2;RCHY1;CDK6;PPM1D;PTEN;CCNB1;CHEK2;APAF1
Long-term potentiation	8	$2.05E-06$	CREBBP;PRKCB;RPS6KA3;CALM2;PPP1CB;CALM1;GRIA1;PRKCA
Colorectal cancer	8	$4.85E-06$	APPL1;FZD3;FZD9;SOS1;DCC;FZD7;APC;FOS
Glioma	7	$8.43E-06$	PRKCB;CALM2;SOS1;CDK6;PTEN;CALM1;PRKCA
ErbB signaling pathway	7	$5.73E-05$	ERBB3;EREG;PRKCB;PAK7;SOS1;ERBB4;PRKCA
Glycosphingolipid biosynthesis-lactoseries	6	$3.65E-07$	B4GALT4;B3GALT2;B4GALT1;FUT9;ST8SIA1;GCNT2
Nonsmall cell lung cancer	6	$3.14E-05$	RASSF5;PRKCB;SOS1;CDK6;STK4;PRKCA
B cell receptor signaling pathway	6	$2.01E-04$	VAV2;NFAT5;PRKCB;SOS1;DAPP1;FOS
Adherens junction	6	$2.67E-04$	CREBBP;LMO7;SSX2IP;TJP1;BAIAP2;PTPRB
TGF-beta signaling pathway	6	$4.50E-04$	ACVR2B;BMPR1A;CREBBP;SMAD7;BMPR2;BMPR1B
GnRH signaling pathway	6	0.001	PRKCB;MAP3K2;CALM2;SOS1;CALM1;PRKCA
Leukocyte transendothelial migration	6	0.002	RASSF5;VAV2;PRKCB;VASP;ITGB1;PRKCA
Natural killer cell mediated cytotoxicity	6	0.005	VAV2;NFAT5;PRKCB;SOS1;SH2D1A;SH2D1A;PRKCA
Cytokine-cytokine receptor interaction	6	0.076	ACVR2B;BMPR1A;TNFSF13B;KITLG;BMPR2;BMPR1B
Basal cell carcinoma	5	$3.79E-04$	FZD3;FZD9;WNT5A;FZD7;APC
Cholera-infection	5	$5.26E-04$	PRKCB;TJP1;ATP6V0A2;KDELR2;PRKCA
Adipocytokine signaling pathway	5	$9.43E-04$	ACSL6;PCK1;IRS2;PPARGC1A;ACSL5
PPAR signaling pathway	5	0.001	ACSL6;PCK1;FABP3;ME1;ACSL5
ECM-receptor interaction	5	0.003	CD47;COL1A1;DAG1;ITGB1;ITGB3
Gap junction	5	0.004	PRKCB;MAP3K2;SOS1;TJP1;PRKCA
T cell receptor signaling pathway	5	0.008	VAV2;NFAT5;PAK7;SOS1;FOS
Renal cell carcinoma	5	0.001	CREBBP;EPAS1;PAK7;SOS1;EPAS1;VHL
Small cell lung cancer	5	0.003	ITGB1;CDK6;TRAF6;PTEN;APAF1
Alzheimer's disease	5	0.049	CALM2;UQCRB;CALM1;APAF1;NDUFA4

Here, we only list important pathways that involve at least 5 target mRNAs of hsa-miR-548 gene family.

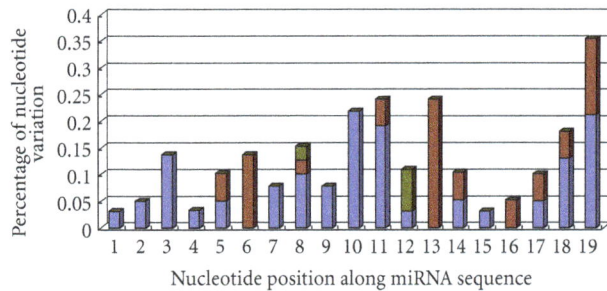

(a)

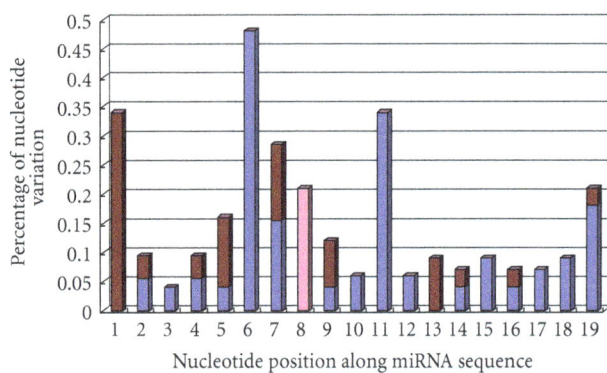

(b)

FIGURE 3: Mutational profiles of hsa-miR-548-5p (a) and hsa-miR-548-3p (b) populations. Products of hsa-mir-548 gene family are estimated their mutational profiles according to each nucleotide position along miRNA sequence. Although these miRNA sequences have higher sequence similarity, higher level of nucleotide variation can be detected between miR-548-5p and miR-548-3p populations, especially the latter.

to each other and formed miRNA:miRNA duplex with 5′-/3′-overhangs. The duplex between miRNAs was similar to typical miRNA:miRNA* duplex, although miRNA-miRNA interaction was not involving a loop structure (Figure S3). Strikingly, due to multicopy pre-miRNAs, the miRNA pairs were detected on chromosomes 8 and 17, respectively (Table 2). The miRNA-miRNA interaction not only existed in sense/antisense miRNAs, and also was detected between other miRNAs, such as hsa-miR-548 h-4-3p, and hsa-miR-548c-5p/hsa-miR-548o-2-5p/hsa-miR-548am-5p, although these miRNA pairs were located on different genomic regions or even different chromosomes (Table 2). The functional interaction networks of these miRNA pairs were reconstructed based on the top 20 predicted target mRNAs. Although miRNA:miRNA interaction could be found between miRNA pairs, they might coregulate biological processes (Figure S4).

3. Discussion

According to the known data, mir-548 gene family mainly expresses in primates. The members of this family are poorly conserved in their sequences, and higher levels of nucleotide divergence exist in some of the members, or even in multicopy pre-miRNAs (Figure S1). Among 69 hsa-mir-548 genes, 1–5 multicopy pre-miRNAs for a specific miRNA can be detected, although adjacent nucleotides are inconsistent and consist of nucleotide variations (Figure S1). The phylogenetic tree based on hsa-mir-548 population shows that some of the members have larger genetic distances than the others (Figure S2). Although this gene family is widely located in almost all the human chromosomes, distribution bias can be found, especially for chromosomes 6, 8, and X (Figure 1). Given that the larger and poorly conserved gene family is originated from the transposable elements (TEs) [28, 32], such universe distribution is probably derived from the evolutionary origin. The evolutionary process also leads to the diversity of miRNA sequences with nucleotide substitution, insertion, or deletion (Figures 2 and 3). Simultaneously, the dynamic evolution further strengthens the functions of the poorly conserved gene family, including their versatile biological roles in complex regulatory network.

Although homologous miRNA sequences are found among different members, "seed shifting" events can be detected (Figure 2). The alternation of 5′ ends of miRNA sequences leads to the variety of seed sequences, and even generates the new target mRNAs. The variety of seed sequences implicates their diverse roles in multiple biological processes, including regulating important signaling pathways and human tumorigenesis (Table 1). Indeed, accumulated evidence show that multiple isomiRs, as well as various sequences, can be generated from a given miRNA locus due to the alternative and imprecise Drosha and Dicer cleavage during pre-miRNA processing [33–38]. This fact also partly contributes to the complexity of the regulatory network, especially when mentioning those isomiRs with novel 5′ ends and seed sequences. Taken together, the "seed shifting" events occurring among different members of miR-548 gene family and among multiple isomiRs in a given miR-548 locus are, respectively, derived from the dynamic evolutionary process and miRNAs processing mechanism, and strengthen the biological roles of miRNA in gene regulatory network. Furthermore, the functional variety may be the adaption to the complex biological processes, and simultaneously implicates the evolutionary trend. Interestingly, except for the seed shifting events in miR-548, higher level of nucleotide divergence can also be detected, especially for the miR-548-3p sequences (Figures 3 and 4). Although both arms of hsa-mir-548 yield to mature miRNA sequences (miR-548-5p and miR-548-3p), they show inconsistent mutational profiles and evolutionary patterns (Figures 3 and 4). The nucleotide divergences in the miRNA sequences lead to the various target mRNAs and biological roles (Table 1). The evolutionary trends and patterns are mainly driven by the functional selection pressure from the complex cellular environment and are based on the repetitive elements and transposable elements.

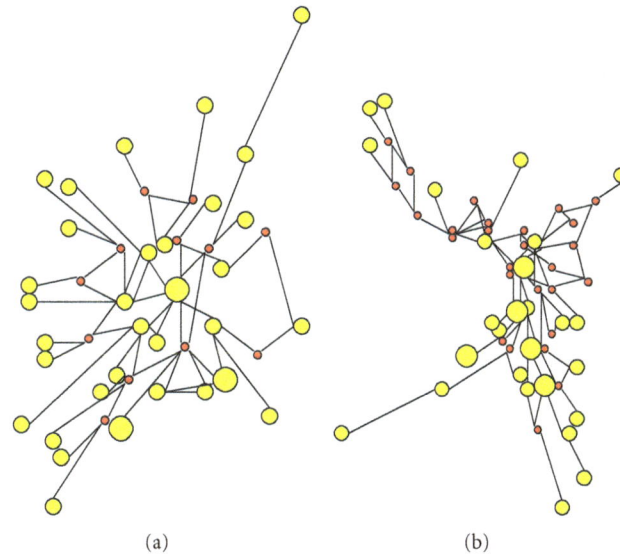

(a)   (b)

FIGURE 4: Phylogenetic networks of hsa-miR-548-5p (a) and hsa-miR-548-3p (b). 37 sequences of miR-548-5p and 33 sequences of miR-548-3p are reconstructed, respectively. More median vector (red circles) are found in network of miR-548-3p.

TABLE 2: miRNA:miRNA interaction in hsa-mir-548 gene family.

Sense/antisense miRNA	miRNAs	Pre-miRNA on chromosome
mul-pre-1: miR-548aa/548d-5p	miR-548aa	8(+): 124360274-124360370
	miR-548d-5p	8(−): 124360274-124360370
mul-pre-2: miR-548aa/548d-5p		17(+): 65467605-65467701
		17(−): 65467605-65467701
miR-548c/548z	miR-548c	12(+): 65016289-65016385
	miR-548z	12(−): 65016289-65016385
miR-548h-4-3p/548c-5p	miR-548h-4-3p	8(−): 26906370-26906480
	miR-548c-5p	12(+): 65016289-65016385
miR-548h-4-3p/548o-2-5p	miR-548o-2-5p	20(+): 37145206-37145275
miR-548h-4-3p/548am-5p	miR-548am-5p	X(−): 16645135-16645208
miR-548t-3p/548d-1-3p	miR-548t-3p	4(+): 174189311-174189384
	miR-548d-1-3p	8(−): 124360274-124360370
miR-548t-3p/548d-2-3p	miR-548d-2-3p	17(−): 65467605-65467701
miR-548t-3p/548ad	miR-548ad	2(+): 35696471-35696552
miR-548z/548o-2-5p	miR-548z	12(−): 65016289-65016385
	miR-548o-2-5p	20(+): 37145206-37145275
miR-548z/548am	miR-548z	12(−): 65016289-65016385
	miR-548am	X(−): 16645135-16645208

Except for hsa-miR-548aa/548d-5p and hsa-miR-548c/548z are endogenous sense/antisense miRNAs, other pairs are located on different chromosomes.

Based on the larger gene family, we also found that some of the members have potential interaction and form miRNA:miRNA duplex (Table 2 and Figure S3). The miRNA-miRNA interaction is recently identified in the miRNA world, which restricts the transcription process [11, 39–42]. These miRNA pairs, such as hsa-mir-548aa and hsa-mir-548d, may be the sense/antisense miRNAs in the same genomic region, or have not location correlation (Table 2). Indeed, the miRNA-miRNA interaction is more prevalent and complex than we previously thought. In this sense,

more miRNA members will be discovered and identified, including those active miRNAs* serving as the gene network regulators [14–17]. The abundant miRNA data will indicate more potential miRNA pairs which can complementarily bind to each other and form miRNA:miRNA duplex. In addition, even if the analysis is not so stringent and permits some mismatches and the complementarily binding of nucleotide U and G, the interaction among miRNAs still can be found as a typical structure of the miRNA:miRNA* duplex. The functional interaction of miRNA pairs can be predicted through their target mRNAs, although they have potential interactions by forming a miRNA:miRNA duplex and restricting each enrichment levels (Figure S4). The miRNA-miRNA interaction in the miR-548 gene family may be the results of the inverted repeat transposons during evolutionary history, while these miRNA pairs might have more potential and abundant regulatory roles and contribute to dynamic expression profiles and multiple biological processes.

4. Materials and Methods

All the miRNA members in the mir-548 gene family (http://www.mirbase.org/cgi-bin/mirna_summary.pl?fam= MIPF0000317), including their annotations, miRNA/ miRNA* and pre-miRNA sequences from different animal species, were obtained from the miRBase database (Release 18.0, http://www.mirbase.org/) [43]. Multiple sequence alignment of miRNA and pre-miRNA sequences were aligned with Clustal X 2.0 [44]. Phylogenetic trees of pre-miRNAs based on Neighbor-Joining (NJ) method were reconstructed with MEGA 5.0 [45] by 1,000 bootstrap resampling. Nucleotide diversity and average number of nucleotide differences of miR-548-5p and miR-548-3p populations were estimated in DnaSP version 5 software [46]. Percentage of nucleotide substitution and insertion/ deletion at each position was estimated for miR-548-5p and miR-548-3p without considering gaps/missing sites in the terminus regions. The most abundant nucleotide at each position was selected as the reference nucleotide, which would help estimate substitution trend more precisely.

Further, phylogenetic network of pre-miRNA members was reconstructed in SplitsTree 4.10 [47] by using the Neighbor-Net method [48] based on Jukes-Cantor model. Based on the network, we attempted to reconstruct the evolutionary history and discover potential evolutionary pattern. All the gaps/missing data were deleted in the phylogenetic tree. In order to infer ancestral miRNA members to understand origin of miRNAs, we also reconstructed evolutionary network [49] with Network 4.6.1.0 (http://www.fluxus-engineering.com/) based on the mature miRNA sequences. Due to various 5′/3′ ends, miR-548 sequences (including miR-548-5p and miR-548-3p) were dealt according to their pre-miRNAs based on the core sequences. Some miRNAs that largely deviated from the core miRNA sequences were removed from the analysis.

Based on a great amount of miRNA members in the gene family, we also searched for the phenomenon of interaction between miRNAs according to miRNA and pre-miRNA sequences. If miRNA sequence could be accurately mapped to other pre-miRNAs, potential miRNA:miRNA interaction could be detected. Generally, the miRNA pairs might be located on sense/antisense strands in the same genomic region, or located on different genomic regions. Similar to miRNA:miRNA* duplex, they could form miRNA:miRNA duplex with 5′-/3′-overhangs.

We integrated the predicted Target mRNAs of the prediction software programs Pictar [50], TargetScan [51]

and miRanda programs [52]. These genes were then queried for gene ontology (GO) enrichments by using CapitalBio Molecule Annotation System V4.0 (MAS, http://bioinfo .capitalbio.com/mas3/). We also constructed functional interaction networks using Cytoscape v2.8.2 Platform [53].

Authors' Contribution

T. Liang, L. Guo, and C. Liu are contributed equally to this work.

Acknowledgments

This work was supported by a research Grant from the National Natural Science Foundation of China (2012104GZ30055), the Natural Science Foundation of the Jiangsu Higher Education Institutions of China (12KJB360001), and the Priority Academic Program Development of Jiangsu Higher Education Institutions (PAPD).

References

[1] D. P. Bartel, "MicroRNAs: genomics, biogenesis, mechanism, and function," *Cell*, vol. 116, no. 2, pp. 281–297, 2004.

[2] D. P. Bartel and C. Z. Chen, "Micromanagers of gene expression: the potentially widespread influence of metazoan microRNAs," *Nature Reviews Genetics*, vol. 5, no. 5, pp. 396–400, 2004.

[3] R. H. A. Plasterk, "Micro RNAs in animal development," *Cell*, vol. 124, no. 5, pp. 877–881, 2006.

[4] M. Lagos-Quintana, R. Rauhut, J. Meyer, A. Borkhardt, and T. Tuschl, "New microRNAs from mouse and human," *RNA*, vol. 9, no. 2, pp. 175–179, 2003.

[5] L. P. Lim, M. E. Glasner, S. Yekta, C. B. Burge, and D. P. Bartel, "Vertebrate microRNA genes," *Science*, vol. 299, no. 5612, p. 1540, 2003.

[6] V. N. Kim and J. W. Nam, "Genomics of microRNA," *Trends in Genetics*, vol. 22, no. 3, pp. 165–173, 2006.

[7] R. Yi, Y. Qin, I. G. Macara, and B. R. Cullen, "Exportin-5 mediates the nuclear export of pre-microRNAs and short hairpin RNAs," *Genes and Development*, vol. 17, no. 24, pp. 3011–3016, 2003.

[8] E. Lund, S. Güttinger, A. Calado, J. E. Dahlberg, and U. Kutay, "Nuclear export of microRNA precursors," *Science*, vol. 303, no. 5654, pp. 95–98, 2004.

[9] E. Bernstein, A. A. Caudy, S. M. Hammond, and G. J. Hannon, "Role for a bidentate ribonuclease in the initiation step of RNA interference," *Nature*, vol. 409, no. 6818, pp. 363–366, 2001.

[10] A. Khvorova, A. Reynolds, and S. D. Jayasena, "Erratum: Functional siRNAs and miRNAs exhibit strand bias," *Cell*, vol. 115, no. 4, p. 505, 2003.

[11] L. Guo, T. Liang, W. Gu, Y. Xu, Y. Bai, and Z. Lu, "Cross-mapping events in miRNAs reveal potential miRNA-Mimics and evolutionary implications," *PLoS One*, vol. 6, no. 5, Article ID e20517, 2011.

[12] J. Liu, M. A. Carmell, F. V. Rivas et al., "Argonaute2 is the catalytic engine of mammalian RNAi," *Science*, vol. 305, no. 5689, pp. 1437–1441, 2004.

[13] G. Meister, M. Landthaler, A. Patkaniowska, Y. Dorsett, G. Teng, and T. Tuschl, "Human Argonaute2 mediates RNA

cleavage targeted by miRNAs and siRNAs," *Molecular Cell*, vol. 15, no. 2, pp. 185–197, 2004.

[14] K. Okamura, M. D. Phillips, D. M. Tyler, H. Duan, Y. T. Chou, and E. C. Lai, "The regulatory activity of microRNA* species has substantial influence on microRNA and 3' UTR evolution," *Nature Structural and Molecular Biology*, vol. 15, no. 4, pp. 354–363, 2008.

[15] K. Okamura, A. Ishizuka, H. Siomi, and M. C. Siomi, "Distinct roles for Argonaute proteins in small RNA-directed RNA cleavage pathways," *Genes and Development*, vol. 18, no. 14, pp. 1655–1666, 2004.

[16] L. Guo and Z. Lu, "The fate of miRNA* strand through evolutionary analysis: implication for degradation as merely carrier strand or potential regulatory molecule?" *PLoS One*, vol. 5, no. 6, article e11387, 2010.

[17] G. Jagadeeswaran, Y. Zheng, N. Sumathipala et al., "Deep sequencing of small RNA libraries reveals dynamic regulation of conserved and novel microRNAs and microRNA-stars during silkworm development," *BMC Genomics*, vol. 11, no. 1, article no. 52, 2010.

[18] R. I. Gregory and R. Shiekhattar, "MicroRNA biogenesis and cancer," *Cancer Research*, vol. 65, no. 9, pp. 3509–3512, 2005.

[19] N. C. Lau, L. P. Lim, E. G. Weinstein, and D. P. Bartel, "An abundant class of tiny RNAs with probable regulatory roles in *Caenorhabditis elegans*," *Science*, vol. 294, no. 5543, pp. 858–862, 2001.

[20] K. Chen and N. Rajewsky, "The evolution of gene regulation by transcription factors and microRNAs," *Nature Reviews Genetics*, vol. 8, no. 2, pp. 93–103, 2007.

[21] R. Niwa and F. J. Slack, "The evolution of animal microRNA function," *Current Opinion in Genetics and Development*, vol. 17, no. 2, pp. 145–150, 2007.

[22] L. Guo, B. Sun, F. Sang, W. Wang, and Z. Lu, "Haplotype distribution and evolutionary pattern of miR-17 and miR-124 families based on population analysis," *PLoS One*, vol. 4, no. 11, article e7944, 2009.

[23] B. M. Wheeler, A. M. Heimberg, V. N. Moy et al., "The deep evolution of metazoan microRNAs," *Evolution and Development*, vol. 11, no. 1, pp. 50–68, 2009.

[24] G. M. Borchert, W. Lanier, and B. L. Davidson, "RNA polymerase III transcribes human microRNAs," *Nature Structural and Molecular Biology*, vol. 13, no. 12, pp. 1097–1101, 2006.

[25] E. J. Devor, A. S. Peek, W. Lanier, and P. B. Samollow, "Marsupial-specific microRNAs evolved from marsupial-specific transposable elements," *Gene*, vol. 448, no. 2, pp. 187–191, 2009.

[26] J. Hertel, M. Lindemeyer, K. Missal et al., "The expansion of the metazoan microRNA repertoire," *BMC Genomics*, vol. 7, article no. 25, 2006.

[27] J. Piriyapongsa, L. Mariño-Ramírez, and I. K. Jordan, "Origin and evolution of human microRNAs from transposable elements," *Genetics*, vol. 176, no. 2, pp. 1323–1337, 2007.

[28] J. Piriyapongsa and I. K. Jordan, "A family of human microRNA genes from miniature inverted-repeat transposable elements," *PLoS One*, vol. 2, no. 2, article no. e203, 2007.

[29] J. Piriyapongsa and I. K. Jordan, "Dual coding of siRNAs and miRNAs by plant transposable elements," *RNA*, vol. 14, no. 5, pp. 814–821, 2008.

[30] N. R. Smalheiser and V. I. Torvik, "Mammalian microRNAs derived from genomic repeats," *Trends in Genetics*, vol. 21, no. 6, pp. 322–326, 2005.

[31] Z. Yuan, X. Sun, D. Jiang et al., "Origin and evolution of a placental-specific microRNA family in the human genome," *BMC Evolutionary Biology*, vol. 10, no. 1, article no. 346, 2010.

[32] Z. Yuan, X. Sun, H. Liu, and J. Xie, "MicroRNA genes derived from repetitive elements and expanded by segmental duplication events in mammalian genomes," *PLoS One*, vol. 6, no. 3, Article ID e17666, 2011.

[33] H. A. Ebhardt, H. H. Tsang, D. C. Dai, Y. Liu, B. Bostan, and R. P. Fahlman, "Meta-analysis of small RNA-sequencing errors reveals ubiquitous post-transcriptional RNA modifications," *Nucleic Acids Research*, vol. 37, no. 8, pp. 2461–2470, 2009.

[34] F. Kuchenbauer, R. D. Morin, B. Argiropoulos et al., "In-depth characterization of the microRNA transcriptome in a leukemia progression model," *Genome Research*, vol. 18, no. 11, pp. 1787–1797, 2008.

[35] M. Lagos-Quintana, R. Rauhut, A. Yalcin, J. Meyer, W. Lendeckel, and T. Tuschl, "Identification of tissue-specific microRNAs from mouse," *Current Biology*, vol. 12, no. 9, pp. 735–739, 2002.

[36] R. D. Morin, M. D. O'Connor, M. Griffith et al., "Application of massively parallel sequencing to microRNA profiling and discovery in human embryonic stem cells," *Genome Research*, vol. 18, no. 4, pp. 610–621, 2008.

[37] J. G. Ruby, C. Jan, C. Player et al., "Large-scale sequencing reveals 21U-RNAs and additional microRNAs and endogenous siRNAs in *C. elegans*," *Cell*, vol. 127, no. 6, pp. 1193–1207, 2006.

[38] L. Guo and Z. Lu, "Global expression analysis of miRNA gene cluster and family based on isomiRs from deep sequencing data," *Computational Biology and Chemistry*, vol. 34, no. 3, pp. 165–171, 2010.

[39] K. E. Shearwin, B. P. Callen, and J. B. Egan, "Transcriptional interference—a crash course," *Trends in Genetics*, vol. 21, no. 6, pp. 339–345, 2005.

[40] C. F. Hongay, P. L. Grisafi, T. Galitski, and G. R. Fink, "Antisense transcription controls cell fate in *Saccharomyces cerevisiae*," *Cell*, vol. 127, no. 4, pp. 735–745, 2006.

[41] A. Stark, N. Bushati, C. H. Jan et al., "A single Hox locus in *Drosophila* produces functional microRNAs from opposite DNA strands," *Genes and Development*, vol. 22, no. 1, pp. 8–13, 2008.

[42] E. C. Lai, C. Wiel, and G. M. Rubin, "Complementary miRNA pairs suggest a regulatory role for miRNA:miRNA duplexes," *RNA*, vol. 10, no. 2, pp. 171–175, 2004.

[43] A. Kozomara and S. Griffiths-Jones, "MiRBase: integrating microRNA annotation and deep-sequencing data," *Nucleic Acids Research*, vol. 39, no. 1, pp. D152–D157, 2011.

[44] M. A. Larkin, G. Blackshields, N. P. Brown et al., "Clustal W and Clustal X version 2.0," *Bioinformatics*, vol. 23, no. 21, pp. 2947–2948, 2007.

[45] K. Tamura, D. Peterson, N. Peterson, G. Stecher, M. Nei, and S. Kumar, "MEGA5: molecular evolutionary genetics analysis using maximum likelihood, evolutionary distance, and maximum parsimony methods," *Molecular Biology and Evolution*, vol. 28, no. 10, pp. 2731–2739, 2011.

[46] P. Librado and J. Rozas, "DnaSP v5: a software for comprehensive analysis of DNA polymorphism data," *Bioinformatics*, vol. 25, no. 11, pp. 1451–1452, 2009.

[47] D. H. Huson, "SplitsTree: analyzing and visualizing evolutionary data," *Bioinformatics*, vol. 14, no. 1, pp. 68–73, 1998.

[48] D. Bryant and V. Moulton, "Neighbor-Net: an agglomerative method for the construction of phylogenetic networks," *Molecular Biology and Evolution*, vol. 21, no. 2, pp. 255–265, 2004.

[49] H. J. Bandelt, P. Forster, and A. Röhl, "Median-joining networks for inferring intraspecific phylogenies," *Molecular Biology and Evolution*, vol. 16, no. 1, pp. 37–48, 1999.

[50] A. Krek, D. Grün, M. N. Poy et al., "Combinatorial microRNA target predictions," *Nature Genetics*, vol. 37, no. 5, pp. 495–500, 2005.

[51] B. P. Lewis, I. H. Shih, M. W. Jones-Rhoades, D. P. Bartel, and C. B. Burge, "Prediction of mammalian microRNA targets," *Cell*, vol. 115, no. 7, pp. 787–798, 2003.

[52] B. John, A. J. Enright, A. Aravin, T. Tuschl, C. Sander, and D. S. Marks, "Human microRNA targets," *PLoS Biology*, vol. 2, no. 11, 2004.

[53] M. E. Smoot, K. Ono, J. Ruscheinski, P. L. Wang, and T. Ideker, "Cytoscape 2.8: new features for data integration and network visualization," *Bioinformatics*, vol. 27, no. 3, Article ID btq675, pp. 431–432, 2011.

21

In Silico Expressed Sequence Tag Analysis in Identification of Probable Diabetic Genes as Virtual Therapeutic Targets

Pabitra Mohan Behera,[1] Deepak Kumar Behera,[2] Aparajeya Panda,[2] Anshuman Dixit,[3] and Payodhar Padhi[2]

[1] *Centre of Biotechnology, Siksha O Anusandhan University, Bhubaneswar, Odisha 751030, India*
[2] *Hi-Tech Research and Development Centre, Konark Institute of Science and Technology, Techno Park, Jatni, Bhubaneswar, Odisha 752050, India*
[3] *Department of Translational Research and Technology Development, Institute of Life Sciences, Nalco Square, Bhubaneswar, Odisha 751023, India*

Correspondence should be addressed to Payodhar Padhi; payodharpadhi@gmail.com

Academic Editor: Ying Xu

The expressed sequence tags (ESTs) are major entities for gene discovery, molecular transcripts, and single nucleotide polymorphism (SNPs) analysis as well as functional annotation of putative gene products. In our quest for identification of novel diabetic genes as virtual targets for type II diabetes, we searched various publicly available databases and found 7 reported genes. The *in silico* EST analysis of these reported genes produced 6 consensus contigs which illustrated some good matches to a number of chromosomes of the human genome. Again the conceptual translation of these contigs produced 3 protein sequences. The functional and structural annotations of these proteins revealed some important features which may lead to the discovery of novel therapeutic targets for the treatment of diabetes.

1. Introduction

To understand the behavior and functionality of various biological processes, it is important to get a clear cut idea of genes and gene products involved, evident by the regulatory interactions of DNA, RNA, and proteins. Rapid advancement in technologies like microarray, sequencing, and spectrometry has contributed vast data for analysis and prediction in the light of genomics and proteomics.

Expressed sequence tags (ESTs) are short stretch of nucleotide sequences (200–800 bases) derived from the cDNA libraries. These are capable of identification of the full-length complimentary gene and mostly used for the identification of an expressed gene. The EST generation process involves sequencing of single segments either $5'$ end or $3'$ end of random clones from cDNA library of an organism. A single sequencing reaction and automation of DNA isolation, sequencing, and analysis can generate many ESTs at a time.

Since their original description and involvement as primary resources in human gene discovery [1], ESTs grow exponentially in various public databases, which will continue till there is suitable funding for the sequencing projects. Although the original ESTs were of human origin, a large number of ESTs are also isolated from model organisms like *Caenorhabditis elegans*, *Drosophila*, rice, and *Arabidopsis*. Public databases like dbEST [2], TIGR Gene Indices [3], and UniGene [4–6] now contain ESTs from a number of organisms for research and analysis. In addition, several commercial establishments maintain some privately funded, in-house collections of ESTs which are available for research. At present, ESTs are widely used throughout the genomics, and molecular biology communities for gene discovery, complement genome annotation, mapping, polymorphism analysis, gene prediction, gene structure identification, and expression studies establish the viability of alternative transcripts and facilitate proteome analysis.

TABLE 1: Information content for *Homo sapiens*.

Sl. no.	Database name	Release	Date	Information content
1	dbEST	040112	April 01, 2012	ESTs 8315296
2	TIGR Gene Indices	17.0	July 28, 2006	ESTs 7233257 HTs 234976
3	UniGene	—	December 23, 2011	mRNAs 209412 Models 212 HTC 20115 3′ ESTs 1693253 5′ ESTs 4027153 Unknown ESTs 927242 Total sequences 6877387

TABLE 2: UniGene information on human diabetes (mRNA and ESTs).

Sl. no.	Name of the gene	Source	mRNA	ESTs
1	Glucokinase (GCK)	*Homo sapiens*	12	46
2	Arginine vasopressin receptor 2 (AVPR2)	*Homo sapiens*	14	10
3	Aquaporin 2 (AQP2)	*Homo sapiens*	07	61
4	Islet cell autoantigen 1 (ICA1)	*Homo sapiens*	10	217
5	SRY (sex determining region Y) box 13 (SOX13)	*Homo sapiens*	09	180
6	Ras-related associated with diabetes (RRAD)	*Homo sapiens*	06	160
7	Ankyrin repeat domain 23 (ANKRD23)	*Homo sapiens*	06	141

Diabetes is a metabolic disorder characterized by hyperglycemia, glucosuria, negative nitrogen balance, and sometimes ketonemia. The clinical symptoms associated with it are retinopathy, neuropathy, and peripheral vascular insufficiencies. Overweight populations with sedentary lifestyle are more prone to diabetes. A recent study reveals that it affects 150 million people and almost 300 million more will be diabetic by the year 2025 [7]. Out of the three major types of diabetes, the non-insulin-dependent (type II diabetes or NIDDM) accounts for 90–95% of the diagnosed cases of the disease. There is no single approach to treat this disease and usually a combination therapy is adopted from different approaches. The worldwide epidemic of type II diabetes led the development of new strategies for its treatment. The discovery of nuclear receptor peroxisome proliferator activated receptors (PPARs) heralded a new era in understanding the patho-physiology of insulin receptors and its related complications [8]. PPARs are known to be the receptor for the fibrate class of hypolipidemic agents, while PPAR agonists reduce hyperglycemia without increasing the amount of insulin secretion. Again few other validated targets are protein tyrosine phosphatase-1B (PTP1B) and glycogen synthase kinase-3 (GSK-3). PTP-1B is a cytosolic phosphatase with a single catalytic domain [9]. *In vitro*, it is a nonspecific PTP and phosphorylates a wide variety of substrates. *In vivo*, it is involved in down regulation of insulin signaling by dephosphorylation of specific phosphotyrosine residues on the insulin receptor. GSK-3 is a type of protein kinase, which mediates the phosphorylation of certain serine and threonine residues in particular cellular substrates. This phosphorylation mainly inhibits the target proteins as in

the case of glycogenesis it inhibits glycogen synthase [10–12]. While a lot of research is focused on validated targets like PTP1B, PPARs, and GSKs, this paper intends identification of novel diabetic genes as virtual target(s). The approach is purely *in silico* and by analysis of ESTs available in public databases.

2. Materials and Methods

2.1. Materials. Databases like dbEST, TIGR Gene Indices, and UniGene are most useful resources containing raw and clusters of ESTs for many organisms. The dbEST is the largest repository of EST data maintained by NCBI. The TIGR Gene Indices of DFCI alphabetically list the ESTs of many organisms. NCBI's UniGene contains gene-oriented clusters of transcript sequences obtained by alignments between transcript sequences and genomic sequences originating from the same gene. The current information content of these three databases is represented in Table 1.

To initiate an *in silico* analysis, the UniGene database was searched for human diabetes gene clusters that reported seven gene entries whose mRNA and ESTs information are listed in Table 2.

The ESTs of all seven gene entries were downloaded and only those originating from pancreas and liver tissue were taken for analysis. Only the 5′ ESTs were considered as the ESTs generated from the 3′ end are most error prone because of the low base-call quality at the start of sequence reads. There were no ESTs of pancreatic or hepatic tissue origin for the gene entry "Aquaporin 2 (AQP2)." Thus we found a total of 34 ESTs from six reported gene entries as listed in Table 3.

TABLE 3: ESTs reported from different genes.

Sl. no.	GB accession no.	Description	Tissue type	EST type	Code*
1		Glucokinase (GCK)			
	DA640823.1	Clone LIVER2005873	Liver	$5'$ read	P
	DA637293.1	Clone LIVER2000237	Liver	$5'$ read	P
	DA638310.1	Clone LIVER2002033	Liver	$5'$ read	P
	CK823298.1	Clone IMAGE:6136115	Pancreas	$5'$ read	P
	BM966889.1	Clone IMAGE:6136115	Pancreas	$5'$ read	P
	BM966913.1	Clone IMAGE:6135860	Pancreas	$5'$ read	P
	BQ101045.1	Clone IMAGE:6135541	Pancreas	$5'$ read	P
2		Arginine vasopressin receptor 2 (AVPR2)			
	BG830436.1	Clone IMAGE:4908956	Pancreas	$5'$ read	—
	BI160709.1	Clone IMAGE:5018991	Pancreas	$5'$ read	P
	BI161076.1	Clone IMAGE:5019146	Pancreas	$5'$ read	—
	BI161438.1	Clone IMAGE:5019572	Pancreas	$5'$ read	—
3		Islet cell autoantigen 1 (ICA1)			
	CB134411.1	Clone L14ChoiCK0-18-B12	Liver	$5'$ read	P
	BX497434.1	Clone DKFZp779M2033	Liver	$5'$ read	A
	BX646846.1	Clone DKFZp779C0346	Liver	$5'$ read	P
	AW583029.1	Clone IMAGE:5637830	Pancreas	$5'$ read	P
	CK904151.1	Clone IMAGE:5672417	Pancreas	$5'$ read	A
	BE736046.1	Clone IMAGE:3639903	Pancreas	$5'$ read	P
	BI715368.1	—	Pancreas	$5'$ read	P
	BI962895.1	Clone IMAGE:5671189	Pancreas	$5'$ read	—
	BI966135.1	Clone IMAGE:5672382	Pancreas	$5'$ read	A
	BM021952.1	Clone IMAGE:5672417	Pancreas	$5'$ read	A
	BU579558.1	Clone IMAGE:6121832	Pancreas	$5'$ read	P
	BU951015.1	Clone IMAGE:6132285	Pancreas	$5'$ read	P
4		SRY (sex determining region Y)-box 13 (SOX13)			
	BE563236.1	Clone IMAGE:3689361	Pancreas	$5'$ read	—
	BE904395.1	Clone IMAGE:3898347	Pancreas	$5'$ read	—
	BE905187.1	Clone IMAGE:3901107	Pancreas	$5'$ read	P
5		Ras-related associated with diabetes (RRAD)			
	BG250011.1	Clone IMAGE:4470428	Liver	$5'$ read	—
	BG250978.1	Clone IMAGE:4472119	Liver	$5'$ read	—
	BG252988.1	Clone IMAGE:4474056	Liver	$5'$ read	P
	BM967357.1	Clone IMAGE:6136533	Pancreas	$5'$ read	P
6		Ankyrin repeat domain 23 (ANKRD23)			
	CB159821.1	Clone L18POOL1n1-19-D04	Liver	$5'$ read	—
	BM127096.1	Clone IMAGE:5675155	Pancreas	$5'$ read	—
	BQ227733.1	Clone IMAGE:6018368	Pancreas	$5'$ read	—
	BU073912.1	—	Pancreas	$5'$ read	—

P: presence of similarity to proteins after translation and A: contains a polyadenylation signal.

2.2. Methods

2.2.1. EST Pre-Processing.

The EST sequences are often of low quality because they are automatically generated without verification and thus contain higher error rates. The ESTs are also contaminated by vector sequences during their synthesis because a part of the vector is also sequenced along with the EST sequences. These sequences should be removed from EST to reduce the overall redundancy and to improve efficacy in further analysis. A comparison of ESTs with various nonredundant vector databases identifies the contamination which is deleted prior to analysis, for example. The EMVEC [13, 14] database removes the vector contamination from the EST sequences using NCBI BLAST2 [15, 16]. Using the UniGene clusters in our analysis is obvious as each cluster is generated by combined information from dbEST, GenBank mRNA database, and electronically spliced genomic DNA. Further they are clustered and cleaned from

TABLE 4: The ESTs and their corresponding contigs obtained from CAP3 Server.

Sl. no.	Gene name	ESTs		No. of contigs	
		Liver	Pancreas	Liver	Pancreas
1	GCK	DA640823.1 DA637293.1 DA638310.1	BM966889.1 BM966913.1	1	1
2	AVPR2	—	BI160709.1 BI161076.1 BI161438.1	—	1
3	ICA1	BX497434.1 BX646846.1	BI715368.1 BI962895.1	1	1
4	SOX13	—	BE563236.1 BE904395.1 BE905187.1	—	1
5	RRAD	BG250011.1 BG250978.1 BG252988.1	—	0	—
6	ANKRD23	—	BQ227733.1 BU073912.1	—	0

contamination (either by bacterial vector sequences or by linker sequences).

2.2.2. EST Clustering and Assembly. The purpose behind EST clustering is to collect overlapping ESTs from the same transcript of a single gene into a unique cluster to reduce redundancy. This is important because all the expressed data coming from a single gene are grouped into an index class which represents information of that particular gene. The clustering or assembly is mainly done by pairwise sequence similarity search between sequences and it consists of three major phases. In the first phase, poor regions of both $5'$ and $3'$ reads are identified and removed. Then the overlapping regions between the sequences are calculated and the false overlaps are removed after their identification. In the second phase, reads are joined to form contigs in decreasing order of overlap scores. Then, both forward-reverse constraints are used to make corrections to the resulting contigs. In the third phase, a multiple sequence alignment of reads is constructed and a consensus sequence along with a quality value for each base is computed for each contig. Base quality values are used in computation of overlaps and construction of multiple sequence alignments. The tissue-based ESTs from six reported genes were subjected to cluster analysis by the CAP3 Server [17]. The subjected ESTs along with their gene names and resulting contigs are listed in Table 4.

2.2.3. Database Similarity Searches. The consensus sequences or contigs (putative genes) obtained from clustering are only useful if their functionality are ascertained and it is only possible by database similarity search using some freely available tools like BLASTN and BLASTX. For transcriptome analysis, the ESTs are additionally aligned to the genome

sequence of the organism using specialized programs like BLAT (BLAST like alignment tool) [18] to assist genome mapping and gene discovery. The 6 contigs generated from 4 genes (GCK, AVPR2, ICA1, and SOX13) were subjected to BLAT analysis with parameters reading (genome: human, assembly: Feb. 2009 (GRCh37/hg19), query type: translated DNA, sort output: Score, output type: hyperlink). The outputs are listed in Table 5.

2.2.4. Conceptual Translation of ESTs. The EST sequences or data is informative only when its ontology, structure, and functions are obvious, for this the ESTs are correlated to protein-centric annotations by most accurate and robust polypeptide translations. The fact governing this process is that the polypeptides act as better templates for the identification of domains and motifs to study protein localization and assignment of gene ontology. The translations of ESTs are initiated by identifying the protein-coding regions or ORFs (open reading frames) from the consensus sequences or contigs. Here all 6 reported contigs were threaded to ESTScan2 [19, 20] tool with parameters reading (format: plain text, species: human, insertion/deletion penalty: −50, output: protein). The graphical view of 6 reported proteins is shown in Figure 1 obtained by BioEdit [21]. From these proteins, only 3 long continuous transcripts (GCK liver, GCK pancreas, and ICA1 liver) were selected for further structural and functional annotations.

2.2.5. Functional Annotation. The functionality of a putative polypeptide is predicted by matching against nonredundant databases of protein sequences, motifs, and family; this is because proteins act as better templates for functional annotation implementing multiple-sequence alignment, profile, HMM generation, phylogenetic analysis, domains, and motif

TABLE 5: BLAT output showing the alignment of contigs versus human genome sorted by score.

Query	Score	Start	End	Qsize	Identity	Chromosome	Strand
				Glucokinase (GCK)			
Contig1	567	1	570	570	100.00%	7	−
Contig1	24	206	230	570	100.00%	1	−
Contig1	21	429	449	570	100.00%	4	−
Contig1	20	386	405	570	100.00%	5	+
Contig1	20	105	124	570	100.00%	3	+
Contig2	27	713	740	938	100.00%	3	−
Contig2	21	778	798	938	100.00%	1	−
Contig2	21	860	880	938	100.00%	X	+
Contig2	20	519	538	938	100.00%	4	+
				Arginine vasopressin receptor 2 (AVPR2)			
Contig3	26	434	460	911	100.00%	2	−
Contig3	25	516	541	911	100.00%	2	−
Contig3	21	336	356	911	100.00%	1	−
				Islet cell autoantigen 1 (ICA1)			
Contig4	586	4	593	593	100.00%	7	−
Contig4	21	570	590	593	100.00%	2	+
Contig4	20	104	123	593	100.00%	1	−
Contig4	20	105	124	593	100.00%	5	+
Contig5	963	7	992	1001	99.20%	7	−
Contig5	127	841	1001	1001	91.00%	16	+
Contig5	40	801	850	1001	90.00%	2	−
Contig5	20	545	564	1001	100.00%	1	−
				Sex determining region Y-box 13 (SOX13)			
Contig6	1283	22	1340	1551	99.30%	1	+
Contig6	35	869	905	1551	97.30%	7	−
Contig6	32	869	905	1551	97.10%	6	−

analysis. In our search for a novel diabetic gene, only 3 translated protein sequences (GCK liver, GCK pancreas, and ICA1 liver) obtained from ESTScan2 were subjected to Inter-ProScan 4.8 [22] with parameters program: iprscan, nocrc: false, goterms: true, appl: blastprodom, fprintscan, hmmpir, hmmpfam, hmmsmart, hmmtigr, profilescan, hamap, patternscan, superfamily, signalp, tmhmm, hmmpanther, gene3d. The results are listed in Table 6.

3. Results and Discussion

Current EST analysis includes several steps and a wide range of computational tools are available for each step featuring different strengths and generate vital information systematically. Again there exists some arguments and confusion in selecting the suitable tools for individual steps of EST analysis and subsequent annotations at DNA and protein level. In our EST analysis for identification of novel diabetic genes as virtual targets for type II diabetes, we have followed a much cited procedure described by Nagaraj et al. [23]. From several successful and widely accessed EST databases, the UniGene database was selected as it uses mRNA and other coding sequence data of GenBank [24] as reference sequences for cluster generation. The UniGene clusters are updated weekly

for progressive data management with the ever increasing EST data in GenBank. It stores all gene isoforms in a single cluster and does not generate consensus sequences. After search for the human diabetic gene in UniGene, the EST sequences of pancreatic and hepatic origin were selected due to their all-round association and greater functionality in the onset and continuation of diabetes. Only the $5'$ ESTs of six genes were considered for analysis as the ESTs generated from the $3'$ end are most error prone. After purposeful clustering of specific ESTs of a particular gene, we found out 1 contig each of hepatic and pancreatic origin for GCK, 1 contig of pancreatic origin for AVPR2, 1 contig each of hepatic and pancreatic origin for ICA1, and 1 contig of pancreatic origin for SOX13. The database similarity search by querying these contigs in BLAT against human genome revealed that both the hepatic contig and the pancreatic contig of GCK were showing good matches with chromosomes (1, 3, 4, 5, and 7) and (1, 3, 4, and X), respectively. The pancreatic contig of AVPR2 was showing good matches with chromosomes (1 and 2). Both the hepatic contig and pancreatic contig of ICA1 were showing good matches with chromosomes (1, 2, 5, and 7) and (1, 2, 7, and 16), respectively. The pancreatic contig of SOX13 was showing good matches with chromosomes (1, 6, and 7). The conceptual

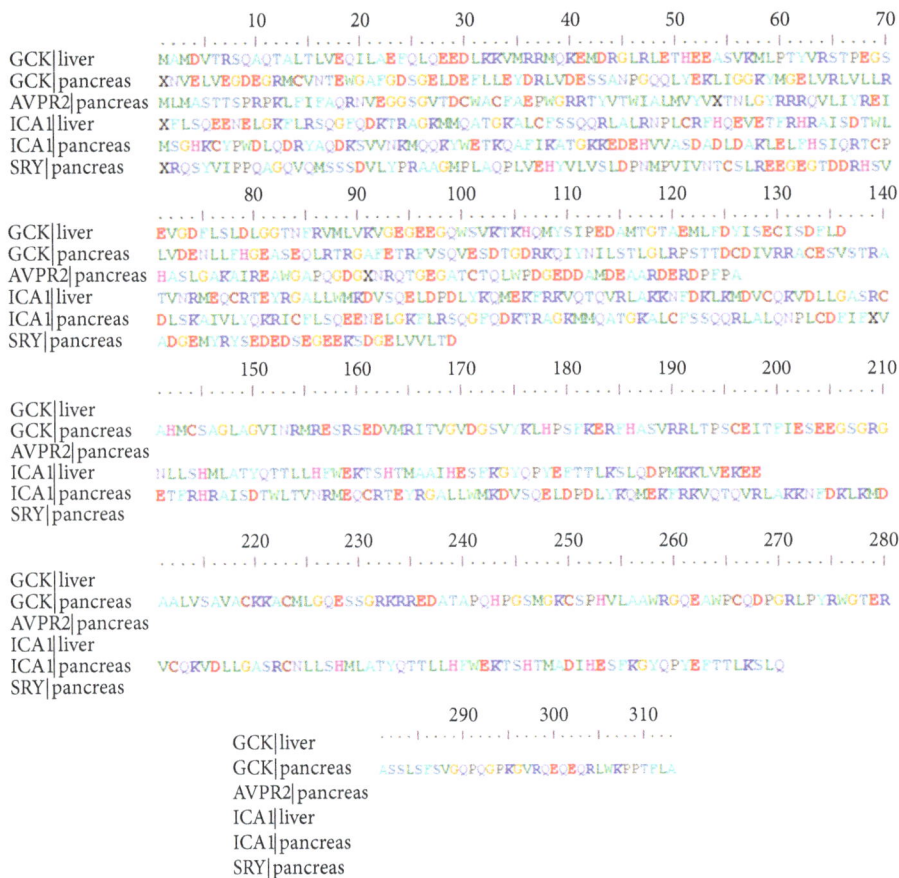

FIGURE 1: Graphical representation of protein sequences obtained from ESTScan2 translations and edited in BioEdit.

translation of these contigs in ESTScan2 provides six protein sequences from which we have considered only three (GCK liver, GCK pancreas, and ICA1 liver) as best for our analysis. The rest three sequences were left due to some erroneous readings (X, which does not code for any amino acid or refers to a stop codon) in their sequence. Thus the three proteins were GCK liver, a protein of 136 amino acids with molecular weight of 15474.87 Daltons; GCK pancreas, a protein of 313 amino acids with molecular weight of 34694.19 Daltons; and ICA1 pancreas, a protein of 270 amino acids with molecular weight of 31690.83 Daltons. These three proteins were named as hypothetical protein 1, hypothetical protein 2, and hypothetical protein 3 for further annotation.

3.1. The Hypothetical Protein 1.

We have reported it from 5' ESTs of liver tissues and it belongs to the hexokinase family of proteins with a distinct N-terminal and C-terminal. It is involved in the primary metabolic process like glycolysis and helps in the ATP-dependant conversion of aldohexose and ketohexose sugars to hexose-6-phosphate. The main function is carbohydrate kinase activity of various metabolic pathways like pentose phosphate pathway, fructose galactose metabolism, and glycolysis. It contains two structurally similar domains represented by PFAM families PF00349 [25]

and PF03727 [26]. In structural classification by CATH, it belongs to the classification lineage of hierarchy 3.30.420.40 featuring 3 (alpha beta), 3.30 (2-layer sandwich), and 3.30.420 (nucleotidyltransferase; domain 5).

3.2. The Hypothetical Protein 2.

We have reported it from 5' ESTs of pancreas tissues and it also belongs to the hexokinase family of proteins with a distinct N-terminal and C-terminal. It is involved in the primary metabolic process like glycolysis and helps in the ATP-dependant conversion of aldohexose and ketohexose sugars to hexose-6-phosphate. The main function is the carbohydrate kinase activity of various metabolic pathways like pentose phosphate pathway, fructose galactose metabolism, and glycolysis. It contains two structurally similar domains represented by PFAM families PF03727 and PF00349. In structural classification by CATH, it belongs to the classification lineage of hierarchy 3.40.367.20 featuring 3 (alpha beta), 3.40 (3-layer (aba) sandwich), and 3.40.367 (hexokinase; domain 1).

Thus both hypothetical protein 1 and hypothetical protein 2 belong to the same family of proteins with common functions, but structurally they have different domains. The hexokinases contain 7 distinct motifs from which the motif 1 encodes the putative ATP-binding domain and motif 2 encodes for the sugar-binding domain. All motifs, except

TABLE 6: The InterProScan annotations for three hypothetical proteins.

Sl. no.	InterProScan applications	Proteins		
		GCK liver	GCK pancreas	ICA1 liver
1	GENE3D	G3DSA: 3.30.420.40	G3DSA: 3.40.367.20	G3DSA: 3.20.1270.60
2	PANTHER	PTHR19443	PTHR19443	PTHR10164
3	PFAM	PF00349	PF03727	PF06456
4	PRINTS	—	PR00475	—
5	PROFILE	—	—	PS50870
6	SMART	—	—	SM01015
7	SUPER FAMILY	SSFS3067	SSFS3067	SSFS3067

motif 6, contain amino acids that project into or near the ATP/sugar-binding pocket. Previously we have assumed that the glucokinase (GCK) gene is expressed and functions irrespective of the tissue types, but now it is obvious that there exit some structural differences although they function as same.

3.3. The Hypothetical Protein 3.

We have reported it from 5' ESTs of liver tissues and it belongs to a family of proteins containing an arfaptin domain with a distinct N-terminal and C-terminal. The arfaptin domain interacts with ARF1 (ADP-ribosylation factor 1), a small GTPase involved in vesicle budding at the Golgi complex and immature secretory granules. The structure of arfaptin shows that, upon binding to a small GTPase, arfaptin forms an elongated, crescent-shaped dimer of three-helix coiled coils [27]. The N-terminal region of ICA69 is similar to arfaptin [28]. It is involved in a neurological system process and secretes several neurotransmitters. It is also involved in a cellular process like cell communication and cell-cell signaling with synaptic transmission. In structural classification by CATH, it belongs to the classification lineage of hierarchy 1.20.1270.60 featuring 1 (mainly alpha), 1.20 (up-down bundle), 1.20.1270 (substrate binding domain of DNAk; chain A; domain 2), and 1.20.1270.60 (Arfaptin, Rac-binding fragment, chain A). Again we also found that the mammalian islet cell autoantigen (ICA69) is a 69 kDa protein [29].

3.4. Molecular Modeling of the Three Hypothetical Proteins.

To initiate the structural annotations of our three hypothetical proteins, we have generated the homology models using Modeller 9.10 [30–33]. The three protein sequences were queried in BLASTP [34] against the PDB [35] database to select their suitable templates. The template for hypothetical protein 1 was 1HKC with 50% sequence identity and a resolution of 2.80 Å. The template for hypothetical protein 2 was 2NZT with 48% sequence identity and a resolution of 2.45 Å. Similarly the template for the hypothetical protein 3 was 1I49 with 34% sequence identity and a resolution of 2.80 Å. After modeling, the three models were subjected to evaluation by development of the Ramachandran plots using PROCHECK NT stand-alone version [36]. The statistical information gathered from the Ramachandran plots revealed that 95.10%, 93.60%, and 95.70% of the residues were in the allowed region for the proteins hypothetical protein 1, hypothetical protein 2, and hypothetical protein 3 at an average resolution of 2.68 Å. Thus the models were perfect for structural annotation. The hypothetical protein 1 had 4 α-helices and 3 β-sheets arranged in a 2-layered sandwich model (Figure 2(a)). The hypothetical protein 2 had 16 α-helices and 2 β-sheets arranged in a 3-layer sandwich model (Figure 2(b)). The hypothetical protein 3 had only 7 α-helices acquiring an up-down bundle model (Figure 2(c)). Therefore the structural annotations by both InterProScan and subsequent modeling were just similar which purposefully validate our work.

In general glucokinase occurs in human liver, pancreas, gut, and brain cells and plays an important role in suitable regulation of carbohydrate metabolism. It works as a glucose sensor and triggers shifts in metabolism or cell function in response to the rising or falling levels of glucose. A mutation of the gene for this enzyme causes several forms of diabetes or hypoglycemia. Human islet cell autoantigen 1 protein is encoded by the ICA1 gene [37, 38]. This protein contains an arfaptin domain and is found in both cytosolic and membrane-bound Golgi complex and immature secretory granules. It also works as an autoantigen in insulin-dependent diabetes mellitus. Our in silico analysis revealed three new proteins from which two (hypothetical protein 1 and hypothetical protein 2) were functionally similar to glucokinase and one (hypothetical protein 3) was functionally similar to the human islet cell autoantigen 1. Due to their association in diabetes, these can be treated as virtual therapeutic targets for treatment of diabetes. There were structural variations among these three proteins and their functional homologues which need further structural analysis and interpretation.

4. Conclusion

The in silico EST analysis of seven reported genes associated with diabetes produced 6 consensus contigs which were annotated functionally and structurally. The functional annotations were similar to the corresponding proteins in which the ESTs were actually categorized. The structural annotations revealed that there is a variation which may be due the differences in source tissue types. This information

P. M. Behera by the University Grants Commission, India, is duly acknowledged.

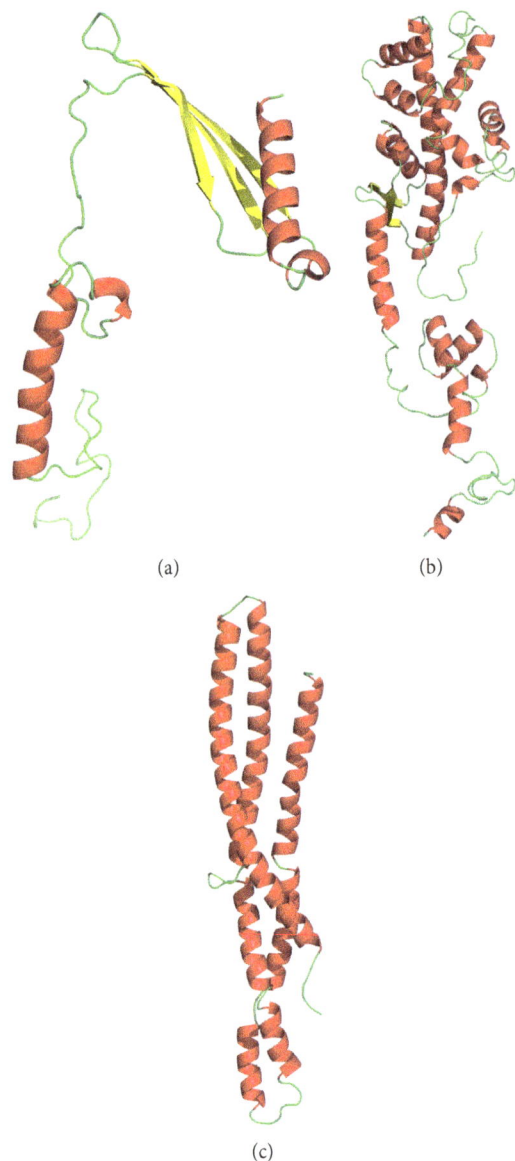

FIGURE 2: 3D representation of homology models of three hypothetical proteins. (a) the homology model of hypothetical protein 1, (b) the homology model of hypothetical protein 2, and (c) the homology model of hypothetical protein 3.

can be used for further structure-based annotations, and new drug designs for the treatment of diabetes.

Acknowledgments

The authors are thankful to the scientific community for their valuable submissions to different public domains from where the data are taken for analysis. Sincere acknowledgements are due to the Vice Chancellor of the Siksha O Anusandhan University and the Chairman of the Hi-Tech Group of Institutions for providing essential facilities in successful research and bringing the paper to the level of publication. The award of the Rajiv Gandhi National Fellowship to P. M. Behera by the University Grants Commission, India, is duly acknowledged.

References

[1] M. D. Adams, J. M. Kelley, J. D. Gocayne et al., "Complementary DNA sequencing: expressed sequence tags and human genome project," *Science*, vol. 252, no. 5013, pp. 1651–1656, 1991.

[2] M. S. Boguski, T. M. J. Lowe, and C. M. Tolstoshev, "dbEST—database for 'expressed sequence tags'," *Nature Genetics*, vol. 4, no. 4, pp. 332–333, 1993.

[3] Y. Lee, J. Tsai, S. Sunkara et al., "The TIGR Gene Indices: clustering and assembling EST and know genes and integration with eukaryotic genomes," *Nucleic Acids Research*, vol. 33, pp. D71–D74, 2005.

[4] J. U. Pontius, L. Wagner, and G. D. Schuler, "UniGene: a unified view of the transcriptome," in *The NCBI Handbook*, National Center for Biotechnology Information, Bethesda, Md, USA, 2003.

[5] D. L. Wheeler, D. M. Church, S. Federhen et al., "Database resources of the National Center for Biotechnology," *Nucleic Acids Research*, vol. 31, pp. 28–33, 2003.

[6] G. D. Sehuler, "Pieces of use puzzle: expressed sequence tags and the catalog of human genes," *Journal of Molecular Medicine*, vol. 75, no. 10, pp. 694–698, 1997.

[7] D. W. Dunstan, P. Z. Zimmet, T. A. Welborn et al., "The rising prevalence of diabetes and impaired glucose tolerance: the Australian diabetes, obesity and lifestyle study," *Diabetes Care*, vol. 25, no. 5, pp. 829–834, 2002.

[8] R. Chakrabarti, P. Misra, R. K. Vikramadithyan et al., "Antidiabetic and hypolipidemic potential of DRF 2519—a dual activator of PPAR-α and PPAR-γ," *European Journal of Pharmacology*, vol. 491, no. 2-3, pp. 195–206, 2004.

[9] Z. Y. Zhang, "Protein tyrosine phosphatases: prospects for therapeutics," *Current Opinion in Chemical Biology*, vol. 5, no. 4, pp. 416–423, 2001.

[10] J. R. Woodgett, "Regulation and functions of the glycogen synthase kinase-3 subfamily," *Seminars in Cancer Biology*, vol. 5, no. 4, pp. 269–275, 1994.

[11] J. R. Woodgett, "Judging a protein by more than its name: GSK-3," *Science's STKE*, vol. 2001, no. 100, p. RE12, 2001.

[12] A. Ali, K. P. Hoeflich, and J. R. Woodgett, "Glycogen synthase kinase-3: properties, functions, and regulation," *Chemical Reviews*, vol. 101, no. 8, pp. 2527–2540, 2001.

[13] W. Baker, A. van den Broek, E. Camon et al., "The EMBL nucleotide sequence database," *Nucleic Acids Research*, vol. 28, no. 1, pp. 19–23, 2000.

[14] T. Etzold, A. Ulyanov, and P. Argos, "SRS: information retrieval system for molecular biology data banks," *Methods in Enzymology*, vol. 266, pp. 114–125, 1996.

[15] S. F. Altschul, T. L. Madden, A. A. Schäffer et al., "Gapped BLAST and PSI-BLAST: a new generation of protein database search programs," *Nucleic Acids Research*, vol. 25, no. 17, pp. 3389–3402, 1997.

[16] E. G. Shpaer, M. Robinson, D. Yee, J. D. Candlin, R. Mines, and T. Hunkapiller, "Sensitivity and selectivity in protein similarity searches: a comparison of Smith-Waterman in hardware to BLAST and FASTA," *Genomics*, vol. 38, no. 2, pp. 179–191, 1996.

[17] X. Huang and A. Madan, "CAP3: a DNA sequence assembly program," *Genome Research*, vol. 9, no. 9, pp. 868–877, 1999.

[18] C. Lottaz, C. Iseli, C. V. Jongeneel, and P. Bucher, "Modeling sequencing errors by combining Hidden Markov models," *Bioinformatics*, vol. 19, no. 2, pp. ii103–ii112, 2003.

[19] C. Iseli, C. V. Jongeneel, and P. Bucher, "ESTScan: a program for detecting, evaluating, and reconstructing potential coding regions in EST sequences," in *Procceedings of the International Conference on Intelligent Systems for Molecular Biology*, pp. 138–148, August 1999.

[20] E. M. Zdobnov and R. Apweiler, "InterProScan—an integration platform for the signature-recognition methods in InterPro," *Bioinformatics*, vol. 17, no. 9, pp. 847–848, 2001.

[21] T. A. Hall, "BioEdit: a user-friendly biological sequence alignment editor and analysis program for Windows 95/98/NT," *Nucleic Acids Symposium Series*, vol. 41, pp. 95–98, 1999.

[22] E. Quevillon, V. Silventoinen, S. Pillai et al., "InterProScan: protein domains identifier," *Nucleic Acids Research*, vol. 33, no. 2, pp. W116–W120, 2005.

[23] S. H. Nagaraj, R. B. Gasser, and S. Ranganathan, "A hitchhiker's guide to expressed sequence tag (EST) analysis," *Briefings in Bioinformatics*, vol. 8, no. 1, pp. 6–21, 2006.

[24] D. A. Benson, I. Karsch-Mizrachi, D. J. Lipman, J. Ostell, and D. L. Wheeler, "GenBank," *Nucleic Acids Research*, vol. 36, no. 1, pp. D25–D30, 2008.

[25] W. S. Bennett Jr. and T. A. Steitz, "Structure of a complex between yeast hexokinase A and glucose. II. Detailed comparisons of conformation and active site configuration with the native hexokinase B monomer and dimer," *Journal of Molecular Biology*, vol. 140, no. 2, pp. 211–230, 1980.

[26] T. A. Steitz, "Structure of yeast hexokinase-B. I. Preliminary X-ray studies and subunit structure," *Journal of Molecular Biology*, vol. 61, no. 3, pp. 695–700, 1971.

[27] C. Tarricone, B. Xiao, N. Justin et al., "The structural basis of Arfaptin-mediated cross-talk between Rac and Arf signalling pathways," *Nature*, vol. 411, no. 6834, pp. 215–219, 2001.

[28] F. Spitzenberger, S. Pietropaololl, P. Verkade et al., "Islet cell autoantigen of 69 kDa is an arfaptin-related protein associated with the golgi complex of insulinoma INS-1 cells," *Journal of Biological Chemistry*, vol. 278, no. 28, pp. 26166–26173, 2003.

[29] J. Cherfils, "Structural mimicry of DH domains by Arfaptin suggests a model for the recognition of Rac-GDP by its guanine nucleotide exchange factors," *FEBS Letters*, vol. 507, no. 3, pp. 280–284, 2001.

[30] N. Eswar, M. A. Marti-Renom, B. Webb et al., "Comparative protein structure modeling using Modeller," in *Current Protocols in Bioinformatics*, Supplement 15, 5.6.1–5.6.30, John Wiley & Sons, New York, NY, USA, 2006.

[31] M. A. Marti-Renom, A. Stuart, A. Fiser, R. Sánchez, F. Melo, and A. Sali, "Comparative protein structure modeling of genes and genomes," *Annual Review of Biophysics and Biomolecular Structure*, vol. 29, pp. 291–325, 2000.

[32] A. Sali and T. L. Blundell, "Comparative protein modelling by satisfaction of spatial restraints," *Journal of Molecular Biology*, vol. 234, no. 3, pp. 779–815, 1993.

[33] A. Fiser, R. K. Do, and A. Sali, "Modeling of loops in protein structures," *Protein Science*, vol. 9, no. 9, pp. 1753–1773, 2000.

[34] S. F. Altschul, W. Gish, W. Miller, E. W. Myers, and D. J. Lipman, "Basic local alignment search tool," *Journal of Molecular Biology*, vol. 215, no. 3, pp. 403–410, 1990.

[35] The Research collaborator for Structural Bioinformatics, Protein Data Bank, http://www.rcsb.org/pdb/home.

[36] R. A. Laskowski, M. W. MacArthur, and J. M. Thornton, "PROCHECK: validation of protein structure coordinates," in *International Tables of Crystallography, Volume F: Crystallography of Biological Macromolecules*, M. G. Rossmann and E. Arnold, Eds., pp. 722–725, Kluwer Academic, Dordrecht, The Netherlands, 2001.

[37] I. Miyazaki, R. Gaedigk, M. F. Hui et al., "Cloning of human and rat p69 cDNA, a candidate autoimmune target in type 1 diabetes," *Biochimica et Biophysica Acta*, vol. 1227, no. 1-2, pp. 101–104, 1994.

[38] M. I. Mally, V. Cirulli, A. Hayek, and T. Otonkoski, "ICA69 is expressed equally in the human endocrine and exocrine pancreas," *Diabetologia*, vol. 39, no. 4, pp. 474–480, 1996.

Permissions

The contributors of this book come from diverse backgrounds, making this book a truly international effort. This book will bring forth new frontiers with its revolutionizing research information and detailed analysis of the nascent developments around the world.

We would like to thank all the contributing authors for lending their expertise to make the book truly unique. They have played a crucial role in the development of this book. Without their invaluable contributions this book wouldn't have been possible. They have made vital efforts to compile up to date information on the varied aspects of this subject to make this book a valuable addition to the collection of many professionals and students.

This book was conceptualized with the vision of imparting up-to-date information and advanced data in this field. To ensure the same, a matchless editorial board was set up. Every individual on the board went through rigorous rounds of assessment to prove their worth. After which they invested a large part of their time researching and compiling the most relevant data for our readers. Conferences and sessions were held from time to time between the editorial board and the contributing authors to present the data in the most comprehensible form. The editorial team has worked tirelessly to provide valuable and valid information to help people across the globe.

Every chapter published in this book has been scrutinized by our experts. Their significance has been extensively debated. The topics covered herein carry significant findings which will fuel the growth of the discipline. They may even be implemented as practical applications or may be referred to as a beginning point for another development. Chapters in this book were first published by Hindawi Publishing Corporation; hereby published with permission under the Creative Commons Attribution License or equivalent.

The editorial board has been involved in producing this book since its inception. They have spent rigorous hours researching and exploring the diverse topics which have resulted in the successful publishing of this book. They have passed on their knowledge of decades through this book. To expedite this challenging task, the publisher supported the team at every step. A small team of assistant editors was also appointed to further simplify the editing procedure and attain best results for the readers.

Our editorial team has been hand-picked from every corner of the world. Their multi-ethnicity adds dynamic inputs to the discussions which result in innovative outcomes. These outcomes are then further discussed with the researchers and contributors who give their valuable feedback and opinion regarding the same. The feedback is then collaborated with the researches and they are edited in a comprehensive manner to aid the understanding of the subject.

Apart from the editorial board, the designing team has also invested a significant amount of their time in understanding the subject and creating the most relevant covers. They scrutinized every image to scout for the most suitable representation of the subject and create an appropriate cover for the book.

The publishing team has been involved in this book since its early stages. They were actively engaged in every process, be it collecting the data, connecting with the contributors or procuring relevant information. The team has been an ardent support to the editorial, designing and production team. Their endless efforts to recruit the best for this project, has resulted in the accomplishment of this book. They are a veteran in the field of academics and their pool of knowledge is as vast as their experience in printing. Their expertise and guidance has proved useful at every step. Their uncompromising quality standards have made this book an exceptional effort. Their encouragement from time to time has been an inspiration for everyone.

The publisher and the editorial board hope that this book will prove to be a valuable piece of knowledge for researchers, students, practitioners and scholars across the globe.

List of Contributors

Oliver Frings, Andrey Alexeyenko and Erik L. L. Sonnhammer
Stockholm Bioinformatics Centre, Science for Life Laboratory, Box 1031, SE-171 21 Solna, Sweden

Oliver Frings and Erik L. L. Sonnhammer
Department of Biochemistry and Biophysics, Stockholm University, SE-106 91 Stockholm, Sweden

Judith E.Mank
Department of Genetics, Evolution and the Environment, University College London, WC1E 6BT, UK

Andrey Alexeyenko
School of Biotechnology, Royal Institute of Technology, SE-171 65 Solna, Sweden

Erik L. L. Sonnhammer
Swedish eScience Research Center, SE-100 44 Stockholm, Sweden

Hiromi Nishida
Agricultural Bioinformatics Research Unit, Graduate School of Agriculture and Life Sciences, The University of Tokyo, Bunkyo-ku, Tokyo 113-8657, Japan

Chun-Tien Chang and Chuan Yi Tang
Department of Computer Science, National Tsing Hua University, Hsin-Chu, Taiwan

Chi-Neu Tsai
Graduate Institutes of Clinical Medical Sciences, Chang Gung University, No. 259 Wen-Hwa, 1st Road, Kwei-Shan, Tao-Yuan 333, Taiwan

Chun-Houh Chen
Institute of Statistical Science, Academia Sinica, Taipei, Taiwan

Jang-Hau Lian, Chi-Yu Hu and Yun-Shien Lee
Department of Biotechnology, Ming Chuan University, Tao-Yuan, Taiwan

Chia-Lung Tsai, Angel Chao, Chyong-Huey Lai and Tzu-HaoWang
Department of Obstetrics and Gynecology, Lin-Kou Medical Center, Chang Gung Memorial Hospital, Chang Gung University, Fu-Hsing Street, Kwei-Shan, Tao-Yuan 333, Taiwan

Chia-Lung Tsai, Tzu-HaoWang and Yun-Shien Lee
Genomic Medicine Research Core Laboratory, Chang Gung Memorial Hospital, No. 5, Fu-Hsing Street, Kwei-Shan, Tao-Yuan 333, Taiwan

Libing Shen, Chao Chen, Hongxiang Zheng and Li Jin
State Key Laboratory of Genetic Engineering and Key Laboratory of Contemporary Anthropology of Ministry of Education, School of Life Sciences, Fudan University, Shanghai 200433, China

Jinsil Kim and Jeffrey C. Murray
Department of Anatomy and Cell Biology, University of Iowa, 500 Newton Road, 2182 ML, Iowa City, IA 52242, USA

Mitchell M. Pitlick, Paul J. Christine, Amanda R. Schaefer and Jeffrey C. Murray
Department of Pediatrics, University of Iowa, 500 Newton Road, 2182 ML, Iowa City, IA 52242, USA

Cesar Saleme
Departamento de Neonatología, Instituto de Maternidad y Ginecología Nuestra Señora de las Mercedes, 4000 San Miguel de Tucumán, Argentina

Belén Comas, Viviana Cosentino and Enrique Gadow
Dirección de Investigación, Centro de Educación Médica e Investigaciones Clínicas (CEMIC), 1431 Buenos Aires, Argentina

Yan Guo, Jiang Li, Chung-I Li and Yu Shyr
Vanderbilt Ingram Cancer Center, Center for Quantitative Sciences, Nashville, TN, USA

David C. Samuels
Center for Human Genetics Research, Vanderbilt University Medical Center, Nashville, TN, USA

Travis Clark
VANTAGE, Vanderbilt University, Nashville, TN, USA

Xiaoling Zhang, Marc E. Lenburg and Avrum Spira
Division of Computational Biomedicine, Boston University School ofMedicine, 72 East Concord Street, E631, Boston, MA 02118, USA

Xiaoling Zhang
Division of Intramural Research, National Heart, Lung and Blood Institute, The NHLBI's Framingham Heart Study, 73 Mt.Wayte Avenue Suite 2, Framingham, MA 01702, USA

Avrum Spira
Pulmonary Center, Boston University Medical Center, 715 Albany Street, Boston, MA 02118, USA

Rajesh Mehrotra, Amit Yadav, Purva Bhalothia and Sandhya Mehrotra
Department of Biological Sciences, Birla Institute of Technology and Science, Pilani, Rajasthan 333031, India

Rajesh Mehrotra and Ratna Karan
Department of Plant Environmentand Soil Sciences, Louisiana State University, Baton Rouge, LA 70894, USA

Lin Liu, Yinhu Li, Siliang Li, Ni Hu, Yimin He, Ray Pong, Danni Lin, Lihua Lu and Maggie Law
NGS Sequencing Department, Beijing Genomics Institute (BGI), 4th Floor, Building 11, Beishan Industrial Zone, Yantian District, Guangdong, Shenzhen 518083, China

J. Beil, L. Fairbairn and T. Buch
Institute for Medical Microbiology, Immunology and Hygiene, Technische Universit¨at M¨unchen, Trogerstraße 30, 81679 Munich, Germany

P. Pelczar
Institute of Animal Laboratory Sciences, VetSuisse Faculty, University of Zurich, Winterthurer Straße 190, 8057 Zurich, Switzerland

Abdullah M. Alzahrani and Hamza Hanieh
Biological Sciences Department, College of Science, King Faisal University, Hofouf 31982, Saudi Arabia

Georgia Ragia and Vangelis G. Manolopoulos
Laboratory of Pharmacology, Medical School, Democritus University of Thrace, Alexandroupolis 68100, Greece

Suping Feng, You Chen and Yaoting Wu
Bioscience and Biotechnology College, Qiongzhou University, Sanya 572200, China

Helin Tong and Jingyi Wang
Institute of Tropical Bioscience and Biotechnology, Chinese Academy of Tropical Agricultural Science, Haikou 571101, China

Yeyuan Chen and Junhu He
Institute of Tropical Crop Variety Resources, Chinese Academy of Tropical Agricultural Sciences, Danzhou 571737, China

Guangming Sun
South Subtropical Crops Research Institute, Chinese Academy of Tropical Agricultural Sciences, Zhanjiang 524091, China

Qiao Zhong, Weidong Xu, YuanjianWu and Hongxing Xu
Department of Laboratory Medicine, Suzhou Municipal Hospital Affiliated Nanjing Medical University, 26 Daoqian Street, Jiangsu, Suzhou 215002, China

Rosa Alduina and Giuseppe Gallo
Department of Science and Molecular and Biomolecular Technology, University of Palermo, Viale delle Scienze, 90128 Palermo, Italy

Hidenori Matsuzaki, Megumi Maeda, Suni Lee, Yasumitsu Nishimura, Naoko Kumagai-Takei, Hiroaki Hayashi, Shoko Yamamoto, Tamayo Hatayama and Takemi Otsuki
Department of Hygiene, Kawasaki Medical School, 577 Matsushima, Kurashiki 7010192, Japan

MegumiMaeda
Department of Biofunctional Chemistry, Division of Bioscience, Okayama University Graduate School of Natural Science and Technology, 3-1-1 Tsushima-Naka, Okayama 7008530, Japan

Hiroaki Hayashi
Department of Dermatology, Kawasaki Medical School, 577 Matsushima, Kurashiki 7010192, Japan

Yoko Kojima, Rika Tabata and Takumi Kishimoto
Research Center for Asbestos-Related Diseases, Okayama Rosai Hospital, 1-10-25 Chikko-Midorimachi, Minami-Ku, Okayama 7028055, Japan

Junichi Hiratsuka
Department of Radiation Oncology, Kawasaki Medical School, 577 Matsushima, Kurashiki 7010192, Japan

Ioannis L. Aivaliotis, Ioannis S. Pateras, Marilena Papaioannou and Christina Glytsou
Molecular Carcinogenesis Group, Department of Histology and Embryology, School of Medicine, National and Kapodistrian University of Athens, 11527 Athens, Greece

Konstantinos Kontzoglou
2nd Department of Propedeutic Surgery, Laikon General Hospital, School of Medicine, National and Kapodistrian University of Athens, 11527 Athens, Greece

Elizabeth O. Johnson
Department of Anatomy, School of Medicine, National and Kapodistrian University of Athens, 11527 Athens, Greece

Vassilis Zoumpourlis
Institute of Biology, Medicinal Chemistry and Biotechnology, National Hellenic Research Foundation, 11635 Athens, Greece

Raquel M. Fernández, Ana Peciña, Maria Dolores Lozano-Arana, Juan Carlos García-Lozano, Salud Borrego and Guillermo Antiñolo
Department of Genetics, Reproduction and Fetal Medicine, Institute of Biomedicine of Seville (IBIS), University Hospital Virgen del Roc´io/CSIC/University of Seville, Avenida Manuel Siurot, s/n, 41013 Seville, Spain

Raquel M. Fernández, Ana Peciña, Salud Borrego and Guillermo Antiñolo
Centre for Biomedical Network Research on Rare Diseases (CIBERER), 41013 Seville, Spain

Brad S. Coates, Craig A. Abel and Thomas W. Sappington
Corn Insects and Crop Genetics Research Unit, ARS, USDA, Ames, IA 50011, USA

Brad S. Coates and Thomas W. Sappington
Department of Entomology, Iowa State University, Ames, IA 50011, USA

Analiza P. Alves, Haichuan Wang, Nicholas J. Miller and Blair D. Siegfried
Department of Entomology, University of Nebraska, Lincoln, NE 68583, USA

Kimberly K. O. Walden and Hugh M. Robertson
University of Illinois, Champaign-Urbana, IL 61801, USA

B. Wade French
North Central Agricultural Research Laboratory, Brookings, ARS, USDA, SD 57006, USA

Mariko Murata and Raynoo Thanan
Department of Environmental and Molecular Medicine, Mie University Graduate School of Medicine, Tsu, 514-8507, Japan

Raynoo Thanan and Shosuke Kawanishi
Faculty of Pharmaceutical Sciences, Suzuka University of Medical Science, Suzuka, 513-8670, Japan

Ning Ma
Faculty of Health Science, Suzuka University of Medical Science, Suzuka, 510-0293, Japan

Tingming Liang and Chang Liu
Jiangsu Key Laboratory for Molecular and Medical Biotechnology, College of Life Science, Nanjing Normal University, Nanjing 210046, China

Li Guo
Department of Epidemiology and Biostatistics and Ministry of Education Key Lab for Modern Toxicology, School of Public Health, Nanjing Medical University, Nanjing 210029, China

Pabitra Mohan Behera
Centre of Biotechnology, Siksha O Anusandhan University, Bhubaneswar, Odisha 751030, India

Deepak Kumar Behera, Aparajeya Panda and Payodhar Padhi
Hi-Tech Research and Development Centre, Konark Institute of Science and Technology, Techno Park, Jatni, Bhubaneswar, Odisha 752050, India

Anshuman Dixit
Department of Translational Research and Technology Development, Institute of Life Sciences, Nalco Square, Bhubaneswar, Odisha 751023, India